# Tunable Evanescent-Mode Filters

# Tunable Evanescent-Mode Filters

## Principles, Implementation, and Applications

*Dimitrios Peroulis*
Purdue University
West Lafayette, IN, USA

*Mohammad Abu Khater*
Philowave
West Lafayette, IN, USA

IEEE*Press*

WILEY

Published by John Wiley & Sons, Inc., Hoboken, New Jersey.
Published simultaneously in Canada.



***Library of Congress Cataloging-in-Publication Data***

Names: Peroulis, Dimitrios author | Khater, Mohammad Abu author
Title: Tunable evanescent-mode filters : principles, implementation, and applications / Dimitrios Peroulis, Mohammad Abu Khater.
Description: Hoboken, New Jersey : Wiley-IEEE Press, [2025] | Includes index.
Identifiers: LCCN 2025020897 (print) | LCCN 2025020898 (ebook) | ISBN 9781394216802 cloth | ISBN 9781394216819 adobe pdf | ISBN 9781394216826 epub
Subjects: LCSH: Microwave filters | Electromagnetic waves
Classification: LCC TK7872.F5 P444 2025 (print) | LCC TK7872.F5 (ebook) | DDC 621.381/31–dc23/eng/20250701
LC record available at https://lccn.loc.gov/2025020897
LC ebook record available at https://lccn.loc.gov/2025020898

Cover Design: Wiley
Cover Images: © Imagery by Mark Hickle, © Tuomas A. Lehtinen/Getty Images

Set in 9.5/12.5pt STIXTwoText by Straive, Chennai, India

*This book is dedicated to my wife, Alina, who with much love, has led us to enjoy many summer mornings into harbors seen for the first time; to our children, Polina, Ann, and Aris, for the treasured gifts they have been in our lives; and to my parents, Aimilios and Katerina, who lovingly and supportively showed me the way.*

*– D. Peroulis*

*To my parents, Mahmoud and Yorseiah, who are my de facto engineering mentors; to my siblings who made that journey enjoyable; and to my incredible wife Ruby, and our wonderful children, Yasmeen and Yazan, who are my daily motivators. For them, and much more, I can only say "Alhamdulillah".*

*– M. Abu Khater*

# Contents

## About the Authors

**Dimitrios Peroulis** is the Senior Vice President for Partnerships and Online at Purdue University. In this role, he leads the offices of industry partnerships, global programs and partnerships, engagement, and online programs. From 2019 to 2023 he served as the Michael and Katherine Birck Head of ECE and special adviser to the Dean of Engineering on online learning. His research interests are focused on the areas of reconfigurable systems, plasma RF electronics, and RF-assisted lyophilization. He received the National Science Foundation CAREER award, is an IEEE and IET Fellow, and has co-authored over 450 journal and conference papers.

**Mohammad Abu Khater** received his PhD from Purdue University. His experience encompasses both academic and industrial positions, and his research work has received several awards and recognitions. He is an IEEE senior member and an active member in the microwave community. Currently, he is the founder and CEO of Philowave, where he is leading the development of advanced wideband interference detection and suppression technologies.

# Preface

On August 1, 2006, the senior author of this book, along with Prof. William Chappell, represented Purdue University in the kickoff meeting of the DARPA Analog Spectral Processors (ASP) program. In this effort we proposed to demonstrate, for the first time, miniaturized tunable filters that could be both high quality and widely tunable. Moreover, they could be implemented with cost-effective industry-standard manufacturing methods. Up until that time, nearly all microwave filters of practical importance in reconfigurable systems either exhibited high quality factors (e.g. static acoustic filters), or wide tuning range (e.g. lumped-element-based tunable filters), but not both. This effort was successful, resulted in a DARPA Phase I award for the Purdue team, and was extended to Phases II, III, and an additional seedling effort for Prof. Peroulis.

After the end of this contract, Prof. Peroulis led or co-led more than 20 additional government and industry tunable filters projects over a period of more than a dozen years. In each project we explored different aspects of this technology including, for example, novel synthesis approaches, advanced filter architectures, micro-mechanical and electronic tuning approaches, optimization methodologies, temperature and vibration-induced challenges, monitoring, control and feedback techniques, low-cost manufacturing technologies, and packaging.

Results from all this work have been published in more than 15 PhD dissertations from Purdue University and dozens of journal and conference papers. In this book we have attempted to organize and present the knowledge gained during these years in an easy-to-follow approach. Our writing style is closer to that of a coach as our goal is to empower and encourage students and engineers with nominal knowledge of RF systems and electromagnetics to dive in the art of designing evanescent-mode devices. This covers basic principles and the physics of operation, implementation and fabrication techniques, and numerous applications addressing state-of-the-art problems.

The book chapters are divided into the three parts mentioned above. Chapters 1 and 2 cover the basic theories of filters and evanescent-mode resonators. Various fabrication processes are then discussed in Chapters 3–5. Filtering applications are discussed in Chapters 6–9, which include advanced design structures. Applications reaching beyond basic filtering functions are in Chapters 10 (passive) and 11 (active). The limitations of evanescent-mode technologies are addressed with monitoring and control techniques in Chapter 12. The contents are supported with numerous visuals that demonstrate simulated and measured performance of physically built prototypes.

We hope you find this book helpful. We would welcome your suggestions for improvements or corrections at dperouli@purdue.edu or mabukhater@ieee.org.

West Lafayette, IN
August 2025

*Dimitrios Peroulis*
*and*
*Mohammad Abu Khater*

## Acknowledgments

This book would not have been possible without the generous financial support of many federal agencies that supported the work performed by the senior author's graduate students over the past two decades. DARPA, ONR, ARL, AFRL, NSF, and NASA supported multiple projects in this area. In the early stages of this work, approximately between 2006 and 2011, Dr. William (Bill) Chappell was a principal or coprincipal investigator and close collaborator in many of these projects. Bill's contributions to this work have been invaluable. Many of the innovative ideas described in this book came from Bill and his graduate students. His leadership, passion, and profound expertise have been instrumental in shaping the development of the field as a whole. We also owe a heartfelt thank-you to Prof. Raafat Mansour. Through his books, research papers and presentations, Prof. Mansour has gifted us with invaluable filter design knowledge and advice. Most importantly, his constant friendship, mentorship, encouragement, and positive energy have been key to our motivation and resilience over the years. We are also grateful to Prof. Gabriel Rebeiz who has made groundbreaking contributions in this area. It is worth noting that in some of the early contracts Prof. Rebeiz's team was also part of one of the performing teams. We have indeed learned so much from his approach to filters and RF engineering in general. Prof. Linda P.B. Katehi and Prof. Kamal Sarabandi have been an inspiration throughout the development of this work. We have learned many of the fundamental RF design principles from their pioneering contributions. Moreover, preliminary filter design/tuning ideas can be found in Prof. Peroulis' dissertation completed under their supervision as well as in other students' doctoral work that showed evanescent-mode substrate-integrated waveguide resonators and filters during the mid-1990s and early 2000s.

We are particularly grateful to the many Purdue University graduate students and researchers whose work has directly or indirectly contributed to the knowledge included in this book. These include, in approximate chronological order, Dr. Himanshu Joshi, Prof. Hjalti Sigmarsson, Prof. Xiaoguang (Leo) Liu,

Dr. Juseop Lee, Dr. Yusha Bey, Prof. Muhammad Shoaib Arif, Prof. Kenle Chen, Prof. Dimitra Psychogiou, Dr. Eric Naglich, Dr. Tsung-Chieh Lee, Dr. Wesley Allen, Dr. Juan Zeng, Dr. Yu-Ting Huang, Dr. Mark Hickle, Dr. Yu Chen Wu, Prof. Mohamed Hagag, Dr. Wei Yang, Dr. ZhengAn (Andy) Yang, Dr. Mahmoud Abdelfattah, and Dr. Michael Sinanis. You may find more details of their contributions in the publications referenced in this book. We also acknowledge that we learned a lot about filter design and implementation from the doctoral works of Dr. Lee Harle and Dr. Andrew Brown. Moreover, we acknowledge that much of the presented work has also been completed at the Purdue University Birck Nanotechnology Center (BRK). The staff members at BRK have provided significant help and technical expertise toward this work.

As we close this short section, we wish to express our most important thanks to our families who are our constant anchors and faithful direction providers. We would not have undertaken this project without their consent and support. I am particularly grateful to my wife, Alina, whose love and support was so critical in even taking the first step. Thank you!

## List of Abbreviations

| Abbreviation | Meaning |
|---|---|
| ABS | acrylonitrile butadiene styrene |
| ABW | absolute bandwidth |
| ADC | analog-to-digital converter |
| ASTM | American Society for Testing and Materials |
| BAL | balanced |
| BPF | bandpass filter |
| BSF | bandstop filter |
| BST | barium strontium titanate |
| CMR | common-mode rejection |
| CMRR | common-mode rejection ratio |
| CNC | computer numerical control |
| CP | charge pump |
| CPW | coplanar waveguide |
| CVD | chemical vapor deposition |
| CW | continuous wave |
| DAC | digital-to-analog converter |
| DRIE | deep reactive-ion etching |
| ECL | emitter-coupled logic |
| EVA | evanescent mode |
| FBW | fractional bandwidth |
| FDD | frequency-division duplexing |
| FEM | finite element method |
| FPFA | field-programmable filter array |
| FPGA | field-programmable gate array |
| GCPW | grounded CPW |
| IL | insertion loss |
| IMN | input matching network |
| LNA | low-noise amplifier |

| | |
|---|---|
| LPF | low-pass filter |
| LTE | long-term evolution |
| MCD | micro-corrugated diaphragms |
| MEMS | micro-electro-mechanical systems |
| MFR | mode frequency ratio |
| NA | network analyzer |
| NRN | non-resonating node |
| OMN | output matching network |
| PA | power amplifier |
| PCB | printed circuit board |
| PFD | phase-frequency detector |
| Qe | external quality factor |
| Qu | unloaded quality factor |
| SE | single ended |
| SIW | substrate-integrated waveguide |
| TMAH | tetramethylammonium hydroxide |
| TZ | transmission zero |
| VCO | voltage-controlled oscillator |
| VNA | vector network analyzer |
| VSWR | voltage standing wave ratio |
| YIG | yttrium–iron–garnet |

# About the Companion Website

This book is accompanied by a companion website:

**www.wiley.com/go/Peroulis/TunableEvanescent-ModeFilters**

This website contains a playlist for 31 videos. The learning objectives are as follows:

- Lecture 1–20:
  Articulate the basic design principles of RF/microwave filters.
  Follow the fundamental design principles to design practical RF/microwave filters.
- Lecture 21–31:
  Articulate the basic design principles based on the coupling matrix method.
  Follow the coupling matrix method to design advanced RF/microwave filters.

# Part I

# Principles

# 1

# Background

## 1.1 Introduction

### 1.1.1 Filters Necessity

Each component in a wireless chain contributes to increasing at least one of the following: power consumption, noise, size, or nonlinearity. It is clearly desirable then to minimize the number of components. Filters are no exception, adding to the size and cost of the system. Nearly all wireless transceivers contain several filter types. This is primarily because filters provide indispensable functions to transceivers. We discuss a few examples below.

Wireless receivers need to accommodate a wide dynamic range of received signal powers. For example, the required adjacent channel leakage in Long Term Evolution (LTE) can be as high as −50 dBm (from a nearby user, for example), while the received signal can be below −100 dBm. To put that into perspective, the sun is approximately 400,000 times brighter than the full Moon. This translates to 56 dB difference in brightness. Such a large input range can cause low-noise amplifiers (LNAs) to become non-linear, which can result in intermodulation or distortion. Channel filters can help by reducing leakage to a level where the LNA can operate with acceptable linearity, ensuring proper reception. This is typically achieved with bandpass fliters (BPFs).

In bidirectional communication systems (e.g., cellular phones), it is desirable to split transmit and receive frequencies. This allows frequency division multiplexing (FDM). Transmitted power is usually much higher than received power. This can cause numerous problems including saturating or even damaging the receiver in some applications. We can limit transmitter leakage to the receiver path by employing diplex filters.

*Tunable Evanescent-Mode Filters: Principles, Implementation, and Applications*,
First Edition. Dimitrios Peroulis and Mohammad Abu Khater.

Companion website: www.wiley.com/go/Peroulis/TunableEvanescent-ModeFilters

Power amplifiers in transmitters usually operate in the nonlinear region to optimize their power efficiency. Nonlinearity at the transmitter can result in out-of-band emissions, which can violate transmission regulations. Those emissions can be suppressed using filters.

Although the examples mentioned above are very common, they are not an exhaustive list. The important role of filters in a communication system cannot be easily replaced by alternatives. Significant research efforts, however, are investigating filterless wireless front-ends. This is discussed in the next section.

### 1.1.2 Alternative Filtering Methods

While tunable and reconfigurable filters have been under development in recent years, other dynamic spectral isolation technologies have also been, and continue to be, investigated. The development of these technologies has been driven by the desire to eliminate filters from RF front-ends. This is particularly important in applications that demand minimal system size, such as consumer electronics, where the cost of off-chip filtering is also a major concern. Moreover, there are high-end applications that require significant isolation, such as the close physical proximity of airborne radars and satellite communication systems.

We discuss below some of the technologies that have been investigated including antenna isolation baffles, mixer-first receiver designs, and self-interference cancellation through beamforming.

**Tunable antenna isolation baffles** have been used to obtain up to 60 dB of narrow-band isolation between co-located antennas and elements of an antenna array [1]. These devices are physically placed between antennas to shift the phase of coupling energy so that it destructively interferes at the receive antenna(s). The phase shift is made tunable through the use of varactors that load the baffles, making them resonant near the frequency of interest. In [1], more than 40 dB isolation was achieved over 10 MHz bandwidths from 3.2 to 3.4 GHz at a variety of scan angles. Such capability can be very useful to increase the transmit-receive isolation in microwave systems. However, due to its reliance on destructive interference, it is difficult to adjust the shape and bandwidth of the high isolation region of the spectrum that results from this technique. Therefore, such a concept could provide valuable supplemental isolation to a reconfigurable filter in diplexing applications, but it remains to be shown that it could replace filtering completely.

**Mixer-first** receiver designs, that do not require RF filtering to operate in some environments, are also under active development [2–6]. These receivers use the impedance-transformation property of passive mixers to implement high-quality-factor filtering by transforming low-quality-factor baseband impedances to RF. Such designs chop a cycle of the clock into multiple pieces with non-overlapping pulses and have distinct advantages over tunable and reconfigurable filters. For example, the center frequency of the passband response is directly controlled by a clock frequency, which is easier to manage than most tunable resonators. In addition, the structures are implemented with switches and capacitors only, enabling them to be linear and designed in integrated circuit technology. Ideally, since no DC current passes through the switches, flicker noise is not a concern. However, receivers that implement mixer-first designs also have some limitations. First, many circuit parameters and performance metrics that are tied to them are in direct competition with each other. Some of these trade-offs are similar to those of classical filters, such as bandwidth versus insertion loss (IL) and selectivity versus noise figure. An increase in selectivity can reduce the level of stopband rejection or decrease the bandwidth over which the system's antenna can be instantaneously impedance matched, as their parameters are linked [7]. In classical filters, stopband rejection is often limited by adjacent coupling paths and is less dependent on selectivity. In addition, there is a theoretical advantage to chopping a clock cycle into numerous short pulses. However, generating multiple non-overlapping pulses that are fractions of the clock frequency is difficult at high frequencies. With current technologies, these systems show excellent performance below 1 GHz and tend to degrade at higher frequencies. Nevertheless, mixer-first designs show great promise for filterless operation in some environments and as isolation supplements to reconfigurable filters.

**Near field cancellation** at receive antenna locations through beamforming has also been recently investigated as a method for increasing isolation between microwave systems. In these systems, antenna excitations are synthesized so that desired far field radiation patterns are maintained while near field patterns destructively interfere at the location of receive antennas. In [8], more than 50 dB isolation improvement was achieved over a 15 MHz bandwidth from 3.1 to 3.6 GHz using beamforming techniques. However, these techniques are limited to narrow bandwidths and require multiple transmit antennas, which may not be possible to implement in some systems.

The techniques discussed above, and others like them, add valuable isolation capability to microwave systems. Nonetheless, their bandwidth limitations, linked parameters and specifications, difficulty/cost in implementation, and/or requirement for multiple antennas or circuit paths lead to capability levels that cannot replace conventional filtering in many applications. While these techniques are important options for supplementing filter technology, reconfigurable filters will likely remain an integral part of dynamic systems, particularly in high-interference environments for the foreseeable future.

## 1.2 Filter Anatomy and Representation

### 1.2.1 The Basic Coupling Matrix (*M* Matrix)

In the early days of filter synthesis (1920s–1970s), nearly all techniques involved the extraction of electrical elements (lumped capacitors and inductors, as well as transmission line lengths) from the polynomials that represented the filter's electrical performance [9]. This method was adequate for the technologies and synthesis demands of the era, but it involved element-by-element extraction of the circuit network and in many cases demanded starting designing from the beginning when a characteristic of the network needed to be changed. As communication systems became more advanced and prolific, new filter synthesis techniques were developed in order to aid designers in meeting increasingly difficult specifications that included innovations like transmission zeros at designed frequencies and group delay equalization across the passband of bandpass filters. One of these advanced filter synthesis methods was developed in 1974 by Atia and Williams [10] to help design filters with challenging specifications associated with satellite communications. The coupling matrix provides a one-to-one correspondence between its elements and the physical resonators and coupling structures of the filter. This is a significant advancement beyond element-by-element extraction because it allows direct modeling of both the resonators (elements) of a filter and all of their couplings, thus enabling faster synthesis of advanced filtering functions. Cameron later developed general techniques to synthesize and generate the coupling matrix in an efficient fashion. This was performed in the low-pass domain, where different topologies may be conveniently obtained using similarity transformations [11, 12].

While the original coupling matrix assumes lossless, dispersion-free, synchronously tuned resonators, and frequency-independent coupling elements, practical considerations can be added to different elements of the coupling matrix. Of particular interest is the ability to analyze and synthesize filters with asynchronously-tuned resonators, or resonators that are individually tuned to different frequencies. This can result in lower complexity and/or cost in tunable filters.

The coupling matrix is a mathematical representation of the relative strengths of coupling between the resonators or elements of a filter. It can be derived from the voltage and current relationships between the elements of a generic equivalent circuit representation of the filter network. Such a generic equivalent circuit representation can be seen in Figure 1.1. The circuit is driven by an open-circuit voltage source $v_S$ with a source resistance of $R_S$ and terminated by a load resistance $R_L$. Individual resonators are composed of 1 Henry (H) inductors and 1 Farad (F) capacitors, producing a normalized resonant frequency of 1 radian/second (rad/s). The current in each resonator is labeled as $i_n$, where $n$ denotes the resonator number. The resonators are coupled to each other, and each coupling is denoted as $M_{m,n}$, where $m$ and $n$ denote resonator numbers. These couplings are set such that the bandwidth of the filter

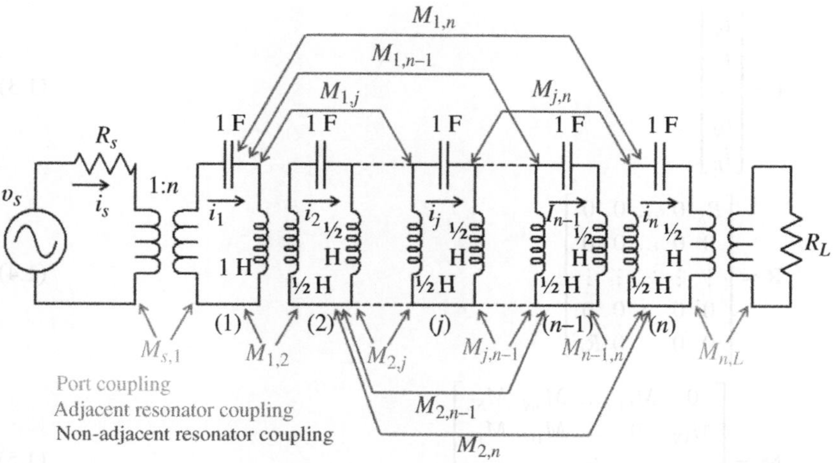


**Figure 1.1** Equivalent circuit model of series- and cross-coupled resonators and marked coupling (*M*) values.

is set to a normalized range from −1 rad/s to 1 rad/s and are assumed to be frequency-independent.

The voltage-current relationship in and between the ports and resonators of the network shown in Figure 1.1 can be expressed in matrix form as shown in equation (1.1),

$$\begin{bmatrix} v_S \\ 0 \\ \vdots \\ 0 \\ 0 \end{bmatrix} = \begin{bmatrix} R_S & jM_{S1} & \cdots & jM_{SN} & jM_{SL} \\ jM_{1S} & j\left(\omega - \frac{1}{\omega}\right) & \cdots & jM_{1N} & jM_{1L} \\ \cdots & \cdots & \ddots & \cdots & \cdots \\ jM_{NS} & jM_{N1} & \cdots & j\left(\omega - \frac{1}{\omega}\right) & jM_{NL} \\ jM_{LS} & jM_{L1} & \cdots & jM_{LN} & R_L \end{bmatrix} \begin{bmatrix} i_S \\ i_1 \\ \vdots \\ i_N \\ i_L \end{bmatrix} \tag{1.1}$$

where $j(\omega - \frac{1}{\omega})$ is the bandpass transformation with center frequency and bandwidth normalized to 1 [13]. The voltage-current relationship in equation (1.1) can be written in simplified form as

$$\mathbf{v} = \mathbf{Z}\mathbf{i} = \left(j\left(\omega - \frac{1}{\omega}\right)\mathbf{U} + \mathbf{R} + j\mathbf{M}\right)\mathbf{i} \tag{1.2}$$

where **v** is the vector matrix on the left-hand side of equation (1.1), **U** is the identity matrix except that $U_{S,S}$ and $U_{L,L}$ are equal to zero, and **i**, **R**, and **M** are

$$\mathbf{i} = \begin{bmatrix} i_S \\ i_1 \\ \vdots \\ i_N \\ i_L \end{bmatrix} \tag{1.3}$$

$$\mathbf{R} = \begin{bmatrix} R_S & 0 & \dots & 0 & 0 \\ 0 & 0 & \dots & 0 & 0 \\ \vdots & \vdots & \ddots & \vdots & \vdots \\ 0 & 0 & \dots & 0 & 0 \\ 0 & 0 & \dots & 0 & R_L \end{bmatrix} \tag{1.4}$$

$$\mathbf{M} = \begin{bmatrix} 0 & M_{S1} & \dots & M_{SN} & M_{SL} \\ M_{1S} & 0 & \dots & M_{1N} & M_{1L} \\ \dots & \dots & \ddots & \dots & \dots \\ M_{NS} & M_{N1} & \dots & 0 & M_{NL} \\ M_{LS} & M_{L1} & \dots & M_{LN} & 0 \end{bmatrix} \tag{1.5}$$

respectively. The **M** matrix in equation (1.5) is called the coupling matrix, and it can be used to completely specify the narrow-band behavior of a filter network, assuming it is composed of synchronously-tuned similar resonators. Here, similar resonators are resonators with the same quality factors, coupling structures, and characteristic impedances. The coupling matrix is a mathematical representation of the coupling values between each of the similar resonators.

It should be noted here that the transmission coefficient ($S_{21}$) can be extracted from the coupling matrix as

$$S_{21}(s) = 2(\mathbf{R} + s\mathbf{U} + j\mathbf{M})^{-1}_{N+2,1} \tag{1.6}$$

where $s$ is the complex frequency. The coupling values are then extracted by equating the numerator and denominator polynomials, such that the desired response (e.g., Chebyshev) is achieved.

Similarly, the value of the reflection coefficient ($S_{11}$) can be calculated from

$$S_{11}(s) = 1 - 2(\mathbf{R} + s\mathbf{U} + j\mathbf{M})^{-1}_{1,1} \tag{1.7}$$

This can be used to find $S_{11}$ once the values of the coupling matrix are found.

### 1.2.2 Coupling-Routing Diagrams

To aid in visualization of the mathematical coupling matrix representation of a given filter, it is common to draw coupling-routing diagrams. Coupling-routing diagrams are a visual representation of the nodes in the circuit that are coupled to each other, along with the relative signs of their coupling values. Nodes of a filter network are the ports, resonators, and non-resonant nodes of the circuit. Relative signs of coupling values and non-resonant nodes are described further in subsections below.

An example of a coupling-routing diagram for a common two-pole series-coupled bandpass filter topology can be seen in Figure 1.2. Such a filter topology can produce a variety of filter responses depending on the coupling values, including Butterworth and Chebychev filter responses. Note that in Figure 1.2, as will be the case in the remainder of this book, the ports of the filter are represented by white circles. Here, $S$ represents the source port, and $L$ represents the load port. It is common to replace $S$ and $L$ with 0 and $N + 1$, where $N$ is the number of resonators in the filter network. Resonators in Figure 1.2 are represented by black circles and are numbered sequentially. Coupling between ports and resonators is represented by solid black lines as all coupling values in this topology are positive. Negative coupling is described in a subsection below. Note that the basic operation of the circuit can be intuitively understood from the coupling-routing diagram. For example, in Figure 1.2, a signal starts at the source ($S$) port. It progresses to resonator 1, which is a series-coupled resonator. A series-coupled resonator is a short circuit at resonance and an open circuit off resonance. Therefore, a signal at the resonant frequency of the resonators progresses through the circuit, while signals at frequencies away from the resonant frequency are reflected back to the source port of the network. This creates a bandpass response, and the specific characteristics of the bandpass response are determined by the coupling ($M$) values.

Similar to Figure 1.2, Figure 1.3 shows the coupling-routing diagram for a common two-pole shunt-coupled bandstop filter (BSF) topology. Note that while labeling and nomenclature are the same for both Figure 1.2 and Figure 1.3, Figure 1.3 represents a different frequency response. There is no direct path between the resonators anymore, and a direct path between the source and load ports of the network was added. Therefore, an input signal at the resonant frequency of the resonators is reflected back to the source due to the resonators in the shunt configuration. Conversely, an input signal at a frequency away from the resonant frequency passes directly from the source to the load, ideally unchanged except for a potential phase shift. Most, if not all, of more complicated resonator-based filters can

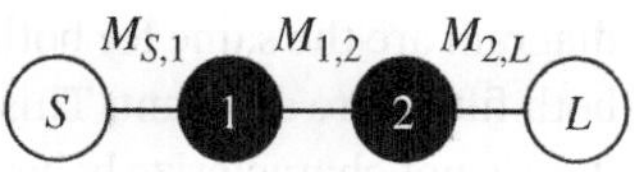


**Figure 1.2** Coupling-routing diagram for a common 2-pole series-coupled bandpass filter topology.

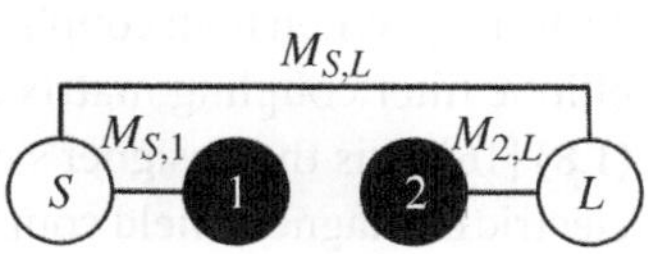


**Figure 1.3** Coupling-routing diagram for a common 2-pole shunt-coupled bandstop filter topology.

be represented with coupling-routing diagrams by adding more resonators to the networks shown in Figures 1.2 and 1.3 along with potentially further slight modifications. Therefore, the coupling matrix and coupling-routing diagram provide a mathematically rigorous and intuitive understanding of a narrow-band filter. Due to their complementary nature, one is almost always shown alongside the other.

### 1.2.3 Additions to the Coupling Matrix for Synthesis of Advanced and Practical Filter Responses

This section is adopted from [168]. While there are expansive tomes and dozens of scholarly articles on useful modifications to the coupling matrix [14, 15], here we focus on four specific additions that are pertinent to the work described in the later chapters of this book. These four additions are positive and negative coupling values, finite resonator quality factors, resonator frequency tuning, and non-resonating nodes. As we will see, these additions enable advanced filtering functions with transmission zeros, accurate estimation of loss and band-edge sharpness, using the off-resonance reactance of a resonator, and accurate representation of resonators that are coupled through multiple coupling structures or over long electrical phase lengths.

#### 1.2.3.1 Positive and Negative Coupling Values

Positive and negative coupling values refer to the relative insertion phase difference between various coupling values. Because of the narrow-band, dispersion-free approximation that is made when using the $M$ matrix method of filter synthesis, all coupling values are assumed to have either −90 electrical degrees or +90 electrical degrees insertion phase. Except in the case of acoustic or other types of non-electrical resonators, all coupling is achieved through magnetic (inductive) or electric (capacitive) fields, or a combination of both. One field type is arbitrarily assigned to a positive coupling, and the other is arbitrarily assigned to negative coupling. If only one of either electric or magnetic field coupling is used in a filter design, that type of coupling is often assigned to be a positive coupling by convention. For example, a Butterworth response filter, represented by a single coupling matrix, can be implemented in two ways: with all electric field coupling or all magnetic field coupling. While the coupling matrix and coupling-routing diagram are the same for both cases, the out-of-band and tuning characteristics of both filters are different. This is a limitation of the basic coupling matrix method; it does not characterize behavior away from the center frequency of the filter very accurately. When both coupling types are present in a filter design, as shown in the elliptic filter coupling matrix and frequency response in Figure 1.4 and equation (1.8) [16], it is the designer's choice whether to assign positive coupling to either electric or magnetic field coupling. The transmission zeros seen in the response in

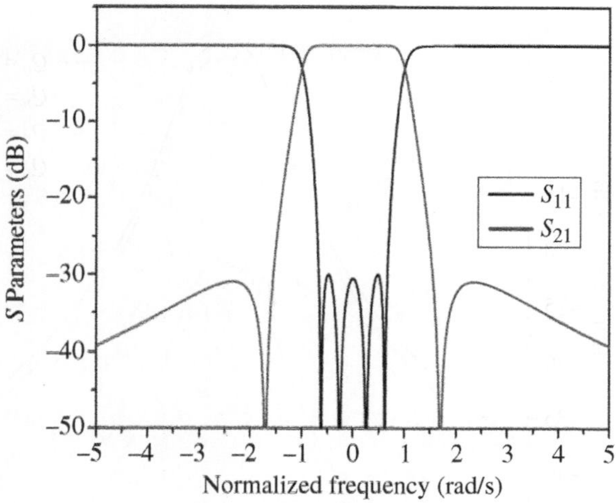


**Figure 1.4** Frequency response of a 4-pole elliptic bandpass filter with transmission zeros at ±1.7 rad/s and 30 dB equi-ripple return loss.

Figure 1.4 are a result of having paths between the source and load ports of the filter utilizing both positive and negative couplings. When these two paths converge at the load port, destructive interference results at the frequency or frequencies where these paths are 180° out of phase, creating the transmission zero(s).

Often, if all electric field coupling values are changed to magnetic field coupling values and all magnetic field coupling values are changed to electric field coupling values, the filter will have the same narrow-band response around its center frequency but different characteristics out of band. In a mixed-coupling topology with both electric and magnetic field couplings, it is common to assign positive coupling values to magnetic field coupling. This convention will be adopted for the remainder of this book.

$$\mathbf{M} = \begin{bmatrix} 0 & 1.27 & 0 & 0 & 0 & 0 \\ 1.27 & 0 & 0.763 & 0 & -0.143 & 0 \\ 0 & 0.763 & 0 & 0.617 & 0 & 0 \\ 0 & 0 & 0.617 & 0 & 0.763 & 0 \\ 0 & -0.143 & 0 & 0.763 & 0 & 1.027 \\ 0 & 0 & 0 & 0 & 1.027 & 0 \end{bmatrix} \tag{1.8}$$

#### 1.2.3.2 Finite Resonator Quality Factors

The basic $M$ matrix filter synthesis method assumes lossless resonators. In any real filter implementation, the resonators will have loss that will affect the insertion loss of the filter and the sharpness of its band edges (Figure 1.5). Additionally, in some filter architectures such as absorptive bandstop filters [17], the unloaded quality factor ($Q_u$) is a design parameter that must be considered. In order to model these effects and filter topologies accurately, $Q_u$ must be added to the $M$ matrix.

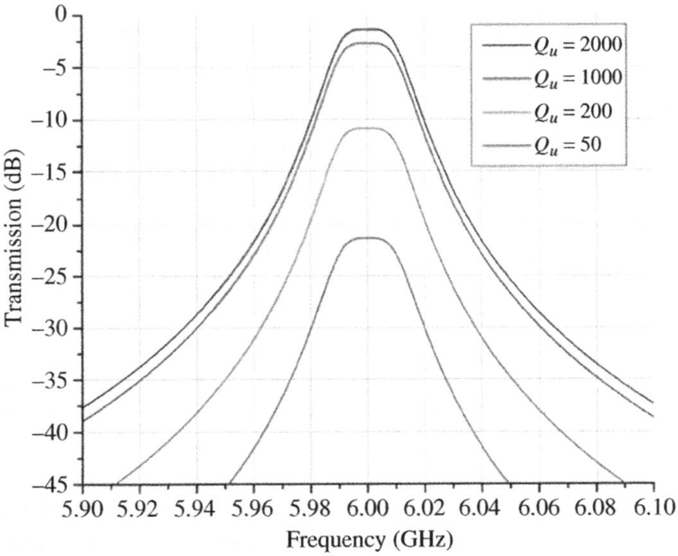


**Figure 1.5** The transmission performance of a filter with a 25 MHz bandwidth and a center frequency of 6 GHz as a function of four different unloaded quality factor values.

It is important to note that $Q_u$-related effects also frequently have a connection to the bandwidth of a filter. For example, a 2% fractional bandwidth $n$-pole filter with a given resonator $Q_u$ will have significantly more insertion loss than a 10% fractional bandwidth $n$-pole filter with the same resonator $Q_u$. However, the $M$ matrix filter synthesis method is normalized such that the filter response is centered at zero rad/s with a bandwidth from −1 rad/s to 1 rad/s. The filter response is scaled at the desired center frequency and bandwidth after the synthesis procedure is complete. In order to account for the $Q_u$-bandwidth relationship during the synthesis procedure, terms are added to the $M$ matrix that depend on both $Q_u$ and the fractional bandwidth ($FBW$) of a filter. For a four-pole series-coupled resonator bandpass filter topology with a finite quality factor, the $M$ matrix can be written as:

$$\mathbf{M} = \begin{bmatrix} 0 & M_{01} & 0 & 0 & 0 & 0 \\ M_{01} & \dfrac{-j}{Q_u\ FBW} & M_{12} & 0 & 0 & 0 \\ 0 & M_{12} & \dfrac{-j}{Q_u\ FBW} & M_{23} & 0 & 0 \\ 0 & 0 & M_{23} & \dfrac{-j}{Q_u\ FBW} & M_{34} & 0 \\ 0 & 0 & 0 & M_{34} & \dfrac{-j}{Q_u\ FBW} & M_{01} \\ 0 & 0 & 0 & 0 & M_{01} & 0 \end{bmatrix} \tag{1.9}$$

where $\frac{-j}{Q_u\ FBW}$ takes the $Q_u$ and $FBW$ into account. Finite $Q_u$ responses can be synthesized using this term.

#### 1.2.3.3 Resonator Frequency Tuning

Of particular interest to the work shown later in this book is the ability to synthesize filter responses composed of resonators that are tuned asynchronously, or tuned to different frequencies. Similar to how the finite quality factor was added to the $M$ matrix in equation (1.9), asynchronous tuning can be added to the $M$ matrix synthesis procedure using the self-coupling terms ($M_{NN}$). A four-pole series-coupled resonator bandpass filter topology with asynchronous resonator tuning has an $M$ matrix of the form:

$$\mathbf{M} = \begin{bmatrix} 0 & M_{01} & 0 & 0 & 0 & 0 \\ M_{01} & M_{11} & M_{12} & 0 & 0 & 0 \\ 0 & M_{12} & M_{22} & M_{23} & 0 & 0 \\ 0 & 0 & M_{23} & M_{33} & M_{34} & 0 \\ 0 & 0 & 0 & M_{34} & M_{44} & M_{01} \\ 0 & 0 & 0 & 0 & M_{01} & 0 \end{bmatrix}. \tag{1.10}$$

Consequently, the actual frequency, due to the self-tuning frequency shift, is given by [18]

$$f_i = f_0\left[\sqrt{1+\left(\frac{M_{ii}\times FBW}{2}\right)^2} - \frac{M_{ii}\times FBW}{2}\right] \tag{1.11}$$

where $f_i$ is the $i$th resonator frequency, $f_0$ is the nominal center frequency of the design, $M_{ii}$ is the self-coupling coefficient of the $i$th resonator. For the general case where $M_{ii} = 0$, the resonant frequency is at $f_0$.

It is important to note that the self-coupling terms ($M_{NN}$) in equation (1.10) are real numbers, whereas in equation (1.9) the self-coupling terms are imaginary. A real self-coupling term denotes frequency tuning of a specific resonator away from the center frequency of the filter response, while an imaginary self-coupling term denotes loss in a particular resonator. Note that the $M$ matrix is multiplied by a factor of $j$ in equation (1.2), reversing the real and imaginary roles of the self-coupling terms in the final voltage-current relationship in line with standard frequency domain analysis. The self-coupling terms can have both a real and imaginary parts (complex) to model both finite $Q_u$ and asynchronous tuning at the same time.

#### 1.2.3.4 Non-Resonating Nodes

Because the $M$ matrix filter synthesis method is a narrow-band approximation, all coupling values have a phase shift of exactly $\pm 90°$. However, in some filter networks, there are coupling sections or transmission lines that have a phase shift that is an integer multiple of $\pm 90°$ as well as coupling sections or transmission lines that have a phase shift that is not an integer multiple of ±90°. Non-resonating nodes are used to model coupling sections or transmission lines that have a phase shift that is an integer multiple of ±90°. Coupling sections and transmission lines that have a phase shift that is not an integer multiple of ±90° cannot be modeled by the $M$ matrix synthesis method alone. Separate and hybrid analysis methods have been developed to address this situation [19]. One separate and one hybrid method are discussed later in this book.

In this book, non-resonating nodes are used primarily in similar-resonator bandstop filters. To understand why they are needed, consider the four-pole lumped element bandstop filter shown in Figure 1.6(a). The filter shown in Figure 1.6(a) is shown in a completely lumped topology, but it has two resonator types. Often, it is most practical for all resonators in a filter to be of the same type. Therefore, it is common to transform the circuit to one with similar resonators using an impedance inverter, which can be implemented using a 90°-transmission line. The resultant topology is shown in Figure 1.6(b). Note that there are three connected 90°-transmission lines that could be viewed as a single 270°-transmission line. In the coupling-routing diagrams of Figures 1.2 and 1.3, coupling values can only span ports and resonators. However, since coupling values in the $M$ matrix method can only have an insertion phase of ±90°, there is no way to represent three 90°transmission lines in series or a 270°-transmission line. While a −90°-transmission line would have the correct insertion phase, the

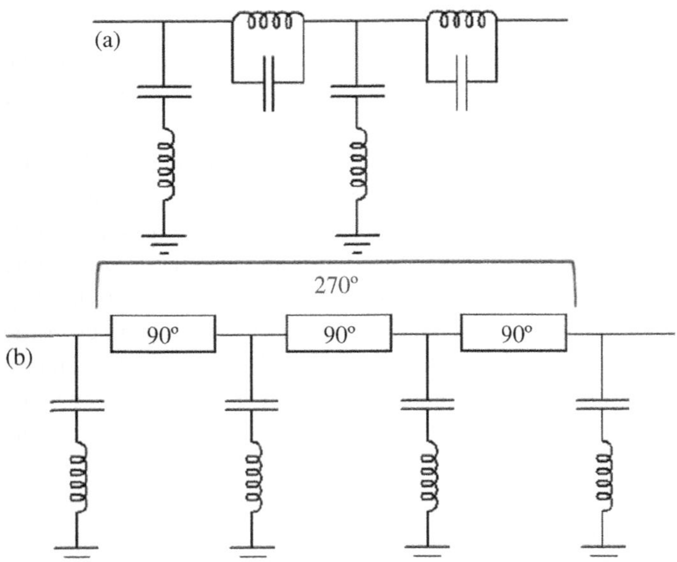


**Figure 1.6** Lumped element bandstop filters. (a) Completely lumped with dissimilar resonators. (b) Lumped element, similar resonators separated by 90°-transmission lines.

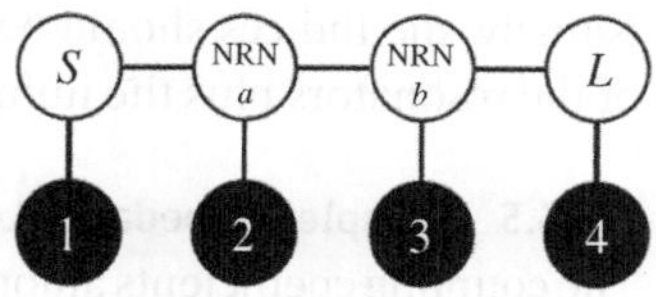


**Figure 1.7** Coupling-routing diagram for the filter topology shown in Figure 1.6(b). NRN = non-resonating node.

topology of the circuit would not be upheld as the four resonators in Figure 1.6(b) are not connected to the same points in the circuit. Non-resonating nodes fill this need. The coupling-routing diagram for the filter topology in Figure 1.6(b) can be seen in Figure 1.7. The non-resonating nodes (NRNs) allow a node for the connection of a 90°-coupling element. The corresponding coupling matrix would have eight rows and eight columns. However, the U matrix from equation (1.2) is no longer the identity matrix with $U_{S,S}$ and $U_{L,L}$ equal to zero when NRNs are used. Instead of producing a frequency-dependent term in each non-port row as before ($j(\omega - \frac{1}{\omega})$) in equation (1.2), the values in the U matrix corresponding to the rows of the NRNs must be changed to zeros as in equation (1.2) where the NRNs are given element numbers (2,2) and (4,4). Note that the NRNs can be given any diagonal element number in the U matrix as long as coupling values in the associated $M$ matrix correspond to the chosen diagonal element numbers. The corresponding $M$ matrix for equation (1.12) can be seen in equation (1.13), where the source is node 0, the load is node 7, and the NRNs are nodes $a$ and $b$. The lack of a resonance term in the chosen diagonal element numbers gives rise to the name "non-resonating node".

$$\mathbf{U} = \begin{bmatrix} 0 & 0 & 0 & 0 & 0 & 0 & 0 & 0 \\ 0 & 1 & 0 & 0 & 0 & 0 & 0 & 0 \\ 0 & 0 & 0 & 0 & 0 & 0 & 0 & 0 \\ 0 & 0 & 0 & 1 & 0 & 0 & 0 & 0 \\ 0 & 0 & 0 & 0 & 0 & 0 & 0 & 0 \\ 0 & 0 & 0 & 0 & 0 & 1 & 0 & 0 \\ 0 & 0 & 0 & 0 & 0 & 0 & 1 & 0 \\ 0 & 0 & 0 & 0 & 0 & 0 & 0 & 0 \end{bmatrix} \tag{1.12}$$

$$\mathbf{M} = \begin{bmatrix} 0 & M_{01} & M_{0a} & 0 & 0 & 0 & 0 & 0 \\ M_{01} & M_{11} & 0 & 0 & 0 & 0 & 0 & 0 \\ M_{0a} & 0 & 0 & M_{a2} & M_{ab} & 0 & 0 & 0 \\ 0 & 0 & M_{a2} & M_{22} & 0 & 0 & 0 & 0 \\ 0 & 0 & M_{ab} & 0 & 0 & M_{b3} & 0 & M_{b7} \\ 0 & 0 & 0 & 0 & M_{b3} & M_{33} & 0 & 0 \\ 0 & 0 & 0 & 0 & 0 & 0 & M_{44} & M_{47} \\ 0 & 0 & 0 & 0 & M_{b7} & 0 & M_{47} & 0 \end{bmatrix} \tag{1.13}$$

The coupling values extraction shown in equation (1.6) is still valid here. The matrix indices used, however, should accommodate the presence of the NRN.

Namely, the indices should be $(N + NRN + 2), 1$, where $N + NRN$ is the number of the resonators plus the number of non-resonating nodes.

#### 1.2.3.5 Complex Impedance Loads

The coupling coefficients among nodes for an extended coupling matrix (including the additions above) are shown in Figure 1.8. It can be shown that the impedance seen from the left-hand side of the NRN looking into the load node is

$$Z' = \frac{1 + jB_L}{M_{LL}^2} \tag{1.14}$$

where $B_{NRN} = 0$, and $M_{LL}$ represents the coupling coefficient between the load node and the NRN. From equation (1.14), we can see that the resistance depends only on $M_{LL}$, while both $M_{LL}$ and $B_L$ determine the reactance of the complex impedance.

The values of $R_S$ and $R_L$ are equal to 1 in equation (1.4), meaning the network system is normalized to 1 Ω. However, the complex source impedance is considered and embedded in the extended coupling matrix.

$M_{LL}$ and $B_L$ can be used to control the impedance value according to equation (1.14). $M_{LL}$ multiplies the numerators of the transmission coefficients by $j$ since $M_{LL}$ provides a $|90°|$ phase shift for $S_{21}$. The characteristic polynomials need to be modified as

$$\begin{aligned} P'(S) &= \pm jP(S)M_{LL}^2 \\ F'(S) &= F(S)M_{LL}^2 \\ E'(S) &= E(S)M_{LL}^2 \end{aligned} \tag{1.15}$$

where the positive sign corresponds to $M_{LL} < 0$ and the negative sign corresponds to $M_{LL} > 0$. The transmission and reflection coefficients are now given by

$$S_{21}(S) = \frac{P'(S)}{E'(S)}, \quad S_{11}(S) = \frac{F'(S)}{E'(S)} \tag{1.16}$$

It is shown in (1.15) and (1.16) that although the coefficients of the characteristic polynomials are scaled, the magnitudes of $S_{21}$ and $S_{11}$ are not affected. Furthermore, the effect of $M_{LL}$ cancels itself in the extended coupling matrix.

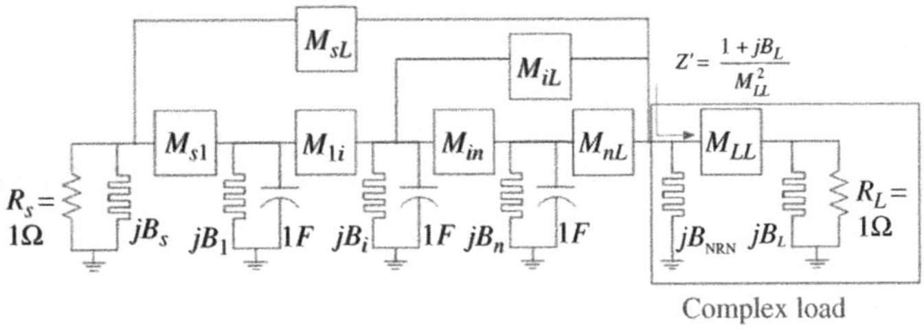


**Figure 1.8** Multi-coupled resonator network includes a complex load impedance in an extended $(n + 3) \times (n + 3)$ coupling matrix. Not all the coupling coefficients are shown.

In conclusion, the main advantages of this approach are

1. Separating the load and the filter using an NRN. The external coupling, $M_{nL}$, is unaffected when a complex impedance is inserted into an extended coupling matrix.
2. The characteristic functions are modified by a real number, $M_{LL}^2$.
3. The coefficient of the highest degree term of $E(S)$ is independent of $B_L$.
4. The port resistances are maintained to be 1 Ω, meaning the complex impedance is incorporated into a 1 Ω system.
5. An analytical solution that reduces complexity in the design process as well as the out-of-band interaction can be accurately described due to the prescribed filter response.

The design procedure is then summarized as follows. With a predetermined complex load, $M_{LL}$ and $B_L$ can be calculated using equation (1.14). The characteristic polynomials should be modified by the known $M_{LL}$ according to equation (1.15). Then the rest of the filter design parameters (coupling coefficients) in the extended coupling matrix can be derived using the well-known processes [14, 15, 20].

**Example 1.1** ***Complex Source Impedance***

To better explain this theory, we include two illustrative examples. First, we will show that the transmission and reflection coefficients derived from the methodology in this section of a 20 dB return loss second-order Chebyshev BPF with 1 Ω source and load impedances are the same as the ones derived from the traditional $(n+2)\times(n+2)$ coupling matrix. Indeed, for $n = 2$, the traditional coupling matrix is

$$\mathbf{M} = \begin{bmatrix} 0 & 1.2247 & 0 & 0 \\ 1.2247 & 0 & 1.6583 & 0 \\ 0 & 1.6583 & 0 & 1.2247 \\ 0 & 0 & 1.2247 & 0 \end{bmatrix}, \tag{1.17}$$

which yields

$$S_{21} = \frac{j4.9745}{4.9996 + 2.9998S + S^2}, \quad S_{11} = \frac{0.5003 + S^2}{4.9996 + 2.9998S + S^2} \tag{1.18}$$

Similarly, for $n = 2$, the extended coupling matrix (including the non-resonating node), for source and load impedances of 1 Ω, is

$$\mathbf{M} = \begin{bmatrix} 0 & 1 & 0 & 0 & 0 \\ 1 & 0 & 1.2247 & 0 & 0 \\ 0 & 1.2247 & 0 & 1.6583 & 0 \\ 0 & 0 & 1.6583 & 0 & 1.2247 \\ 0 & 0 & 0 & 1.2247 & 0 \end{bmatrix} \tag{1.19}$$

which yields

$$\begin{aligned} S_{21} &= \frac{4.9745}{4.9996 + 2.9998S + S^2} \\ S_{11} &= \frac{0.5003 + S^2}{4.9996 + 2.9998S + S^2} \end{aligned} \tag{1.20}$$

Equations (1.18) and (1.20) show the same magnitude frequency responses and validate equations (1.15) and (1.16).

An additional step illustrates the process of designing such a complex source impedance filter, where an arbitrary normalized complex source of $0.87 + j2.28\ \Omega$ is chosen. This complex source corresponds to $M_{LL}$ and $B_L$ of 1.07211 and 2.62069, respectively. In Figure 1.9, a skewed filter response is shown where a traditional Chebyshev BPF with 20 dB return loss (1 Ω system filter) is directly connected to this complex impedance using an extended coupling matrix. A non-traditional filter is needed to match this normalized complex source impedance to 1 Ω and to maintain the filter response. Following the extended coupling matrix design procedure we first derive the conventional characteristic polynomials $P(S)$, $F(S)$, and $E(S)$ of the second-order Chebyshev BPF with 20 dB return loss. These polynomials are then modified according to equation (1.16) to produce $P'(S)$, $F'(S)$, and

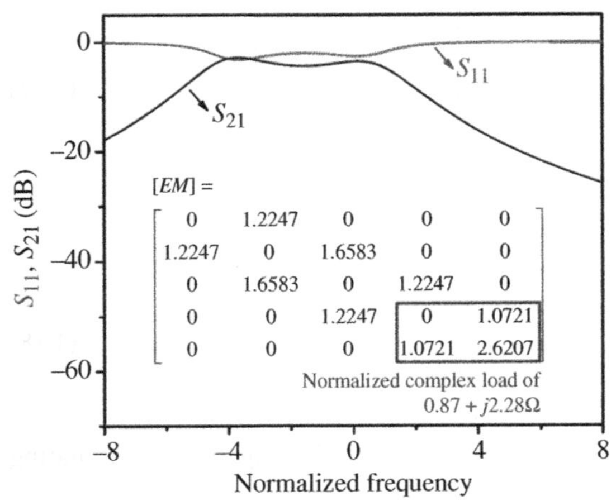


**Figure 1.9** S-parameters response when a Chebyshev BPF with 20 dB return loss is directly connected to a normalized complex source of $0.87 + j2.28\ \Omega$.

$E'(S)$. The $S$-parameters are then given by

$$S_{21} = \frac{j4.9745 \times (-j1.07211^2)}{(4.9996 + 2.9998S + S^2) \times 1.07211^2} \tag{1.21}$$

and

$$S_{11} = \frac{(0.5003 + S^2) \times (1.07211^2)}{(4.9996 + 2.9998S + S^2) \times (1.07211^2)}. \tag{1.22}$$

The remaining coupling coefficients in the extended coupling matrix can be found as

$$\mathbf{M} = \begin{bmatrix} 2.62069 & 1.07211 & 0 & 0 & 0 \\ 1.07211 & 0 & 1.313015 & 0 & 0 \\ 0 & 1.313015 & -3.93075 & 1.6583 & 0 \\ 0 & 0 & 1.6583 & 0 & 1.2247 \\ 0 & 0 & 0 & 1.2247 & 0 \end{bmatrix}. \tag{1.23}$$

The filter responses created by the traditional coupling matrix in equation (1.17), and the extended coupling matrix in equation (1.23), are as seen in Figure 1.10 with the solid and dashed lines, respectively. The traditional matrix is connected

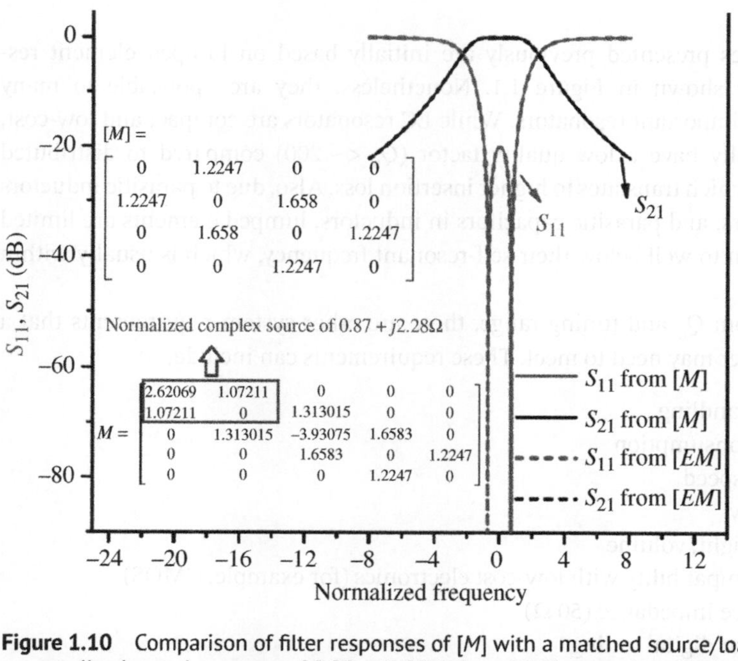


**Figure 1.10** Comparison of filter responses of $[M]$ with a matched source/load and with a normalized complex source of $0.87 + j2.28\ \Omega$ for a 20 dB return loss Chebyshev BPF.

to 1 Ω on both sides while the extended coupling matrix represents the filter with termination ports of $0.87 + j2.28\ \Omega$ on the source side and 1 Ω on the load. The fact that these two traces overlap means that by using the extended coupling matrix, we can design a filter to connect to a complex impedance while maintaining the specified filter response by increasing the external coupling coefficient on the source side from 1.2247 to 1.313015 and detuning the first resonator to −3.93075 in this example.

We would like to note again that this method relies on frequency-invariant reactive elements. As a result, it is narrow-band approximation. A complex load filter bandwidth depends on the relation of the load impedance slope, $\frac{\partial}{\partial\omega}Z_L(\omega)$, and the *FBW*. In other words, if the complex load impedance remains relatively constant within the filter's *FBW*, then the proposed methodology can be applied and will likely yield a practically useful network [21–24]. Combining the techniques described above, a BPF-Bandstop Filter (BSF) cascade can provide good performance in terms of independently tuning the passband and stopband and maximum flexibility of dynamically relocating the transmission zeros to increase the selectivity of the passband and tune the bandwidth.

## 1.3 Tunable Resonators in Filters

The theories presented previously are initially based on lumped element resonators, as shown in Figure 1.1. Nonetheless, they are applicable to many practically-important resonators. While LC resonators are compact and low-cost, they typically have a low quality factor ($Q_u <\sim 200$) compared to distributed elements, which translates to higher insertion loss. Also, due to parasitic inductors in capacitors, and parasitic capacitors in inductors, lumped elements are limited in operation to well below their self-resonant frequency, which is usually within 1–10 GHz.

Apart from $Q_u$ and tuning range, there are other system requirements that a tunable filter may need to meet. These requirements can include,

1. Power handling
2. Power consumption
3. Tuning speed
4. Linearity
5. Size, weight, volume
6. Cost, compatibility with low-cost electronics (for example, CMOS)
7. Reference impedance (50 Ω)
8. Analog or digital tuning
9. Immunity to vibration, shock, temperature, noise on tuning voltage

Numerous alternative technologies exist to build resonators with high-$Q_u$ at high frequencies. Below are a few of the most common tunable resonators.

### 1.3.1 Planar Tunable Resonators

RF/Microwave filters based on planar resonators, such as microstrip, stripline, and coplanar waveguide (CPW) resonators, are widely used in wireless communication systems due to their simple structure and low cost in high-volume manufacturing. Tunable resonators can be made by loading them with tuning elements to change their effective electrical lengths. Figure 1.11 shows several filter architectures using planar tunable resonators. The "diode" symbol represents a general tuning element, which can be implemented with a variety of technologies, such as semiconductor varactors, ferroelectric varactors, RF micro-electromechanical systems (MEMS) switches/varactors, etc.

Semiconductor varactors, including PIN diodes and GaAs Schottky diodes, are popular tuning elements because of their availability, low price, and high tuning range. The very fast response time (nanosecond range) of semiconductor varactors is another big advantage where high tuning speed is required. Many planar tunable filters have been demonstrated with semiconductor varactors [25, 26]. The major disadvantages of semiconductor varactors are their low $Q_u$ (<150), limited linearity, and low power handling.

RF MEMS is another promising candidate for making tuning elements in planar resonators. Due to their high $Q_u$ and simple biasing requirement, RF MEMS

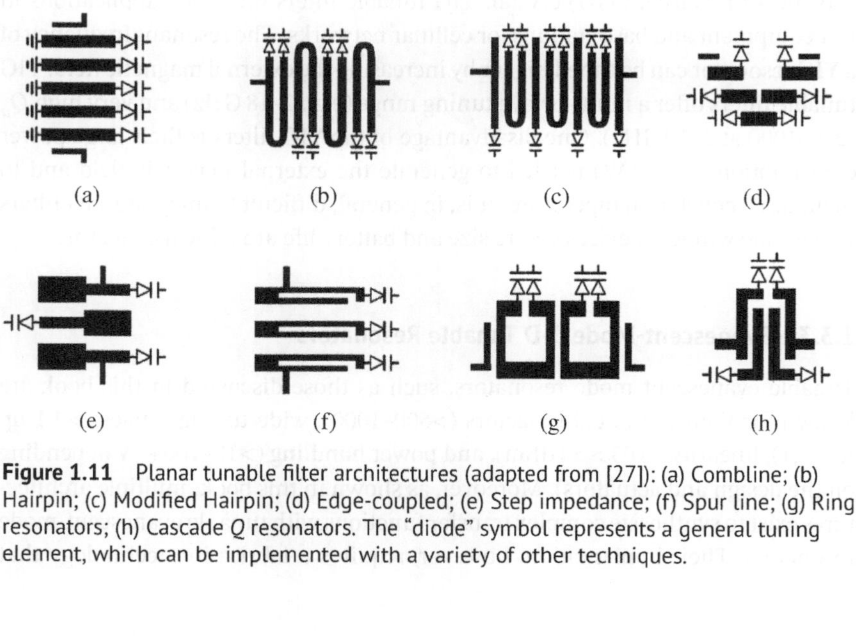


**Figure 1.11** Planar tunable filter architectures (adapted from [27]): (a) Combline; (b) Hairpin; (c) Modified Hairpin; (d) Edge-coupled; (e) Step impedance; (f) Spur line; (g) Ring resonators; (h) Cascade $Q$ resonators. The "diode" symbol represents a general tuning element, which can be implemented with a variety of other techniques.

varactors make it possible to design more complex tunable filters with lower loss. Many RF MEMS enabled tunable filters have been demonstrated in the literature [28–30]. Excellent tuning performances have been achieved over wide frequency ranges. However, the $Q_u$ of planar resonators is generally still less than 400.

Ferroelectric materials, such as Barium Strontium Titanate (BST), can also be used to make varactor at RF/microwave frequencies [31–34]. The permittivity of the ferroelectric materials can be tuned with an externally applied electric field. Therefore, voltage tunable capacitors can be readily made by sandwiching a thin film of ferroelectric material between two mettalic electrodes. Material loss is a particular problem for ferroelectric varactors. Although a lot of effort has been spent on perfecting the deposition and fabrication processes, this type of varactor still suffers from a relatively low $Q_u$ (50–100 at 1–10 GHz). Temperature sensitivity is also a known issue for ferroelectric varactors.

In general, planar tunable resonators are well suited for low-cost RF/microwave systems that require broad frequency coverage with $Q_u$ of generally less than 250. The insertion loss of these tunable filters is primarily limited by either the low $Q_u$ of the varactors or the low $Q_u$ of the resonators themselves. In order to make even lower loss tunable filters, intrinsically high-$Q_u$ resonators need to be used for making tunable filters.

### 1.3.2 Ferrimagnetic Tunable Resonators

Many practical tunable filters are based on ferrimagnetic resonators, such as the Yttrium-Iron-Garnet (YIG) crystal. YIG tunable filters find wide applications in test equipment and base stations for cellular networks. The resonant frequency of a YIG resonator can be tuned higher by increasing the external magnetic field. YIG tunable filters offer a multi-octave tuning range (e.g., 2–18 GHz) and very high $Q_u$ (200–1000 at 2–10 GHz). One disadvantage of the YIG filters is their large power consumption (0.75–3 W) needed to generate the external magnetic field and to maintain a constant temperature. It is, in general, difficult to integrate YIG filters in portable wireless devices where size and battery life are critical concerns.

### 1.3.3 Evanescent-Mode 3-D Tunable Resonators

Tunable evanescent mode resonators, such as those discussed in this book are known for their high-quality factors (>500–1000), wide tuning ranges (>2:1 up to >3:1), linearity (IIP3 > 50 dBm), and power handling (>10–100+ W depending on the design and actuators). Moreover, as shown in this book, multiple architectures exist to synthesize complex transfer functions with tunable evanescent-mode resonators. Their implantation, however, requires a fabrication technology that permits substrate-integrated waveguides and their associated tuners (mechanical or electronic).

The following bullet points concisely summarize the advantages and drawbacks of tunable evanescent-mode resonators and compare them with planar tunable resonators.

Tunable Evanescent-Mode Resonators:

- **Size:** These resonators are generally compact due to their resonator design, which is beneficial for space-constrained applications. However, they require a process that enables substrate integrated waveguides with appropriate tuning elements.
- **Loss:** They exhibit low insertion loss, particularly in narrow-band applications, which is crucial for maintaining signal integrity.
- **Linearity:** Mechanical tuning contributes to better linearity, making them ideal for high-performance RF systems where signal quality is paramount.
- **Power handling:** Their 2.5D nature and associated tuning elements provide greater design flexibility when high power handling is needed.
- **Tuning speed:** Tuning speed depends on the employed tuning elements. Faster tuning speeds often require solid-state components that may limit other loss or linearity of the filter.

Planar Tunable Resonators:

- **Size:** Planar resonators tend to be compact and they offer ease of integration with other planar components.
- **Loss:** They tend to have higher insertion loss compared to evanescent-mode filters.
- **Linearity:** They tend to be lower in linearity, particularly when solid-state tuning elements are employed.
- **Power handling:** Power handling is limited by the 2D nature and tuning elements of the resonator.
- **Tuning speed:** The same above-mentioned considerations apply here.

Overall, tunable evanescent-mode resonators tend to excel in performance and design flexibility, making them suitable for high-performance RF applications where signal quality cannot be compromised. Tunable planar resonators offer greater ease of integration and lower complexity, which can be advantageous in certain systems. Both types of resonators play pivotal roles in the evolution of wireless communication technologies.

Table 1.1 summarizes the performance measures of available tunable resonator technologies in general. Evanescent-mode resonators show a feasible compromise

**Table 1.1** Performance comparison between tunable resonator technologies.

| Technology | Active | Ferrimagnetic | Ferroelectric | Microstrip | Evanescent-mode |
|---|---|---|---|---|---|
| Size | Sub-mm | 10 s $cm^3$ | mm-cm | mm-cm | **mm to cm** |
| Quality factor | Low (1–50) | Very High (100–1000) | Low (10 s) | Medium (10–400) | **High (500–1000)** |
| Power consumption | ≃mW | 1 to 10 W | ≃0 | ≃0 | **≃0** |
| Power handling | Low (mW) | High (10s W) | Medium (< 1 W) | Medium (< 10 W) | **High (10–100+ W)** |
| Maximum Operating Frequency | 10s GHz | 10s GHz | <3 GHz | < 10 GHz | **≥ 100 GHz** |
| Frequency tunability | > octave | > octave | 20–30% | 20–30% | **> octave** |

between size, $Q_u$, power, and tunability. They are also relatively low-cost to produce. These features, along with their versatility in coupling types for complex filter structures, make them ideal for advanced reconfigurable filters serving numerous practical applications.

For the remainder of this book, the principle of operation, fabrication technologies, and design procedures of evanescent-mode resonators are discussed in detail.

# 2

# Evanescent-Mode Resonators

## 2.1 Physical Structure

Traditional cavity resonators are briefly discussed first as an introduction to evanescent-mode resonators.

A cavity resonator can be formed when a conductive sheet (e.g., copper) forms the surface of a cuboid. The minimum resonant frequency for the cavity resonator is dictated by its largest dimension, which should be $>\lambda/2$, where $\lambda$ is the wavelength. The resulting $Q_u$ is typically high (1,000s). Figure 2.1 shows a cavity resonator at its lowest resonant frequency. In the dimensions shown in the figure, the resonant frequency is approximately 21 GHz. While this can still be useful in many applications, the majority of communication systems operate within the DC-7 GHz regime. This means that for most practical applications the corresponding resonator needs to be at least three times bigger than this one.

Another limitation of conventional cavity resonators is their limited possible tuning. If the dimensions were tunable (using some form of mechanical actuators), doubling the frequency would require halving the size, which is a significant motion to be covered in a reasonable volume and speed.

To operate a resonator below its fundamental mode (or at the evanescent mode), a conductive post can be added as shown in Figure 2.2. The loading post is typically placed in the middle of the resonator (which is the reason those resonators are sometimes called coaxial cavity resonators). The loading post is in contact with one end of the resonator and is in close proximity to its opposite end, creating capacitive loading (which is the reason those resonators are sometimes called capacitively-loaded or highly-loaded resonators). With external dimensions similar to the cavity resonator discussed earlier, the evanescent-mode resonator has a significantly lower resonant frequency (~4 GHz), with a reasonable penalty on $Q_u$.

*Tunable Evanescent-Mode Filters: Principles, Implementation, and Applications,*
First Edition. Dimitrios Peroulis and Mohammad Abu Khater.

Companion website: www.wiley.com/go/Peroulis/TunableEvanescent-ModeFilters

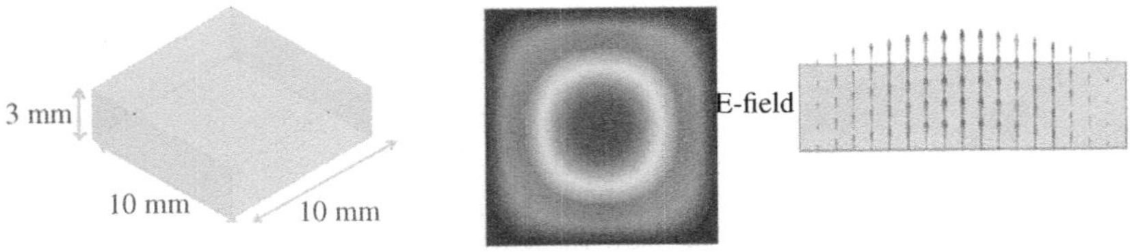


**Figure 2.1** Traditional cavity resonator along with its electric field distribution when it is resonating at the $TE_{101}$ mode (fundamental mode).

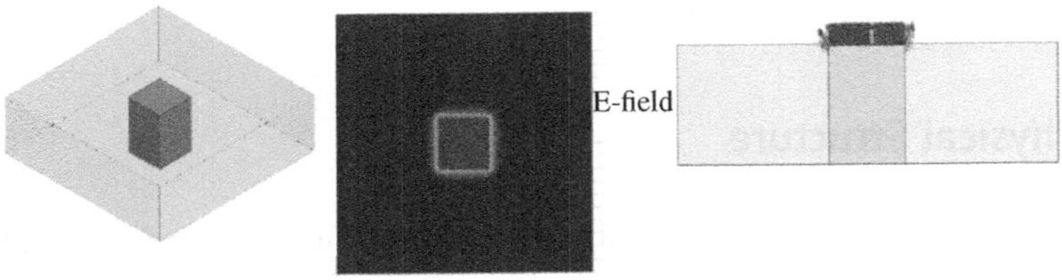


**Figure 2.2** Evanescent-mode resonator along with its electric field distribution at resonance.

The resulting quality factor is still in the several 100s and can reach over 2,000 depending on the design and fabrication technology.

The presence of the post confines the electric field between the top wall of the resonator and the surface of the post. This makes its resonance sensitive to the effective loading capacitance. As a result, changing the distance between the post and the cavity ceiling by as little as 10 s of micrometers can result in nearly an octave tuning. Detailed analysis of the frequency and the quality factor are discussed in the following section.

It is important to notice that the structure of evanescent-mode resonators is not necessarily cuboid. Depending on the fabrication technology, it can be a tapered cuboid, cylinder, or tapered cylinder.

## 2.2 Analysis

### 2.2.1 Coaxial Cable Approximation

We can employ a simple model to analyze a coaxial evanescent-mode resonator (Figure 2.3). First, the electric field distribution between the post and the cavity ceiling is practically uniform. Consequently, we can represent the area above the post as a lumped capacitor. Similarly, the concentrically circular magnetic field surrounding the post implies that the electrically-small post and cavity sidewall can be modeled as a shorted coaxial line, effectively functioning as an inductor.

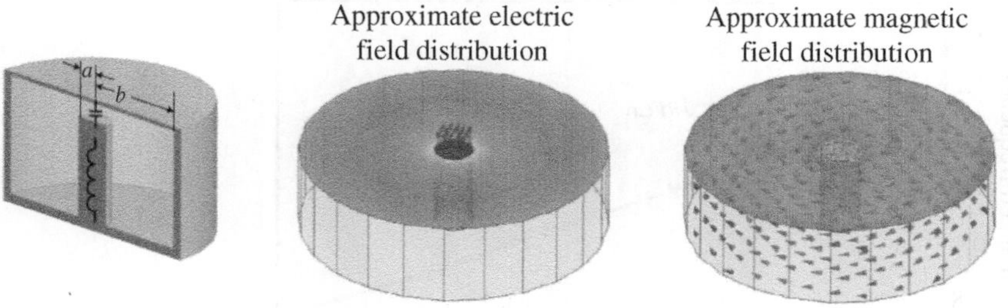


**Figure 2.3** Cross section of cylindrical (coaxial) evanescent-mode resonator along with its electric and magnetic fields.

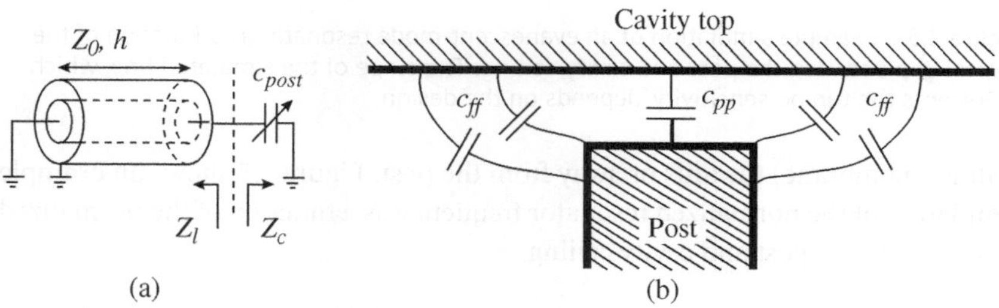


**Figure 2.4** (a) Coaxial cable approximation of evanescent-mode resonators. (b) $C_{post}$ consists of parallel-plate capacitance and fringing capacitance.

This approximation is illustrated in Figure 2.4(a). As a result, at resonance,

$$Z_L + Z_C = 0 \tag{2.1}$$

$$jZ_0 \tan \beta h + \frac{1}{j2\pi f C_{post}} = 0 \tag{2.2}$$

$$Z_0 \tan\left(2\pi f \frac{h}{c}\right) = \frac{1}{2\pi f C_{post}} \tag{2.3}$$

The symbols above are as follows: $c$ is the speed of light, $f$ is the design resonant frequency, $C_{post}$ is the post capacitance, $h$ is the length (often referred as height) of the post-formed transmission-line section, and $Z_0$ is the characteristic impedance of this line, which can be controlled by the parameters $a$ (post radius) and $b$ (cavity radius) (Figure 2.3).

Although the parallel plate capacitance ($C_{pp}$) could provide a reasonable first-order approximation of the post capacitance, a more careful full-wave simulation that also captures the fringing-field capacitance ($C_{ff}$) is often required (Figure 2.4).

To tune the evanescent-mode resonator, it is necessary to alter $C_{post}$ in equation (2.3). A mechanical actuator often achieves this by bending the often-thin cavity

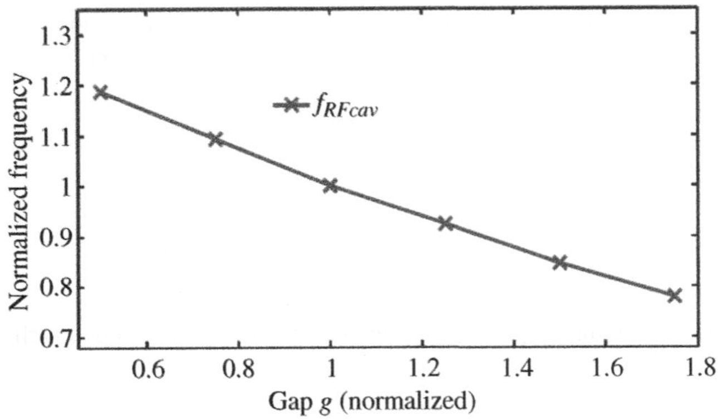


**Figure 2.5** Example simulation of an evanescent-mode resonator as a function of the critical gap between the post and cavity ceiling. The slope of the simulated line, which represents the tuning sensitivity, depends on the design.

ceiling (membrane) towards or away from the post. Figure 2.5 shows an example simulation of the normalized resonator frequency as a function of the normalized gap between the post and cavity ceiling.

#### 2.2.1.1 Unloaded Quality Factor

The unloaded quality factor $Q_u$ is an essential figure of merit for resonators which represents loss. A low $Q_u$ can translate to inferior filter performance, in terms of insertion loss for example. In this section, we derive an approximate closed-form expression for $Q_u$, which will be used to optimize the structure of evanescent-mode resonators. The analysis below is for the cylindrical evanescent-mode resonator.

In order to find $Q_u$, a resonator is modeled as an RLC circuit. In the model in Figure 2.6, the capacitance and inductance are as described in Section 2.2.1. The resistances stem from the finite conductance of the sheet (or $R_s \neq 0$). Once

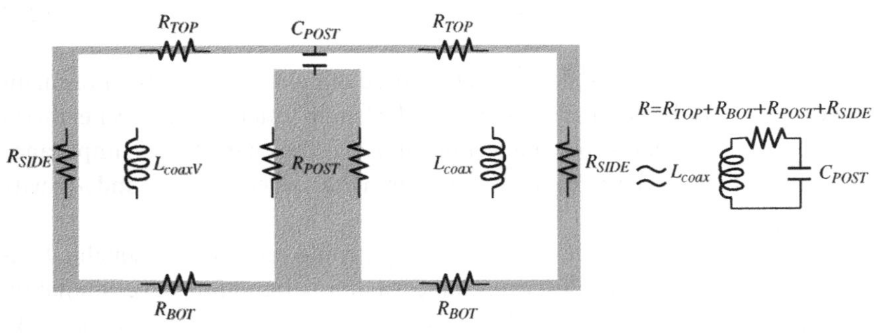


**Figure 2.6** Cross section of an evanescent-mode resonator along with its approximate RLC circuit model.

the values of the approximate RLC circuit are known, the quality factor can be found using

$$Q_u = \frac{\omega_0 L_{coax}}{R} \tag{2.4}$$

where $\omega_0$ is the design resonance, $L_{coax}$ is the equivalent inductance, and $R$ represents the effective resistance.

To find the inductance, in (2.2), the structure of the coaxial cavity is approximated as a short-length shorted transmission line. This yields

$$|Z_L| = \omega L_{coax} = Z_0 \tan\left(2\pi f \frac{h}{c}\right) \approx Z_0 2\pi f \frac{h}{c} \quad for\ h \ll \lambda \tag{2.5}$$

From which

$$L_{coax} \approx Z_0 \frac{h}{c} = \underbrace{\frac{1}{2\pi}\sqrt{\frac{\mu}{\epsilon}}\ln\left(\frac{b}{a}\right)}_{Z_0} h \underbrace{\sqrt{\mu\epsilon}}_{1/c} = \frac{1}{2\pi}\mu h \ln\left(\frac{b}{a}\right) H \tag{2.6}$$

where $\mu$ and $\epsilon$ are the permeability and permittivity of the cavity dielectric material, $b$ and $a$ are the radii of the resonator and the post, respectively.

The equivalent resistance is found next. The post and the sidewall resistances represent a sheet with a length of $h$ and a width of $2\pi a$ and $2\pi b$. Consequently,

$$R_{POST} = R_s \frac{h}{2\pi a} \tag{2.7}$$

and

$$R_{SIDE} = R_s \frac{h}{2\pi b} \tag{2.8}$$

where $R_{POST}$ and $R_{SIDE}$ are as defined in Figure 2.6. For $R_{TOP}$ and $R_{BOT}$, the resistance is calculated by integrating the resistances of infinitesimally thin rings. For simplicity, the tip of the post is assumed to have the same radius as the post itself $a$ (no additional area is added, as discussed in later chapters)

$$R_{TOP} = R_{BOT} = \int_a^b R_s \frac{1}{2\pi r} \cdot dr = \frac{R_s}{2\pi} \ln\left(\frac{b}{a}\right) \tag{2.9}$$

From (2.7), (2.8), and (2.9), the total resistance in the model is

$$R = \frac{R_s}{2\pi}\left(\frac{h}{a} + \frac{h}{b} + 2\ln\left(\frac{b}{a}\right)\right) \tag{2.10}$$

These results can now be used with (2.4) to find $Q_u$

$$Q_u = \omega_0 \frac{\mu}{R_s} \frac{\ln\left(\frac{b}{a}\right)}{\left(\frac{1}{a} + \frac{1}{b} + \frac{2}{h}\ln\left(\frac{b}{a}\right)\right)} \tag{2.11}$$

The expression in (2.11) gives several insights on how to increase $Q_u$. Naturally, having lower sheet resistance is desired, but this is typically dictated by the available materials and fabrication process. Increasing the height of the resonator also increases $Q_u$. The height cannot be increased arbitrarily it is often limited by the substrate thickness for a given fabrication technology. Also, other resonant modes might appear with a large $h$. While increasing $a$ and $b$ (simultaneously) increases $Q_u$, their effects independently are not easily comprehended from (2.11). Increasing both of them increases the size of the resonator, which can be limited by the form factor of the filter. As a result, we will assume that the external dimensions $b$ and $h$ are constant and will investigate the value of $a$ that will produce the largest $Q_u$. To do that, we set

$$\frac{\partial Q_u}{\partial a} = \omega_0 \frac{\mu}{R_s} \frac{\frac{-1}{a}\left(\frac{1}{a}+\frac{1}{b}+\frac{2}{h}\ln\left(\frac{b}{a}\right)\right) + \ln\left(\frac{b}{a}\right)\left(\frac{1}{a^2+\frac{2}{ha}}\right)}{\left(\frac{1}{a}+\frac{1}{b}+\frac{2}{h}\ln\left(\frac{b}{a}\right)\right)^2} = 0 \quad (2.12)$$

which results in

$$a = be^{-W\left(\frac{1}{e}-1\right)} \quad (2.13)$$

where $W$ is the product logarithm function (or Lambert $W$-function). Numerically, this translates to

$$a = 0.27846\, b \quad (2.14)$$

It should be no surprise that the result in (2.14) corresponds to $Z_0 \approx 76.6\Omega$, which is the minimum loss impedance of a coaxial transmission line.

In Figure 2.7, the quality factor values are found for a constant $b = 5$ mm and a variable $a$. The maximum value of $Q_u$ corresponds to $a = 0.27846\, b,\ b = 1.39$ mm, regardless of the sheet resistance.

It should also be noted that the result in (2.14), while insightful, should be used carefully since it implies several assumptions mentioned earlier. Furthermore, this result was based on optimizing $Q_u$ only, regardless of other performance measures such as tuning range or manufacturability, which are discussed later.

### 2.2.2 Tapered Resonator Model

For relatively small-sized resonators (below ~5 mm), fabrication of tapered cylindrical structure can be more compatible with microfabrication techniques (more on that in Chapter 4). Figure 2.8 illustrates a two-dimensional view of the tapered resonator along with an approximate equivalent model. The tapered transmission

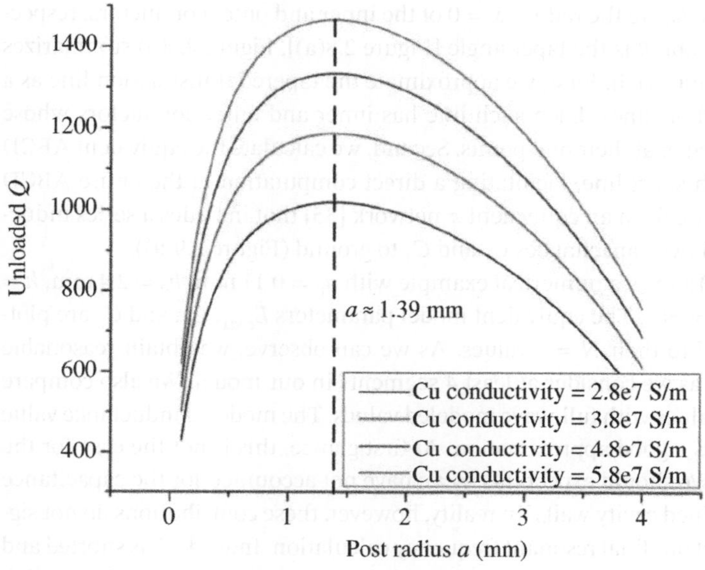


**Figure 2.7** $Q_u$ versus post radious $a$ at different conductivities (sheet resistances). The dimensions are $b = 5$ mm, $h = 3$ mm, the simulations are centered at 4.5 GHz by adjusting the capacitive gap.

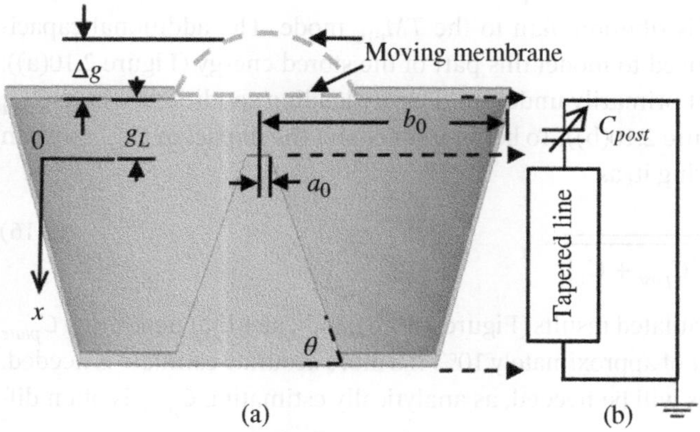


**Figure 2.8** Schematic of a tapered evanescent-mode resonator with most critical dimensions. A two-dimensional view (a) and an approximate equivalent circuit (b) are shown.

line can be modeled through its impedance $Z_0$ based on the following equation that gives its value at every position $x$,

$$Z_0(x) = \frac{1}{2\pi}\sqrt{\frac{\mu_0}{\epsilon}}\ln\left(\frac{b_0 - \dfrac{x}{\tan(\theta)}}{a_0 + \dfrac{x}{\tan(\theta)}}\right) \tag{2.15}$$

in which $a_0$ and $b_0$ are the radii at $x = 0$ of the inner and outer conductors, respectively, at $x = 0$, and $\theta$ is the taper angle [Figure 2.8(a)]. Figure 2.9(a) summarizes our modeling approach. First, we approximate the tapered transmission line as a cascade of $N$ short lines. Each such line has inner and outer conductors whose radii are calculated at their mid points. Second, we calculate the equivalent ABCD matrix for each short line, facilitating a direct computation of the entire ABCD matrix. Third, we form an equivalent $\pi$ network [35] that includes a series inductance $L_{coax}$ and two capacitances $C_1$ and $C_2$ to ground (Figure 2.9(a)).

Figure 2.9(b) shows a numerical example with $a_0 = 0.11$ mm, $b_0 = 2.4$ mm, $h = 1.5$ mm, and $\theta = 60°$. The equivalent model parameters $L_{coax}$, $C_1$, and $C_2$ are plotted normalized to their $N = 1$ values. As we can observe, we obtain reasonable results as long as we consider at least 4 segments in our model. We also compare our modeled values with full-wave modeled values. The modeled inductance value $L_{coax}$ represents a good approximation. At first glance, this is not the case for the capacitances. We should expect this as we have not accounted for the capacitance due to the inclined cavity walls. In reality, however, these contributions do not significantly affect our final resonant frequency calculation. Indeed, $C_2$ is shorted and so it does not play a role. Moreover, $C_1$ is added to $C_{post}$ as they are both in parallel. However, since it is significantly smaller than $C_{post}$, an accurate estimation of $C_1$ is not needed.

In addition to the TEM mode, a portion of the electric field energy is retained in an electric field distribution akin to the $TM_{010}$ mode. The additional capacitance $C_{plate}$ is introduced to model this part of the stored energy (Figure 2.10(a)). This mode is relevant primarily under mild cavity loading conditions where $C_{post}$ is not dominant (Figure 2.10(b)). To better understand the impact of $C_{plate}$, we can calculate $\omega_0$ by ignoring it, as

$$\omega_0 \approx \frac{1}{\sqrt{L_{coax}\left(C_{post} + C_1\right)}} \tag{2.16}$$

Comparison with simulated results (Figure 2.10(b)) indicates that neglecting $C_{plate}$ may result in an error of approximately 10%. If a more accurate estimate is needed, full-wave simulations will be needed, as analytically estimating $C_{plate}$ is often difficult.

In conclusion, $C_{post}$, $C_1$, and $C_{plate}$ (defined in Figure 2.10) form the total resonator capacitance as

$$C_{resonator} = C_{post} + C_1 + C_{plate} \tag{2.17}$$

In practice, we often try to load an evanescent-mode resonator very heavily to minimize its volume and maximize its tuning range. This can lead to critical gaps in the order of just a few micrometers (although it can be challenging to achieve them from an implementation technology point of view). In such cases, $C_{post}$ tends

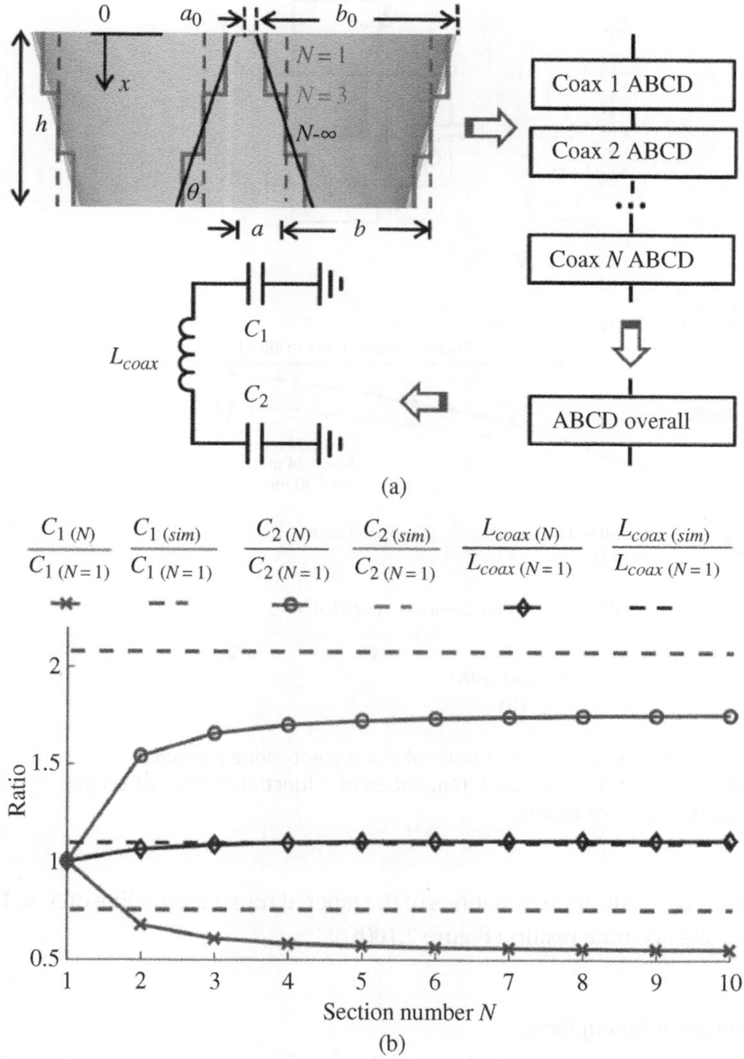


**Figure 2.9** (a) Modeling approach of the tapered evanescent-mode resonator. The tapered line can be approximated as a cascade of short uniform sections, each modeled by its equivalent ABCD matrix. (b) A simulated example for $a_0 = 0.11$ mm, $b_0 = 2.4$ mm, $h = 15$ mm, and $\theta = 60°$. The graph shows that the proposed equivalent circuit converges after four or more transmission line segments are used to model the tapered section.

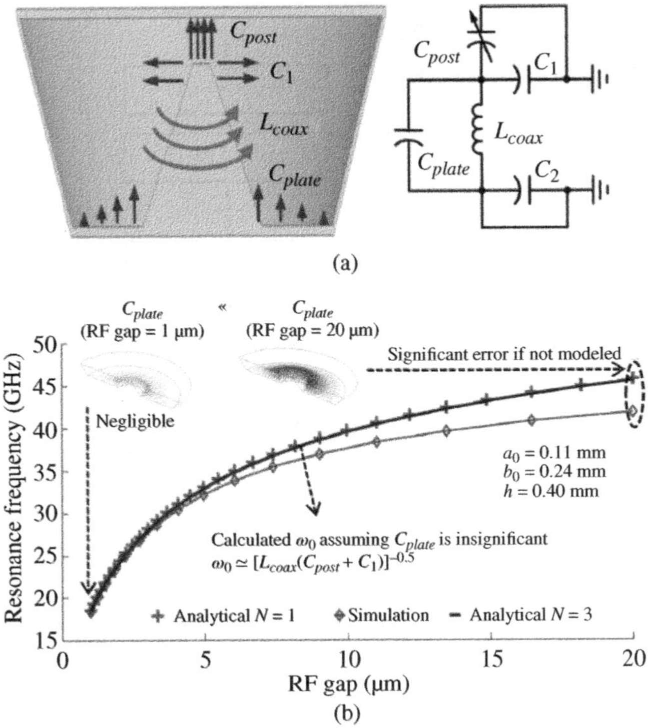


**Figure 2.10** (a) Equivalent circuit of a tapered evanescent-mode resonator. (b) Calculated and simulated resonance frequencies as a function of the critical gap between the post and cavity ceiling.

to dominate. Even crude approximations of the tapered transmission line of $N = 1$ or 2 tend to yield accurate results (Figure 2.10(b)).

#### 2.2.2.1 Frequency Tuning Ratio

Following the previous analysis and assuming that $C_{post} >> C_{plate} + C_1$, we can compute the resulting tuning ratio as

$$\frac{\omega_{0H}}{\omega_{0L}} \approx \left( \frac{g_L}{g_L + \Delta g_{max}} + \frac{C_1 + C_{plate}}{C_{postL}} \right)^{-\frac{1}{2}} \tag{2.18}$$

in which $g_L$ and $\Delta g_{max}$ are defined in Figure 2.8, $\omega_{0H}$ and $\omega_{0L}$ are the high and low frequencies, respectively. Notice that we have used $C_{postL}$ to denote the capacitance that corresponds to $\omega_{0L}$, which is realized when the minimum gap $g_L$ is achieved.

This can be found using the parallel plate approximation

$$C_{postL} \approx \frac{\epsilon_0 \pi a_0^2}{g_L} \tag{2.19}$$

The selected tuning method governs the first part of (2.18). On the other hand, the cavity geometry largely determines its second part. A smaller starting gap $g_L$ results in a higher tuning range. It is therefore common practice to fix $g_L$ to the smallest value $g_{Lmin}$ possible as primarily determined by the manufacturing process/tolerances and the needed RF power handling [36]. Once this critical gap is determined, $C_{postL}$ only depends on $a_0$. Additionally, the second part of (2.18) demonstrates that a higher tuning range is correlated with a bigger $C_{postL}$. Moreover, we should remember that once we choose $a_0$ (and therefore $C_{postL}$), there is a set of $h$ and $h$ that minimizes $C_1 + C_{plate}$ in (2.18) resulting to the maximum tuning range. The following equation from the coaxial cable analysis can be used to approximate the capacitance $C_1$ as

$$C_1 \approx \frac{2\pi\epsilon}{\ln\left(\frac{b}{a}\right)} h \tag{2.20}$$

Using (2.6) and (2.19) for $\omega_0 \approx 1/L_{coax}\, C_{postL}$ yields

$$h = \frac{2g_L}{\omega_{0L}^2 a_0^2 \epsilon_0 \mu \ln\left(\frac{b}{a}\right)} \tag{2.21}$$

This can be a useful expression for $h$ when we attempt to calculate $C_1$ through (2.20).

To further clarify this design process, we will demonstrate it with a specific numerical example and the resonator with vertical walls depicted in Figure 2.11. For this resonator, we will select a very small initial gap of $g_L = 1$ μm. We should keep in mind though that such a gap is often hard to achieve in practice. Moreover, it limits the maximum power handling of the resonator. We will also assume that the maximum gap control available from the tuning method is $\Delta g_{max} = 19$ μm. The initial resonant frequency is assumed to be $\omega_{0L} = 20$ GHz. Moreover, we assume that the post radius is $a_0 = 90$ μm. Based on these assumptions, we can use (2.21) to identify a range of values for $h$ and $b$ for the cavity. Each set of these values yields a different tuning range. The highest tuning range is obtained when $h = 0.6$ mm and $b = 0.83$ mm (Figure 2.11(a)). It is interesting to observe from this figure and analysis that, in general, lower aspect rato cavities result in higher tuning ranges.

Figure 2.11(b) plots all critical capacitances at the high end of the frequency range of this resonator. We have used the above-mentioned equations for $C_1$ and $L_{coax}$. Full-wave simulations have been used for $C_{post}$. The most important conclusion from these capacitance calculations/simulations is that there is an outer

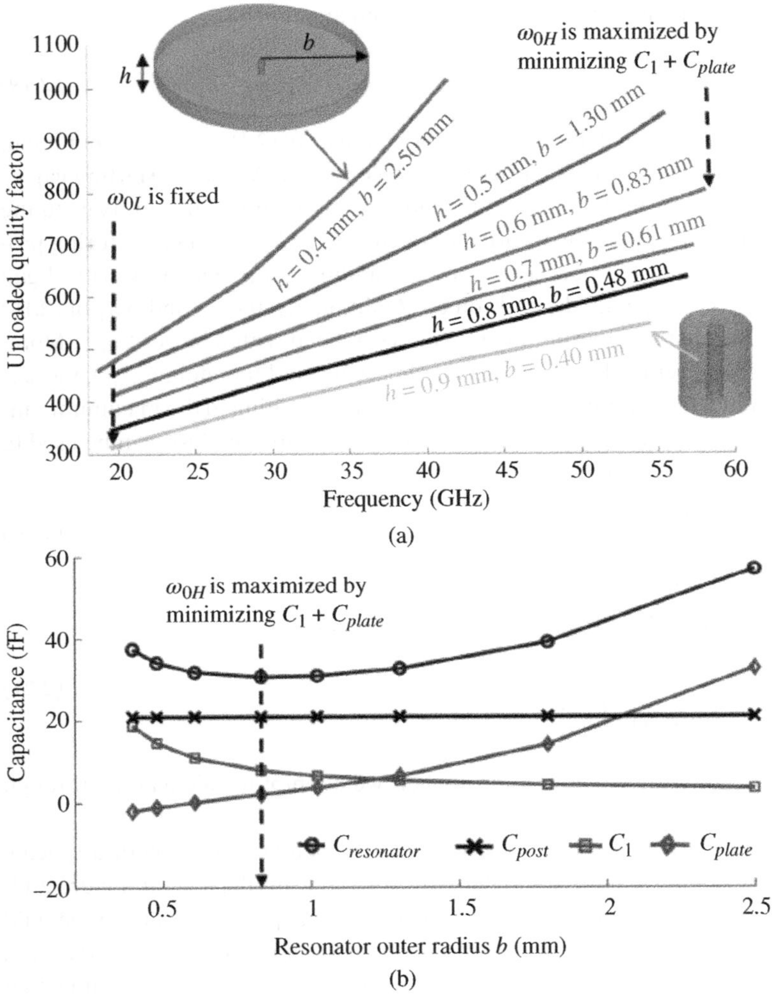


**Figure 2.11** (a) Unloaded quality factor as a function of frequency for an example resonator with minimum gap of 1 μm, gap change of 19 μm, and post radius of 90 μm. Notice that there is a set of values for $h$ and $b$ that yield the maximum possible tuning range. (b) Calculated and simulated capacitances of the resonator at the high end of its tuning range as a function of its outer radius. Note the existence of an optimal outer radius $b$ that minimizes $C_1 + C_{plate}$ and maximizes the highest achievable frequency and thus the resulting tuning range.

radius $b$ that minimizes the value of $C_1 + C_{plate}$ and, as a result, maximizes the highest achievable frequency $\omega_{oH}$. Consequently, this optimal radius $b$ results in the highest tuning range.

### 2.2.3 Tuning Range and Quality Factor Co-optimization

In Section 2.2.1.1, we derived the quality factor expression in (2.11), and its geometric optimization in (2.14), assuming constant external dimensions $b$ and $h$. Also, this optimization targeted increasing $Q_u$, regardless of the achievable frequency tuning range. In this section, we discuss an alternative design approach to accommodate for both, $Q_u$ and tuning range.

From (2.4), we can conclude that $Q_u$ is directly proportional to $L_{coax}$. Increasing $L_{coax}$, however, decreases $C_{post}$ according to (2.16), which decreases the tuning range as concluded from (2.18). As a result, a tradeoff exists between $Q_u$ and the tuning range. We have summarized below a practically-useful methodology that satisfies a desired tuning range while maintaining an optimal quality factor. The suggested design steps are:

1. Select the minimum critical gap $g_{Lmin}$ of the resonator that will define the low end of its tuning range. As mentioned before, this is often dictated by the fabrication process/tolerances as well as the required power handling.[1]
2. Set the post radius close to the minimum value $a_{0\,\min}$ allowed by the fabrication process. This choice will minimize $C_{postL}$ and maximize $L_{coax}$, yielding the highest quality factor.
3. Optimize the resonator cavity dimensions (Figure 2.11) to obtain the highest tuning ratio.
4. If step 3 does not yield an acceptable tuning range, it will be necessary to increase $a_0$ and repeat the process. This will result in a lower quality factor but increased tuning range. On the other hand, if the calculated tuning range exceeds the target, it may be advisable to relax some of the constraints to make fabrication easier. For example, we can increase $g_{Lmin}$ and repeat the process until we achieve the required tuning range.

To demonstrate this process, we have optimized a cavity resonator in the 20–40 GHz range as shown in Figure 2.12. The primary design constrains are shown in the figure. It is interesting to note than no designs with $a_0 < 100\ \mu\text{m}$ satisfy the tuning range requirements. The solid line in Figure 2.12 shows the optimized solution that corresponds to $a_0 = 100\ \mu\text{m}$. This design achieves the needed tuning range with the highest possible quality factor. Designs with $a_0 > 100\ \mu\text{m}$ result in higher tuning ranges, but also exhibit lower quality factors.

---

1 High power results in a high voltage across the capacitive gap, which can cause a breakdown.

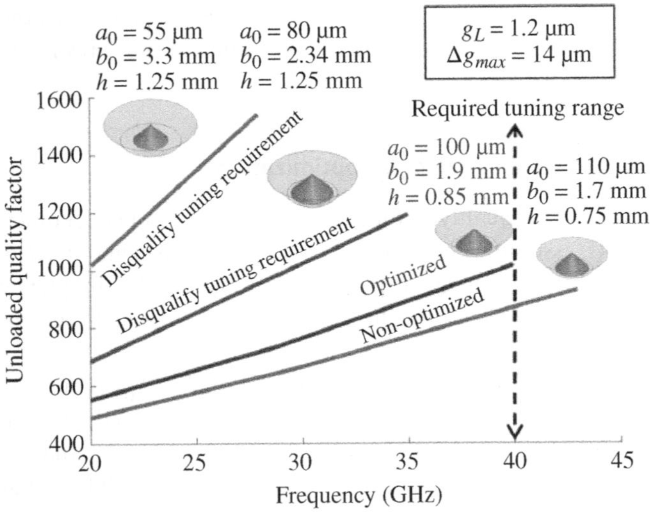


**Figure 2.12** A numerical example summarizing the presented design methodology for a 20–40 GHz resonator.

## 2.3 Coupling Techniques

In Section 1.2.2, the coupling routing diagram values and diagrams were presented abstractly. In this section, we will take a detailed look at how they are physically implemented in the evanescent-mode technology.

Coupling means that energy is transferred either from/to an external source/load or within resonators. An interesting feature of the discussed evanescent-mode resonator is that nearly all electric field energy is concentrated on top of its post (Figure 2.2). Hence it is often easier to couple resonators through their magnetic fields. External coupling is discussed first.

### 2.3.1 External Coupling

External coupling can be realized in various ways, some of which depend on whether the filter is a bandpass or a bandstop. Figure 2.13 shows a few of the common ones.

The aperture coupling is typically formed by an arc-shaped opening beneath the transmission line. In a bandpass filter (BPF), the transmission line is typically shorted to the ground beyond the aperture. This increases the current and, hence, the coupled magnetic field. The main advantage of aperture coupling is simplicity in fabrication and design. The out-of-band frequency response, however, is relatively limited since the aperture structure itself can resonate. Radiation losses are also higher compared to other sealed coupling structures.

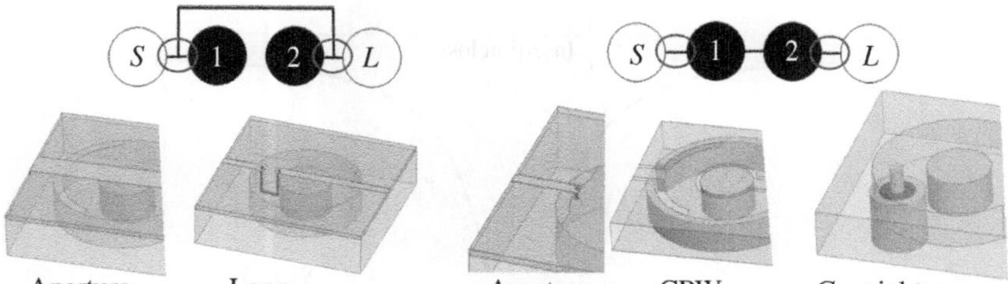


**Figure 2.13** Examples of common external coupling methods in evanescent-mode BPFs and BSFs.

Coplanar waveguide (CPW) coupling is also common in BPFs. Its structure is similar to that of aperture coupling, except that it does not require an extra metal layer for the microstrip in the aperture structure.

Tapped coupling is also common in BPFs. Here, the RF signal is injected from the top or the bottom of the resonator through the center conductor of a coaxial cable. This maintains the resonator sealed, minimizing radiation losses, which is desirable in high-frequency designs. This advantage comes at the cost of complicated fabrication.

Loop coupling can be thought of as the tapped coupling of bandstop filters (BSFs). The structure inside the resonator has to return to sustain the source-load coupling. We sometimes prefer such a technique as it has a much higher self-resonance frequency compared to aperture coupling. This means that a filter can achieve a large spurious-free operating range.

While the external coupling methods mentioned above are very common, many others exist. More examples are discussed in Part III of the book.

In Chapter 1, we noticed that in order to get a specific frequency response from a filter, well-defined coupling values ($M$) are required. It is obvious that the geometry of the coupling structure determines the coupling value. But how can we tell what that actual value is? The method to determine the coupling value depends on the filter type.

**BSF External Coupling Extraction:**

Similar to BPFs, evanescent-mode resonators are designed and optimized first. Then, the external coupling is determined (as discussed below). This process can repeat until both design criteria, for the resonator and the coupling, are satisfied.

In a BSF, a resonator is connected in a 'shunt' configuration. As a result, a two-port test can determine the value of the external coupling. The external coupling value depends on the value of the unloaded quality factor of the resonator (which is the intrinsic quality of the resonator itself). The value of $Q_u$ can be found in the two-port response, shown in Figure 2.14(a), using

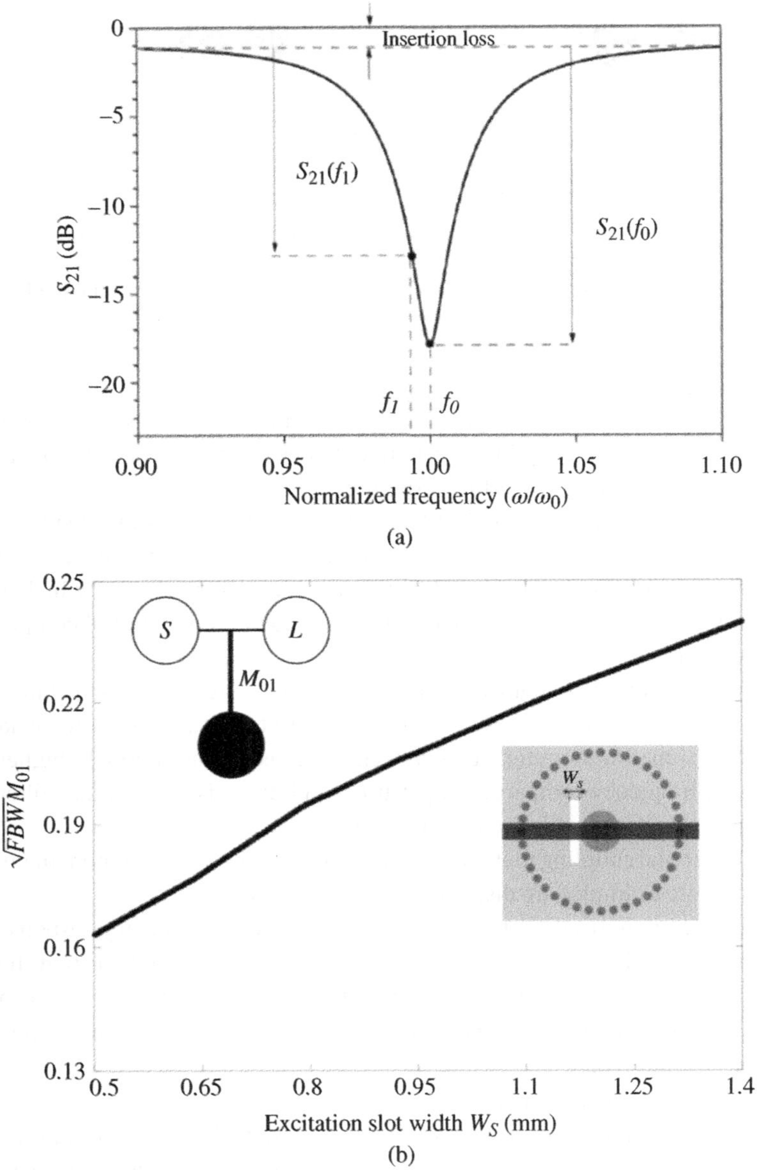


**Figure 2.14** (a) Frequency response of a single resonator in a BSF configuration. (b) A top view of an evanescent-mode filter with its simulated external quality factors in a bandstop coupling structure configuration. Designers will often optimize the coupling geometry to achieve the needed external coupling.

$$Q_u = \frac{f_0 f_1}{|f_0^2 - f_1^2|} \sqrt{\frac{\frac{{S_{21}}^2(f_1)}{{S_{21}}^2(f_0)} - 1}{1 - {S_{21}}^2(f_1)}} \tag{2.22}$$

where the frequencies $f_0$ and $f_1$ represent the resonant frequency and a frequency close to $f_0$ respectively. It is important to notice that the values of $S_{21}$ are the absolute values, rather than the dB values. Once $Q_u$ is known, the external coupling can be found as

$$M_{01} = \frac{1}{\sqrt{FBW}} \sqrt{\frac{2(1 - S_{21}(f_0))}{Q_u S_{21}(f_0)}} \tag{2.23}$$

Simulated results of these equations for an aperture coupling are shown in Figure 2.14(b).

This method does not account for non-ideal effects, including passband insertion loss and transfer function asymmetry resulting from passband mismatch. Non-zero insertion loss can be accounted for by subtracting the passband loss from the measured attenuation levels $S_{21}(f0)$ and $S_{21}(f1)$, as illustrated in Figure 2.14. In cases of elevated reflection within the passband, the precision of this method may be enhanced by performing multiple calculations of $Q_u$ and $M_{01}$, followed by averaging the results, utilizing frequencies for $f_1$ that are both above and below $f_0$.

**BPF External Coupling Extraction:**

Once the resonator is designed at the desired frequency and optimized, an external coupling structure is chosen. The single-port response is tested or measured, typically using an FEM simulation. Specifically, the group delay is measured, which is indicative of the coupling value. The external coupling depends on the external quality factor $Q_e$. This is defined as the ratio of the peak energy held in a resonator during a cycle of oscillation to the energy dissipated per radian to the external circuit, excluding the internal losses of the resonator. The external quality factor can be found as [14]

$$Q_e = \frac{2\pi f_0 \tau_{11}(f_0)}{4} \tag{2.24}$$

where $\tau_{11}(f_0)$ is the reflection group delay at resonance. Once we find $Q_e$, the external coupling for a bandpass structure can be calculated using

$$M_{01} = \frac{1}{\sqrt{FBW \times Q_e}} \tag{2.25}$$

For example, once a coupling structure is determined (assuming the *FBW* is known), the structure is then simulated and the coupling value is found using

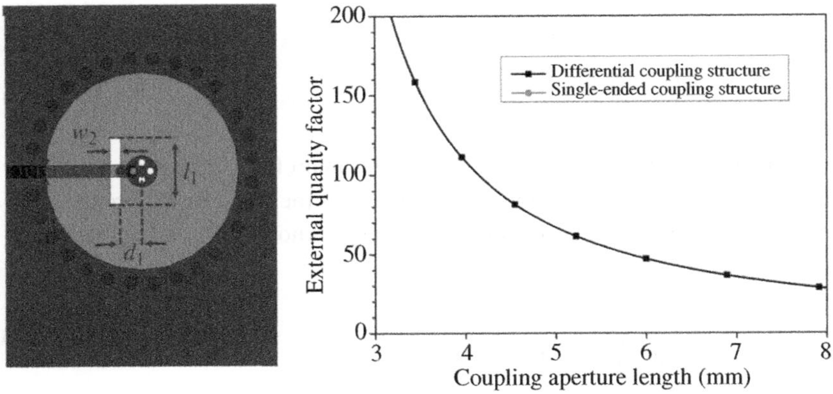


**Figure 2.15** A top view of an evanescent-mode filter with its simulated external quality factors for a bandstop coupling configuration. We can achieve the required external quality factor by optimizing the coupling geometric parameters.

(2.25). The designer can then determine whether this value needs to be increased or decreased, to achieve the sought $M_{01}$ value. Figure 2.15 shows how external coupling structure dimensions affect the external quality factor.

An alternative two-port method to calculate the external coupling for BPFs can be found in [14].

It is important to notice that having an external coupling in a resonator will shift the resonance frequency and affect $Q_u$. Consequently, designs typically go through several iterations between optimizing the resonator and getting the correct coupling value.

To find $Q_u$ in a BPF structure, a weakly coupled single resonator response is measured/simulated as shown in Figure 2.16. The 3-dB bandwidth dictates the value of $Q_u$ by

$$Q_u = \frac{f_0}{f_2 - f_1} \tag{2.26}$$

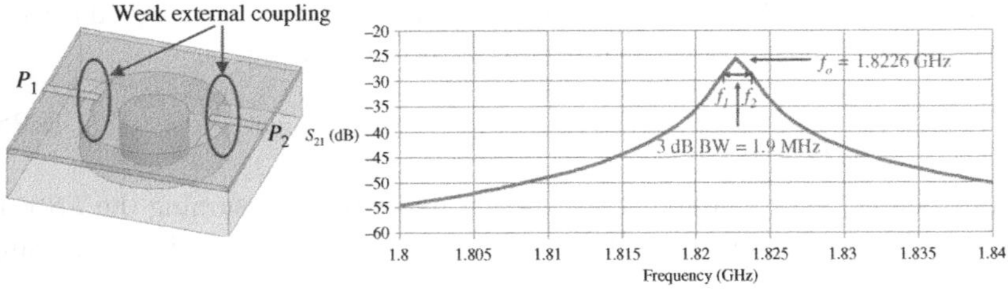


**Figure 2.16** Weak external coupling is used to fund $Q_u$ in a BPF structure.

### 2.3.2 Inter-Resonator Coupling

Resonator-to-resonator coupling, often indicated as inter-resonator coupling, can also be realized in many ways. The key point is that energy needs to couple from one resonator to the other, typically in the form of magnetic mutual coupling. For example, Figure 2.17(a) shows how an opening between two evanescent-mode resonators creates mutual coupling. If additional coupling is required, a structure similar to aperture external coupling can be added between the two resonators (Figure 2.17(b)).

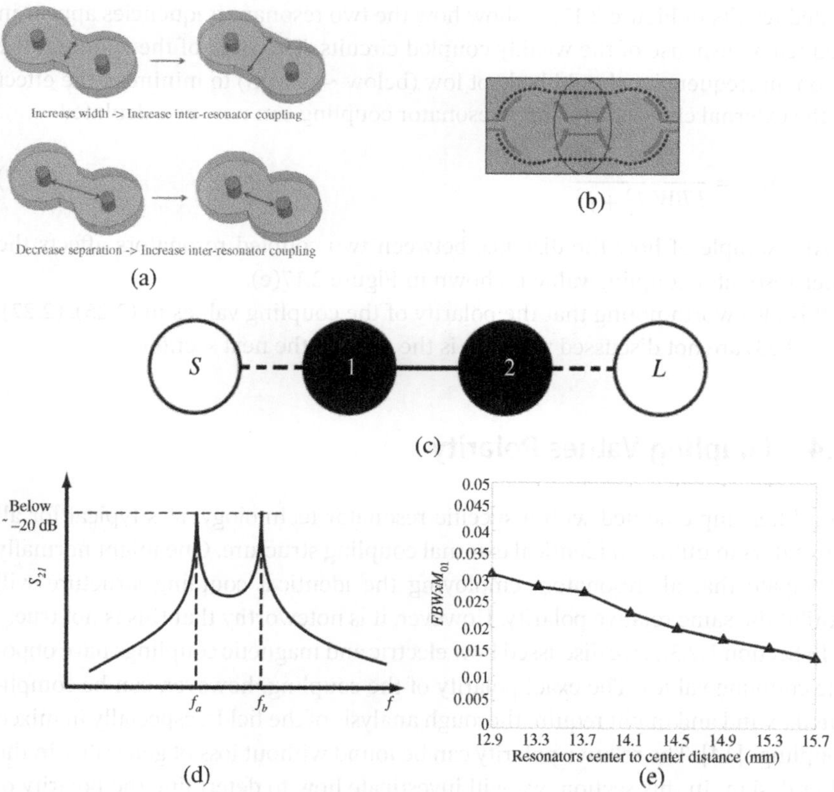


**Figure 2.17** (a) Coupling energy from one resonator to the next, often referred to as inter-resonator coupling, can be accomplished by an appropriate opening between the resonators. This opening allows for magnetic field coupling. (b) To strengthen the inter-resonator coupling, an additional aperture between the resonators can be utilized. (c) Inter-resonator coupling test requires weak external coupling, resulting in peaks in $M_{21}$ below 20 dB. (d) The odd and even resonances should have a peak below ~ −20 dB. (e) An example of how the distance between two resonators dictates the inter-resonator coupling.

Inter-resonator coupling is typically required in BPFs, but sometimes they exist in BSFs to expand their functionalities. Extracting the coupling value, however, is similar in both cases.

**Inter-Resonator Coupling Extraction:**

The structure for testing/simulating inter-resonator coupling values is shown in Figure 2.17(c). The two-port measurement setup requires two weak external couplings (large $Q_e$). The two resonators, while tuned to the same frequency, interact with each other and end up resonating in even and odd modes (detailed analysis in [14]).

The two resonant frequencies determine the inter-resonator coupling. The simulated results in Figure 2.17(d) show how the two resonant frequencies appear in frequency response of the weakly coupled circuits. The peak of the values at the resonant frequencies should be kept low (below ~−20 dB) to minimize the effect of the external circuits. The inter-resonator coupling can now be calculated

$$M_{12} = \frac{1}{FBW}\frac{f_a^2 - f_b^2}{f_a^2 + f_b^2} \tag{2.27}$$

An example of how the distance between two coupled resonators affects the inter-resonator coupling value is shown in Figure 2.17(e).

It is also worth noting that the polarity of the coupling values in (2.25), (2.27), and (2.23) are not discussed yet. This is the topic of the next section.

## 2.4 Coupling Values Polarity

In a filter implemented with a specific resonator technology, it is typical for all resonators to utilize an identical external coupling structure. One might normally anticipate that all resonators employing the identical coupling structure will exhibit the same relative polarity. However, it is noteworthy that this is not true.

In Section 1.2.3.1, we discussed how electric and magnetic couplings have opposite coupling values. The exact polarity of the coupling, however, can be complicated to find and might require thorough analysis of the fields, especially in mixed couplings [14]. The relative polarity can be found without loss of generality in the filter design. In this section, we will investigate how to determine the polarity of coupling structures.

### 2.4.1 External Coupling Polarity

Let's examine the filter depicted in Figure 2.18. It comprises two evanescent-mode cavity resonators that are connected to a source-to-load microstrip transmission line through coupling apertures located in the shared ground plane. Assuming an

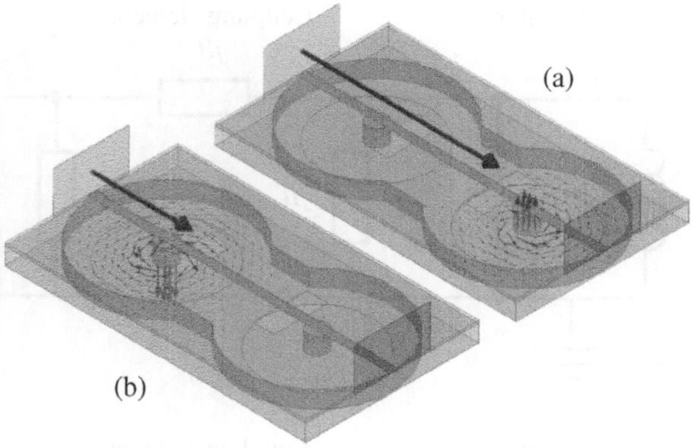


**Figure 2.18** Two resonators are coupled to a source-to-load line. The two couplings have opposite signs because of their coupling apertures locations with respect to their posts.

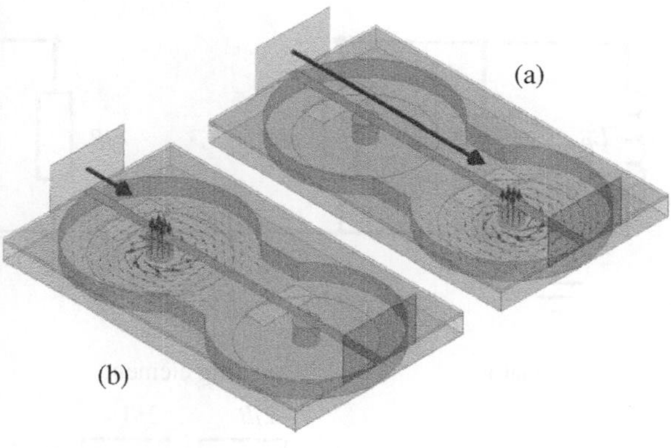


**Figure 2.19** A seemingly minor difference of the relative location of the coupling apertures with respect to Figure 2.18 results in the same coupling signs for both resonators.

energy flow from left to right, notice that the microstrip line excites the left cavity *after* it passes over its post, while it excites the right cavity *before* it reaches its post. The two coupling apertures achieve opposite coupling polarities as a result. This is shown in the figure by plotting the relative directions of magnetic and electric fields. For these plots, each resonator is excited independently while the other resonator is shorted out. The reference plane of the excitation is de-embedded up to the midpoint of the coupling aperture.

Now compare this with the configuration illustrated in Figure 2.19. In this case, the miscrostrip line couples energy to both cavities *before* it passes their posts. In this case, both cavities exhibit the same sign of external coupling. The orientation of the electric fields in this figure is indicative of their identical polarity.

### 2.4.2 Inter-Resonator Coupling Polarity

In Figure 2.20(a), an impedance inverter is used to couple resonators. The shown circuit model of the inverter can be implemented with capacitors ($B = \omega C_c$) or

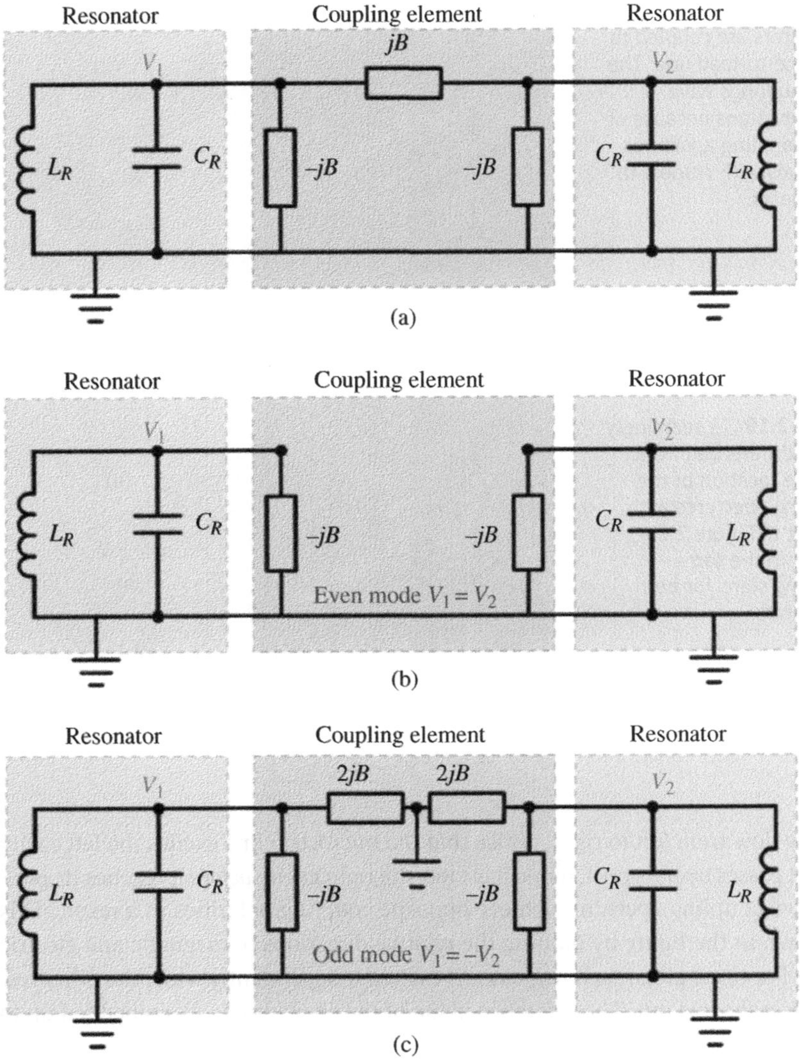


**Figure 2.20** (a) Coupled resonators with a circuit model of an impedance inverter. (b) The coupled resonator in even-mode oscillation. (c) The coupled resonator in odd-mode oscillation.

inductors $[B = -1/(\omega L_c)]$. The values of $C_c$ and $L_c$ can be positive or negative. Since evanescent-mode filters typically couple magnetically, the analysis below assumes inductive $B$.

The resonant frequencies of the circuit in Figure 2.20(a) are found using even/odd analysis. For the even case, shown in Figure 2.20(b),

$$\omega_e = \sqrt{\frac{1}{C_R}\left(\frac{1}{L_R} - \frac{1}{L_c}\right)} \mid V_1 = V_2 \tag{2.28}$$

It is important to notice that since the resonant frequency is still dominated by the resonator itself, it is safe to assume that $L_c > L_R$.

As for the odd mode, we get

$$\omega_o = \sqrt{\frac{1}{C_R}\left(\frac{1}{L_R} + \frac{1}{L_c}\right)} \mid V_1 = -V_2 \tag{2.29}$$

Assuming $L_c > 0$ (resulting in a negative $B$, typically considered negative coupling), $\omega_e < \omega_o$. This causes a −90° insertion phase.

Assuming $L_c < 0$ (resulting in a positive $B$, typically considered positive coupling), $\omega_e > \omega_o$. This causes a +90° insertion phase.

As a result, in order to tell whether an inter-resonator coupling is positive or negative, the resonator voltage polarity needs to be detected at the lower resonant frequency. If the polarities are the same (even), then the coupling is negative and vice versa. The polarities can be determined using FEM simulations.

To demonstrate the coupling polarity concept, Figure 2.21 shows the electromagnetic fields of coupled evanescent-mode resonators at the lower resonant frequency. It can be seen that the standard coupling in Figure 2.21(a) results in a similar electric field polarity, which translates to negative coupling. Figure 2.21(b)

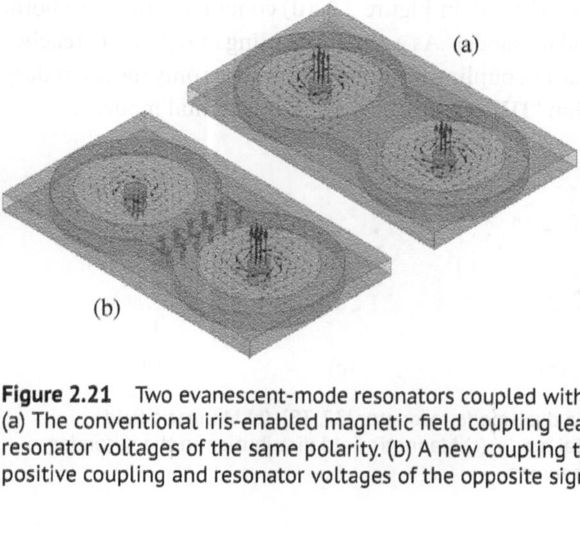


**Figure 2.21** Two evanescent-mode resonators coupled with two different techniques. (a) The conventional iris-enabled magnetic field coupling leads to negative coupling and resonator voltages of the same polarity. (b) A new coupling technique that results in positive coupling and resonator voltages of the opposite signs.

depicts a different coupling technique that results in a positive coupling sign, which can be concluded from the fact that the fields are opposing each other in the two resonators.

## 2.5 Advanced Evanescent-Mode Structures

The evanescent-mode resonators discussed thus far are sealed structures, with a single post in the middle. While this structure is the most common one, others exist with various properties. Below are some examples.

### 2.5.1 Dual-Mode Resonators

Dual-mode resonators differ from the traditional ones by having two posts. They are typically symmetric around the center. A simplified structure is shown in Figure 2.22(a). (Compare this structure to the one in Figure 2.3.)

The presence of two posts creates two distinct resonant modes, typically at two different frequencies. This can be very beneficial for multi-band devices/systems.

Those two resonances are the fundamental or even one [Figure 2.22(b)], and the secondary or odd one [Figure 2.22(c)]. In the fundamental resonance, the two posts behave as one, with practically no magnetic field between them. The secondary resonance shows the two posts having opposite polarities, resulting in a strong magnetic field between them.

The field distribution is important in order to couple to the intended mode. For example, the region marked with "I" in Figure 2.22(d) contains fields from both, the fundamental and secondary modes. As a result, coupling in region "I" reaches both resonant modes. Similarly, coupling in region "II" reaches only the secondary mode, and coupling in region "III" reaches only the fundamental mode.

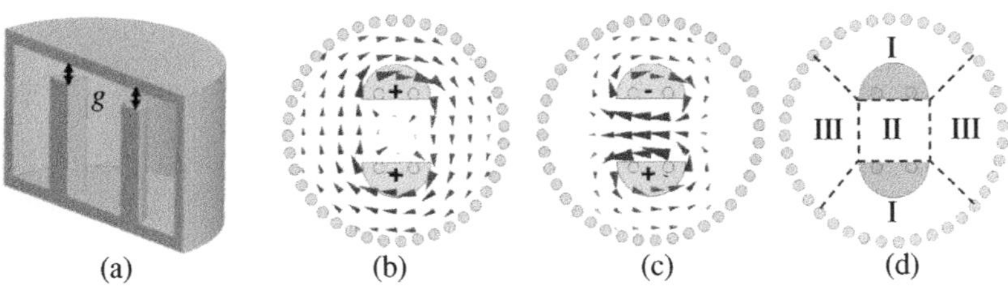


**Figure 2.22** (a) Structure of a dual-mode resonator [37, 38]. (b) Magnetic field distribution in the fundamental mode. (c) Magnetic field distribution in the secondary mode. (d) Available coupling regions.

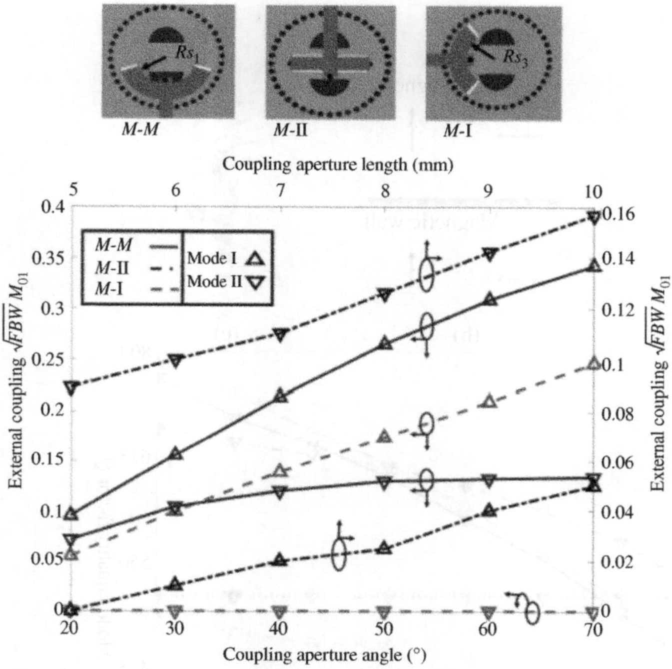


**Figure 2.23** Simulated external coupling coefficient $M_{01}$ of a dual-mode resonator coupled in different regions that can be realized with various aperture lengths and angles.

The effect of different coupling locations is shown in Figure 2.23. For example, while coupling in region "I" has a finite value for both modes, coupling in region "III" has a finite value for the fundamental mode only, and a zero in the secondary mode.

It is important to notice that higher-order resonators (e.g., quad-mode with four posts) are also achievable.

### 2.5.2 Fractional Mode Resonators

In Figure 2.24(a), we can see the magnetic field in a traditional evanescent-mode resonator. The direction of the magnetic field allows for bisecting the resonator along one of its symmetry planes, which is equivalent to a magnetic wall (Figure 2.24(b) details this concept). This results in a half-mode evanescent resonator, as shown in Figure 2.24(c). Compared to the circuit model in Figure 2.3, this bisection doubles the value of the inductance and halves the value of

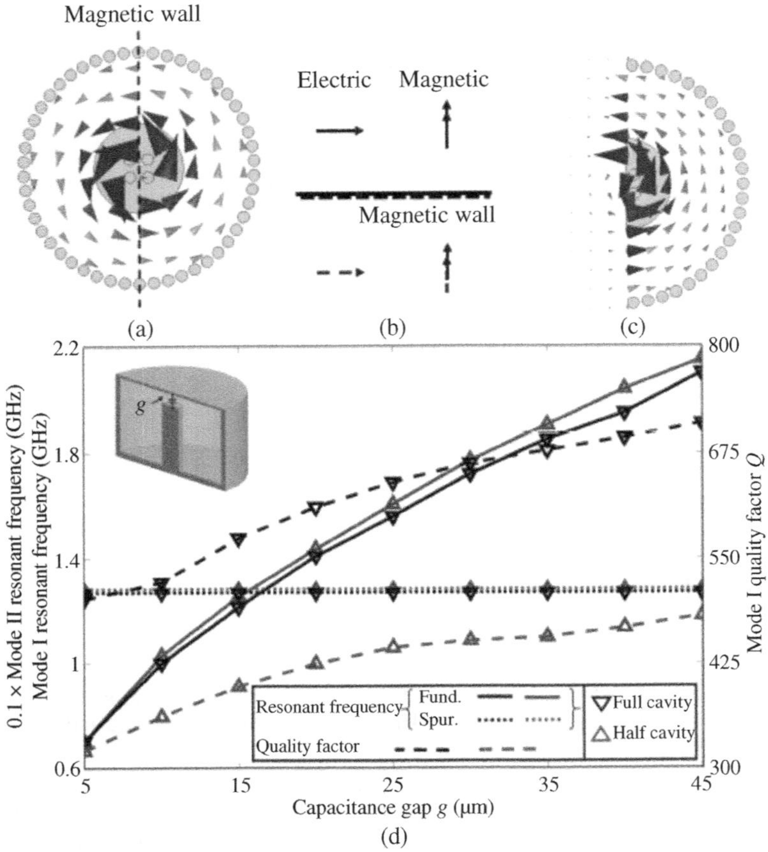


**Figure 2.24** Basic idea of a half-mode resonator. (a) Simulated magnetic field of a conventional full-wave evanescent-mode resonator. (b) The basic image theory ideas over a magnetic wall [39]. (c) Simulated magnetic field of a half-mode evanescent-mode resonator. Notice the similarity to (a). (d) Simulated fundamental and first spurious frequencies for the full- and half-cavity resonators. These example simulations are conducted for resonators implemented on a 100-mil TMM3 Rogers substrate. The post radius is 3.1 mm and the cavity radius is 8 mm.

capacitance, resulting in ideally the same resonant frequency at half the volume. This serves in reducing the size.

Figure 2.24(d) shows the performance of the half-mode resonator in comparison with the full-mode resonator. The fundamental frequencies of both resonators are approximately equal, with no effect on the first spurious resonance mode (which typically dictates the upper limit on the usable frequency of a resonator). Due to the reduced volume of the half-mode resonator, less energy is stored, resulting in a lower $Q_u$. Radiation losses also contribute to the reduction of $Q_u$ since the fields in the half-mode resonator are no longer contained within a closed volume.

While the structure in Figure 2.24 is for a half-mode resonator, quarter modes are also achievable.

## 2.6 Filter Examples

We will discuss two examples of evanescent-mode filters, a BPF and a BSF. The exact details of the fabrication, however, will be discussed in later chapters.

### 2.6.1 Chebyshev Bandpass Filter

The coupling diagram in Figure 1.2 represents an example of a second-order BPF. The first response is dictated by the inter-resonator and external coupling values. In this case, it is a Chebyshev response with 15-dB reflection ripple and 2.7% fractional bandwidth around 900 MHz. The frequency-normalized response is described in (1.6). The Chebyshev coefficients for this response are represented in the following coupling matrix

$$\mathbf{M}_{BPF} = \begin{bmatrix} 0 & 1.037 & 0 & 0 \\ 1.037 & 0 & 1.287 & 0 \\ 0 & 1.287 & 0 & 1.037 \\ 0 & 0 & 1.037 & 0 \end{bmatrix} \tag{2.30}$$

In other words, the external couplings have the value of 1.037, and the inter-resonator coupling is 1.287.

First, the dimensions of a single resonator are chosen. The external diameter is set at 19 mm; the inner post and the tip of the post radii were chosen such that the resonant frequency is at the required value with a maximized $Q_u$.

Second, the external coupling is designed. For a BPF, a CPW coupling is typically convenient. The size of the external coupling structure is swept until (2.24) and (2.25) are satisfied.

The inter-resonator coupling is then realized by combining two similar resonators. The required inter-resonator coupling is not easily achieved by just an opening between the resonators. As a result, an additional CPW inter-resonator structure is employed to strengthen the coupling. The coupling dimensions are also swept until (2.27) is satisfied. Fine tuning is required since the external and inter-resonator couplings will affect the resonant frequency, in addition to each other. Figure 2.25 summarizes the structure and dimensions of this filter.

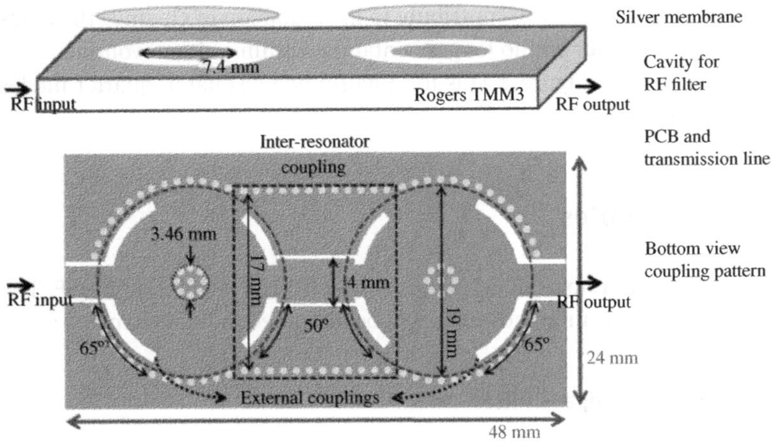


Figure 2.25 Example evanescent-mode second-order BPF.

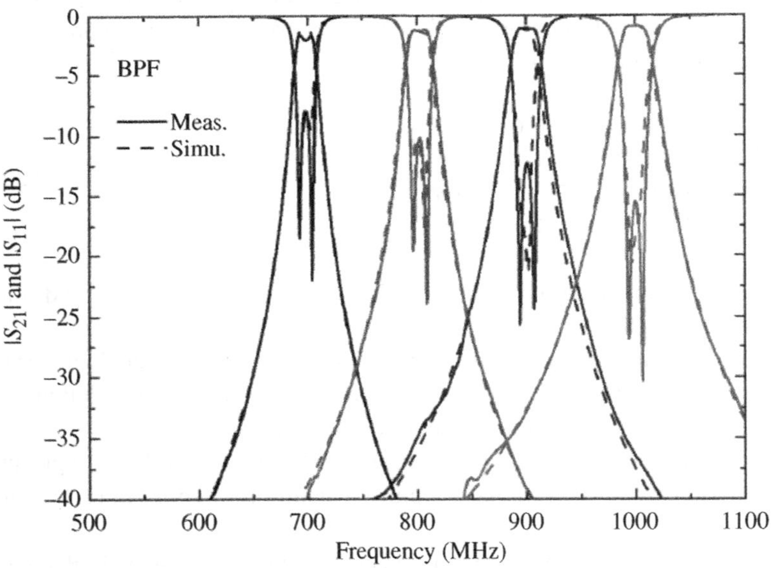


Figure 2.26 Measured and simulated frequency responses of the example tunable filter.

The top of the resonator is flexible to bend, to allow tuning. Various tuning methods are discussed in Chapters 3 and 4. Figure 2.26 shows the results from this BPF. Since the design is a second-order Chebyshev, the reflection coefficient shows two notches within the passband.

### 2.6.2 Butterworth Bandstop Filter

Figure 1.3 shows the coupling diagram of a second-order BSF. The two external coupling values are chosen such that the response follows a maximally flat Butterworth response.

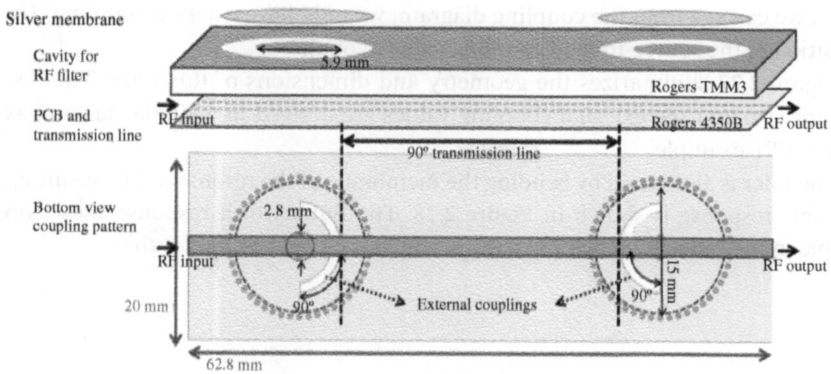


**Figure 2.27** Example evanescent-mode second-order tunable BSF.

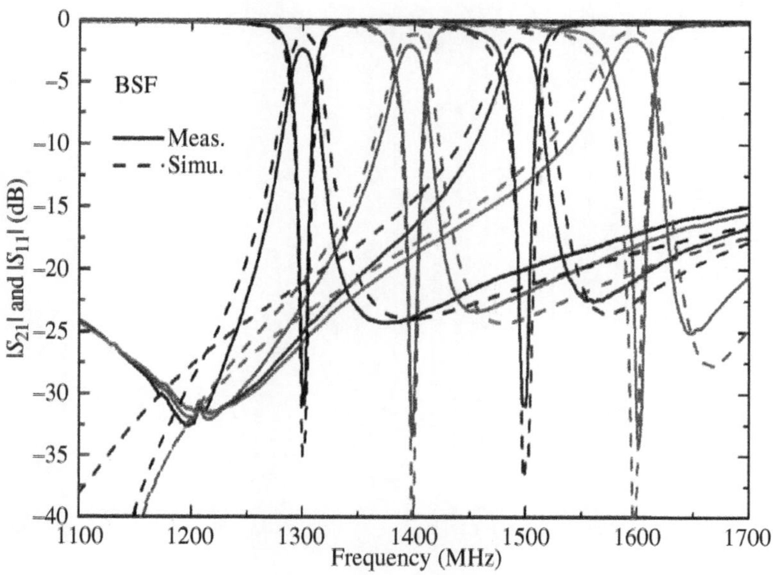


**Figure 2.28** Frequency response of the example evanescent-mode second-order BSF.

The filter is designed at 1.5 GHz, with a 25 MHz bandwidth (1.67%). As a result, the coupling matrix coefficients that deliver this response are

$$\mathbf{M}_{BSF} = \begin{bmatrix} 0 & 1.189 & 0 & 1 \\ 1.189 & 0 & 0 & 0 \\ 0 & 0 & 0 & 1.189 \\ 1 & 0 & 1.189 & 0 \end{bmatrix} \tag{2.31}$$

As we can see from the coupling diagram, we only have external couplings (in addition to the source-to-load coupling value of one).

Figure 2.27 summarizes the geometry and dimensions of this filter. The resonator size and the external coupling values are derived in a similar fashion as in the BPF example.

The filter is also tuned by bending the membrane of the resonator. The resulting tunable response is shown in Figure 2.28. The Butterworth response shows no ripple in the reflection coefficient ($S_{11}$), which is evident in the results.

# Part II

# Implementation

# 3

# Printed Circuit Board Technology

## 3.1 Evanescent-Mode Resonator Structure

The cylindrical part of an evanescent-mode resonator is compatible with the standard printed circuit board (PCB) fabrication process. The resonator structures in bandpass and bandstop filters are slightly different, primarily due to the required external couplings or source-load coupling.

The BPF will be discussed first since it requires a single PCB substrate. The fabrication process starts with a bare PCB substrate as shown in Figure 3.1(a). A bare PCB substrate is clad with thin copper layers (typically 17 or 34 μm thick). The first step is to drill the vias that define the cylindrical structures, and then plate the vias to ensure a good conduction between the top and the bottom layers. The result of this step is shown in Figure 3.1(b).

The top and bottom metal layers are now ready to be patterned. This can be done by etching or milling the areas where the copper is not needed. In Figure 3.1(c), the top layer shows the coplanar waveguide (CPW) coupling into the resonators. On the bottom layer, Figure 3.1(d), the loading post is also patterned. A tunable membrane can now be attached to the finished PCB, which is discussed in the next section.

For a BSF, the fabrication process is very similar to that of the BPF as shown in Figure 3.1(a)–(d). The main difference is that an additional PCB substrate is required for the microstrip as a part of the external and source-to-load couplings as shown in Figure 3.1(e) and (f). The bottom layer of the microstrip substrate and the top layer of the cavity substrate have similar patterns since they will be laminated to deliver the BSF.

It is important to note that this fabrication process, while generally accurate, is not unique. Filters might require different coupling structures as shown in

*Tunable Evanescent-Mode Filters: Principles, Implementation, and Applications*,
First Edition. Dimitrios Peroulis and Mohammad Abu Khater.

Companion website: www.wiley.com/go/Peroulis/TunableEvanescent-ModeFilters

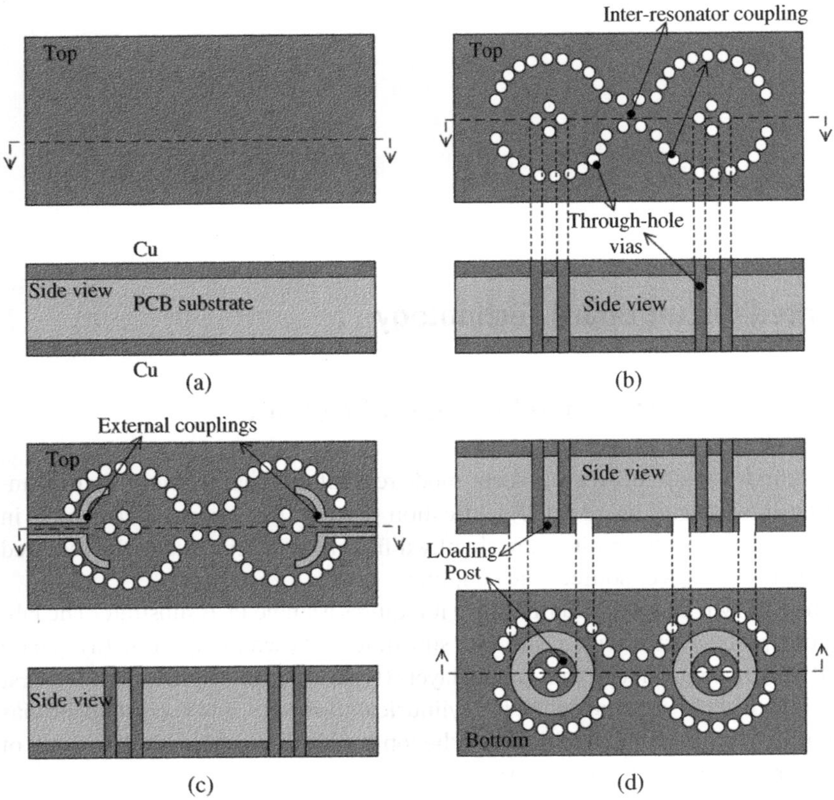


**Figure 3.1** BPF fabrication process in a PCB technology. (a) Bare PCB substrate. (b) PCB substrate after drilling and plating the vias that define the shape of evanescent-mode resonators. (c) Top and (d) bottom views of the PCB substrate after patterning the shapes that define the external coupling in a BPF and the center loading post of the resonator.

Figure 2.13. It is also customary to plate the PCB with gold to prevent copper corrosion and oxidation.

### 3.1.1 Practical Considerations

The through-hole vias shown in Figures 3.1 and 3.2 should, ideally, be placed as close as possible to each other, resembling a cylinder. The spacing between the vias, however, cannot be arbitrarily reduced and is dependent on the board thickness and the via diameter. The density of the vias then should be studied carefully in the simulation stage against the specifications of the PCB fabrication.

The substrate of the filter is typically required to have several properties, including (1) low loss tangent, (2) appreciable thickness (frequency dependent),

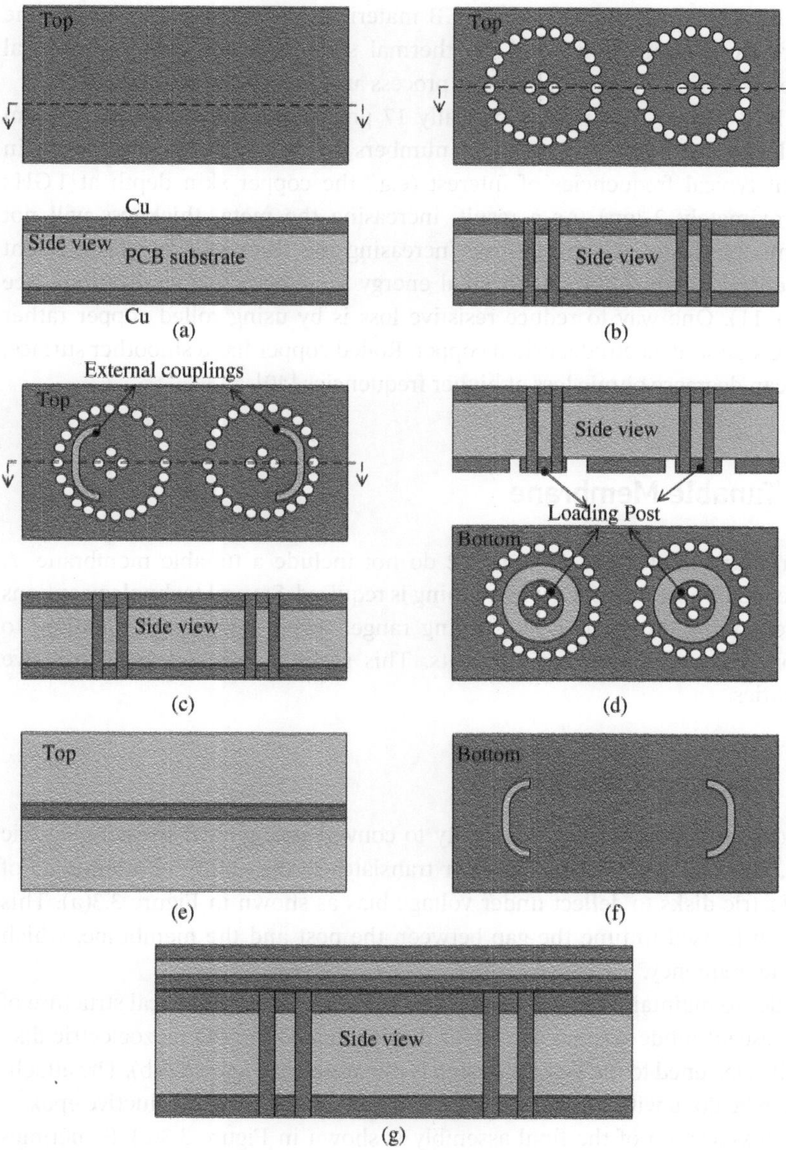


**Figure 3.2** BSF fabrication process in a PCB technology. (a) Bare PCB substrate. (b) PCB substrate after drilling and plating the vias that define the shape of evanescent-mode resonators. (c) Top and (d) bottom views of the PCB substrate after patterning the shapes that define the external coupling in a BPF, and the center loading post of the resonator. (e) Top and (f) bottom view of the signal substrate which is laminated on the cavity substrate to form the BSF. (g) Final structure of the BSF after laminating the substrates in (c) and (e).

(3) multilayer compatibility (some PCB materials are too brittle to handle the pressure from lamination), and (4) thermal stability. Consequently, material selection is a critical part of the design process and should not be overlooked.

The PCB copper thickness is typically 17 μm before processing, and 34 μm after plating the vias. Both of these numbers are much larger than the skin depth at typical frequencies of interest (e.g., the copper skin depth at 1 GHz is approximately 2 μm). As a result, increasing the metal thickness will not substantially reduce its ohmic loss. Increasing the thickness, however, might be beneficial in spreading out thermal energy in some active applications (see Chapter 11). One way to reduce resistive loss is by using rolled copper rather than the standard electrodeposited copper. Rolled copper has a smoother surface, which can decrease ohmic loss at higher frequencies [40].

## 3.2 Tunable Membrane

The structures in Figures 3.1 and 3.2 do not include a tunable membrane. A tunable membrane is needed when tuning is required. Several technology options exist depending on the required tuning range, speed, and power handling to mention a few common requirements. This section discusses two attractive possibilities.

### 3.2.1 Piezoelectric Disk Tuners

A piezoelectric material has the ability to convert mechanical stress to electric charge, and vice versa.[1] This behavior translates to the ability of a laminate of piezoelectric disks to deflect under voltage bias as shown in Figure 3.3(a). This deflection is used to tune the gap between the post and the membrane, which tunes the frequency.

In order to maintain a sealed conductive surface on the cylindrical structure of an evanescent-mode resonator, a silver disk is attached to the piezoelectric disk before it is fastened to the PCB. This step is illustrated in Figure 3.3(b). The attachment can be done with a standard soldering method, or using conductive epoxy.

The cross section of the final assembly is shown in Figure 3.3(c). Sometimes it is necessary to electroplate outside the post with thicker copper to center the piezoelectric tuner in the middle of the gap. The thickness of the conductive epoxy, with the small required deflections, typically makes this extra electroplating step not necessary.

---

1 A more accurate definition of piezoelectricity exists in the physics communities. The given definition, however, is sufficient for filter applications. Piezoelectric materials are very common in spark generators and inkjet printers. The phenomenon is observed in many materials including table sugar.

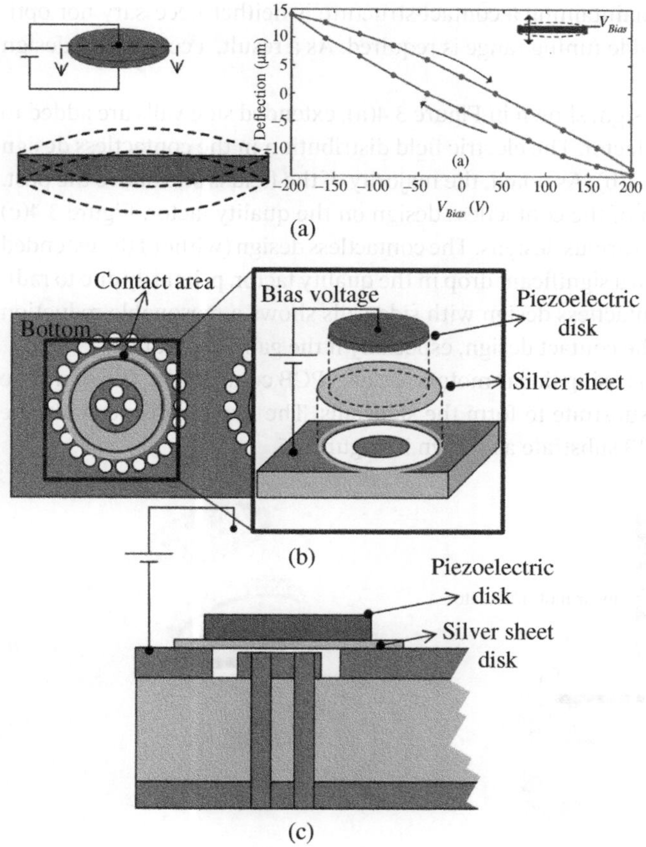


**Figure 3.3** (a) Piezoelectric laminate disks can bend under voltage bias. (b) The piezoelectric disk is attached to a silver disk, and the silver disk is attached to the evanescent-mode PCB, creating a tunable resonator. (c) A cross section of the assembled filter.

In order to prevent an electric short between the post and the silver disk, an insulating material is deposited on the post. The most common example is Parylene using chemical vapor deposition (CVD).

While piezoelectric devices require relatively high voltages (10 s to 100 s V), they consume practically no current. As a result, their power consumption is limited to charging and discharging their equivalent capacitance, which is often negligibly small.

### 3.2.2 Contactless Mechanical Actuators

Recent advancements in micro linear actuators, particularly in size and accuracy, have enabled their use as tuning elements in filters. Due to their ability to move over a wide range, maintaining a contact structure is neither necessary nor optimal, especially if a wide tuning range is required. As a result, a contactless design was adopted.

In a contactless design, shown in Figure 3.4(a), extended sidewalls are added to enhance the quality factor. The electric field distribution in the contactless design is shown in Figure 3.4(b). As noted, the majority of the field is still above the post.

To study the effect of the contactless design on the quality factor, Figure 3.4(c) shows the results for various designs. The contactless design (without the extended sidewall) suffers from a significant drop in the quality factor, primarily due to radiation losses. The contactless design with sidewalls shows a reasonable reduction in $Q_u$ compared to the contact design, especially if the gap $d_2$ is reduced.

Contactless designs using linear motors are also PCB compatible. The structure requires additional substrate to form the sidewalls. The moving disk can also be fabricated from a PCB substrate as shown in Figure 3.5.

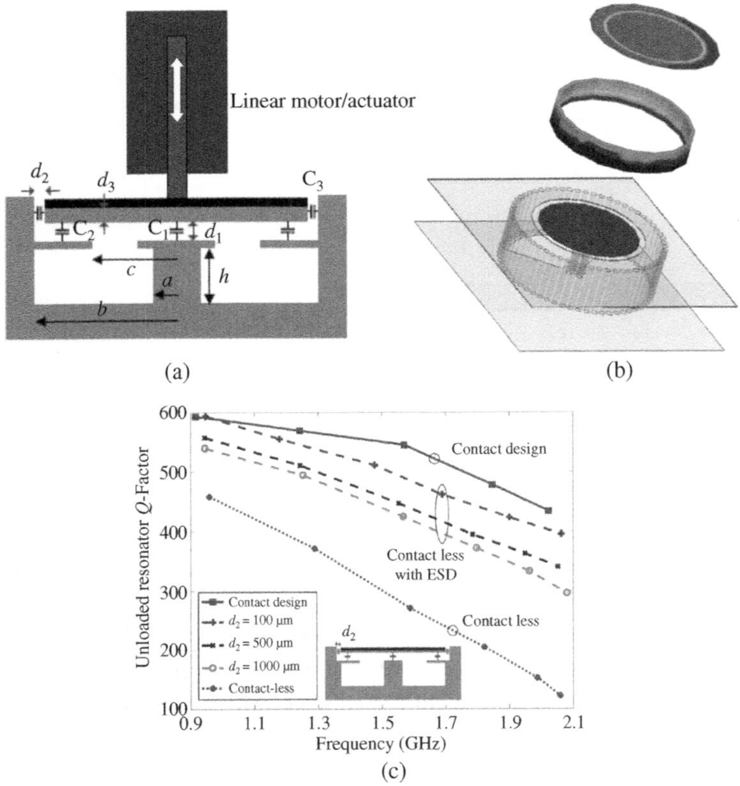


**Figure 3.4** (a) Contactless evanescent-mode resonators with extended sidewall structure. (b) Exploded view with the electric field overlaid. (c) Quality factor comparison.

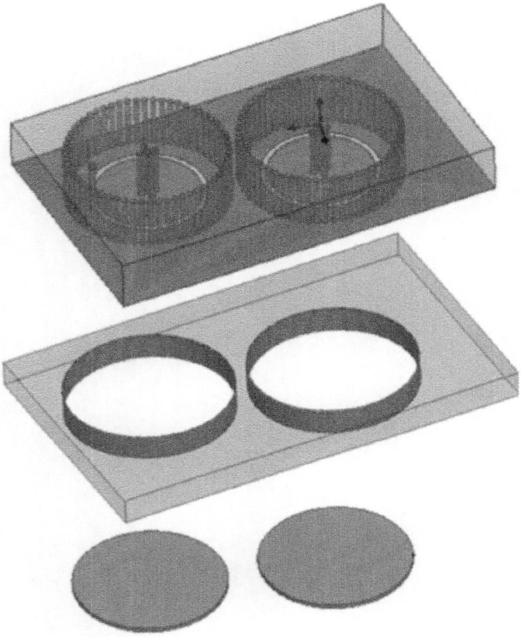

**Figure 3.5** Contactless evanescent-mode resonator structure in PCB technology.

**Table 3.1** Tuning mechanism comparison.

| | **Piezoelectric** | **Mechanical actuators** |
|---|---|---|
| Advantages | • Compact<br>• Low cost<br>• Low power | • High accuracy<br>• Digital interface<br>• Repeatable |
| Disadvantages | • Hysteresis<br>• High voltage | • High cost<br>• Bulky |

Table 3.1 summarizes the advantages and the disadvantages of using the tuning mechanisms discussed in this chapter.

# 4

# Silicon Microfabrication

## 4.1 Generic Structure

An example structure of a silicon microfabricated bandpass filter (BPF) with a MEMS tuner is shown in Figure 4.1. The cavity top ceiling is a silicon substrate sputter-coated with a layer of gold film. This forms two flexible diaphragms (for the two resonators) that control the resonant frequency. A pair of electrodes (gold-coated silicon-etched posts) are placed above the tuning diaphragms through appropriately etched via holes on the back side of the diaphragms. A layer of silicon dioxide electrically isolates the electrodes from the silicon substrate. The bias on the electrodes creates an electric field that generates an electrostatic force, pulling the membrane closer to the electrodes, and further from the post. This allows independent control of each diaphragm. As an additional design detail to be noted, the height of the bias electrode determines the initial electrostatic bias gap, which accordingly defines the maximum applied actuation voltage for a given diaphragm deflection. Two feeding pins are inserted into the filter cavities through the etched via holes and coupled to the cavity resonator by means of magnetic coupling (coaxial tap external coupling in Figure 2.13). It should be noted that the feeding pins always maintain DC contact with the upper ceiling of the cavity for optimal mechanical stability, lower fabrication uncertainty, and stronger magnetic field coupling.

Additional silicon substrate is needed to form the resonators and the inter-resonator coupling. The lamination of these structures forms a BPF as shown in Figure 4.2. For the rest of this chapter, the design aspects of silicon microfabrication and related issues are discussed.

*Tunable Evanescent-Mode Filters: Principles, Implementation, and Applications*,
First Edition. Dimitrios Peroulis and Mohammad Abu Khater.

Companion website: www.wiley.com/go/Peroulis/TunableEvanescent-ModeFilters

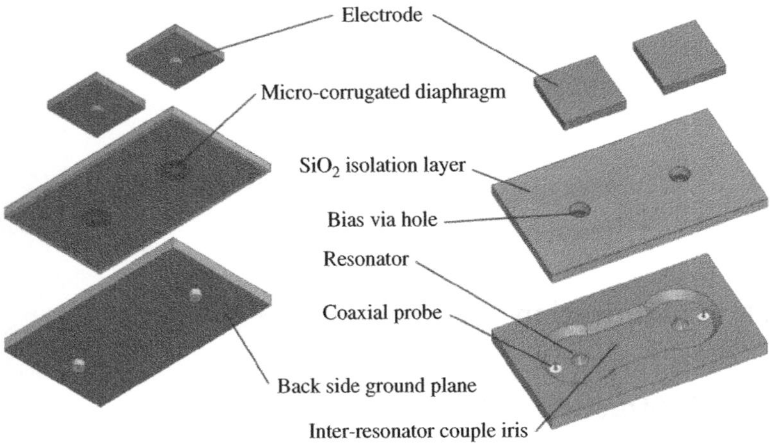


**Figure 4.1** 3-D exploded view of an evanescent-mode tunable BPF filter implemented in silicon microfabrication.

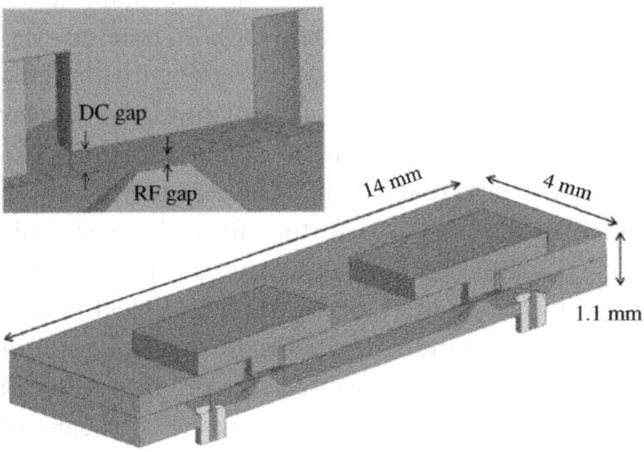


**Figure 4.2** Assembled filter from Figure 4.1.

## 4.2 MEMS Tuner Design

The bending behavior of a flat membrane can be described by:

$$P_{flat} = \frac{4t_d}{R_d^2}\left(\sigma_0 + 1.33\frac{E_d t_d^2}{4R_d^2}\right)\Delta g + 2.83\left(\frac{E_d t_d}{R_d^4}\right)\Delta g^3 \tag{4.1}$$

where $P_{flat}$ is the pressure imposed on the flat diaphragm and $\Delta g$ is the out-of-the-plane deflection. $E_d$, $t_d$, and $R_d$ are the Young's modulus, diaphragm thickness, and diaphragm radius respectively. $\sigma_0$ is the diaphragm residual stress. Scaling such a design into a smaller resonator cavity operating at a higher frequency demands further reduction of $R_d$. This increases the required $P_{flat}$, which results in a higher actuation voltage. Using softer material (lower Young's

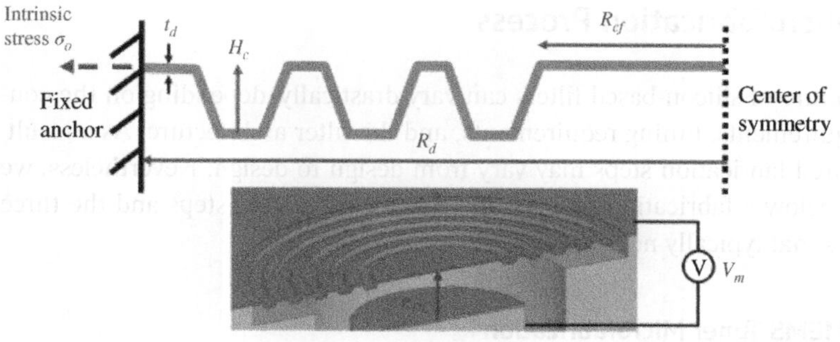


**Figure 4.3** Micro-corrugated diaphragms (MCDs) which reduce the required electrostatic actuation voltage.

modulus) can mitigate the problem. The scaling bottleneck, however, still exists since $R_d$ remains at the highest power among all parameters.

To overcome this problem, an alternative approach employing micro-corrugated diaphragms (MCDs) is used. The MCD exhibits significantly lower tensile rigidity in the radial direction [41]. Instead of stretching the material, bending of an MCD elongates its corrugation sections and effectively reduces the radial tensile stress. As a result, an MCD demands lower bias voltage and suffers less material stretch for the same amount of deflection. This concept is shown in Figure 4.3. The bending behavior of an MCD can be modeled as the superposition of a stress-free corrugated diaphragm and a flat diaphragm with initial stress [41], which yields:

$$P_{MCD} = \left( A_p \frac{E_d t_d^3}{R_d^4} + \sigma_0 \frac{4B_p \left(1 - \mu^2\right) t_d}{2.83 R_d^2} \right) \Delta g + B_p \frac{E_d t_d}{R_d^4} \Delta g^3 \tag{4.2}$$

In the above equation, $P_{MCD}$ represents homogeneous pressure on the corrugated diaphragm. The small deflection stiffness coefficient $A_p$ and large deflection stiffness coefficient $B_p$ are both related to the corrugation profile factor, which is dominated by the ratio of $t_d$ to corrugation depth $H_c$. The electrostatic pressure induced by two parallel DC biased electrodes is given by (assuming the dimensions of the electrodes are much larger than the DC gap)

$$P_{MCD} = -\frac{\frac{1}{2}\epsilon_0 V^2}{g_{DC0} - \Delta g} \tag{4.3}$$

where $V$, $\epsilon_0$, $g_{DC0}$, are the bias voltage, vacuum permittivity, and initial gap between the electrode and diaphragm, respectively. Equating (4.2) and (4.3) yields the relationship between bias voltage and diaphragm deflection. Solving the equation $d(V) = d(\Delta g) = 0$ yields the pull-down voltage and the continuous tuning range.

## 4.3 Microfabrication Process

The structure of silicon-based filters can vary drastically, depending on the coupling requirements, tuning requirements, and the filter architecture. As a result, the required fabrication steps may vary from design to design. Nevertheless, we describe below a fabrication process that includes the basic steps and the three substrates that typically need to be processed.

### 4.3.1 MEMS Tuner Microfabrication

The following steps assume a double-sided polished silicon (Si) wafer whose crystal orientation is <1 0 0>. The first step defines the corrugations using photolithography followed by a buffered-oxide-etch (BOE) process. This makes the top silicon dioxide layer a hard mask of the corrugation pattern. Wet-etching the silicon in a tetramethylammonium hydroxide (TMAH) bath forms trapezoid corrugation profiles. Additional materials such as Triton X-100 surfactant retain the corrugation geometry and surface smoothness. The outcome of this step is shown in Figure 4.4(a).

In the second step, the remaining sacrificial silicon oxide is etched with a photolithography and BOE etching process, resulting in the cross section shown in Figure 4.4. It is important to keep the silicon oxide outside the membrane since it acts as a DC insulator in the third step, the membrane metal is sputtered on top of the corrugations. This metal is typically gold or a gold alloy, due to its chemical stability and relatively low tensile stiffness (Young's modulus). As can be seen in Figure 4.5(a), this metal layer takes the shape of the corrugations formed earlier in the silicon. The metal sputtering is typically preceded with a thin (~1–10 nm) seed layer of titanium, to enhance adhesion.

The sputtered metal is then patterned with photolithography and etching in the fourth step. The etching removes a large area of gold on the die boundary as well as small releasing holes in the diaphragm area. The release holes are used later to access the silicon beneath the metal layer (the last step in the MEMS tuner fabrication). They also help in reducing the air drag when tuning the membrane.

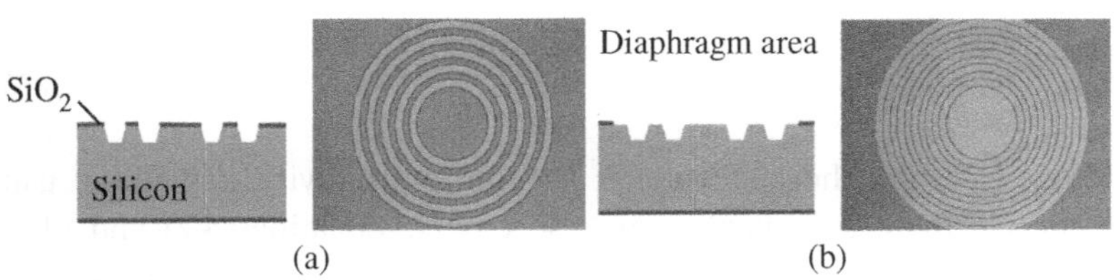


**Figure 4.4** MCD fabrication (a) step 1: Silicon wet-etch to pattern the corrugations. (b) step 2: Oxide-etch removes oxide under MCD area.

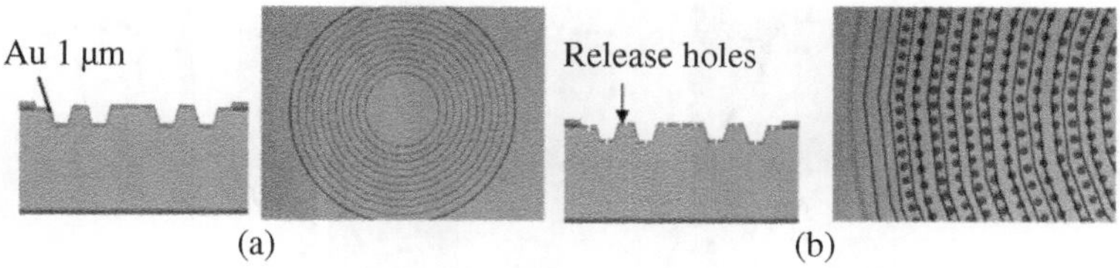


**Figure 4.5** MCD fabrication (a) step 3: Metal sputtering (Au in this case). (b) step 4: Patterning the metal sputtered in step 3.

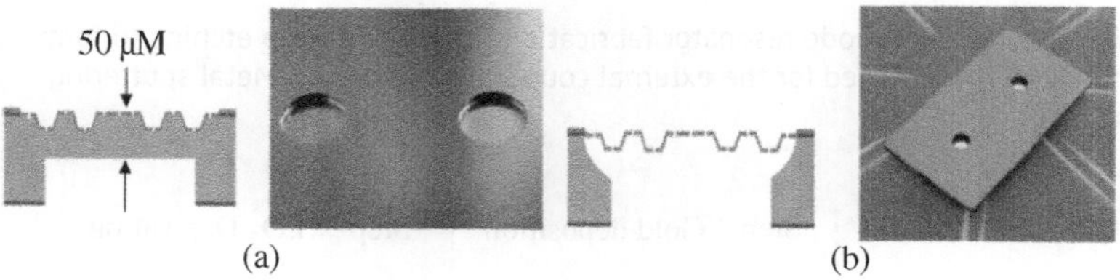


**Figure 4.6** MCD fabrication (a) step 5: DRIE is used to etch the back side of the wafer with vertical walls. (b) step 6: Final step. The membrane is released with Xenon difluoride silicon etching process.

Additional etching might be required to remove the Ti seed layer. The outcome of this step is shown in Figure 4.5(b).

In the fifth step, a recess is formed on the back side of the wafer, beneath the membrane. This is done with deep reactive-ion etching (DRIE). This eases the next and final step in fabricating the MEMS tuner, which is etching the remaining silicon beneath the membrane metal. This is done with a Xenon difluoride silicon etching process. At this point, the membrane is released and is bendable, as shown in Figure 4.6.

### 4.3.2 Evanescent-Mode Resonator Microfabrication

Starting on a double-sided polished silicon wafer with crystal orientation <1 0 0>, a TMAH etching technique creates a cavity with a post in the middle. This is typically preceded with photolithography. The RF feed holes (external coupling) at the bottom of the cavity are etched with the DRIE process. Subsequently, BOE etching removes all remaining silicon dioxide. Metal is then sputtered on the top and bottom sides of the cavity, forming the resonator structure. These steps are summarized in Figure 4.7.

### 4.3.3 Bias Electrode Microfabrication

The electrode is also fabricated on a double-sided polished silicon wafer. First, a DRIE process on the front side of the wafer creates the biasing posts. Metal

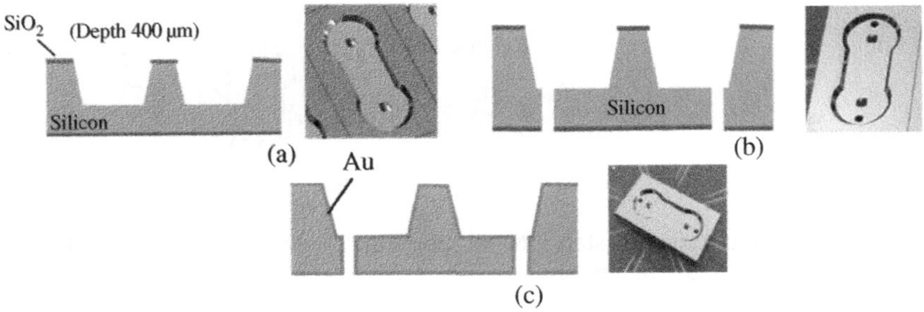


**Figure 4.7** Evanesceent-mode resonator fabrication. (a) TMAH silicon etching to form the resonator. (b) DRIE is used for the external coupling structure. (c) Metal sputtering (Au in this case).

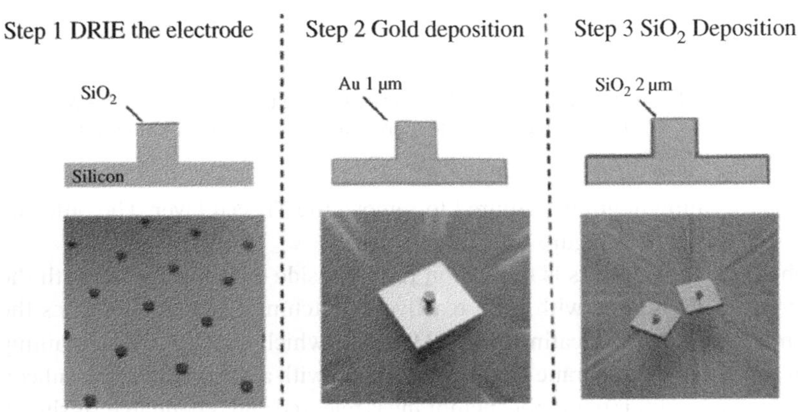


**Figure 4.8** Bias electrode fabrication.

sputtering then coats the electrode post and side walls. Finally, the gold is covered with silicon dioxide to protect against accidental short between the bias electrode and the MEMS membrane. Figure 4.8 demonstrates these steps along with sample photos.

### 4.3.4 Filter Assembly

Each one of the three wafers discussed above is diced with the corresponding dimensions, and then cleaned to prepare them for the assembly of an evanescent-mode filter structure.

First, the MEMS tuner is laminated on the resonator. Typically, the MEMS release process (last step in Section 4.3.1) is done after the lamination, to minimize the mechanical stress on the membrane. One common method to laminate the diced pieces is based on thermal compression. Conventional thermal compression, however, requires relatively high temperatures (>300 °C) and pressures that can damage the components. As a result, surface activation is typically used, where plasma irradiation roughens the surface slightly [42]. This can reduce the required temperature to approximately 150 °C.

Lastly, the bias electrode is fitted on top of the laminated structure, resulting in the tunable evanescent-mode resonator shown in Figure 4.9.

As we can see in Figure 4.9, no source-to-load coupling exists. This is an indication that this structure is for a BPF. A BSF structure would require an additional substrate for a transmission line structure to couple source and load. An example for such a stack up is shown in Figure 4.10. We would like to note here that the BSF in Figure 4.10 combines the MEMS tuner and the bias electrode processes into a single wafer. The fabrication steps involved for this, however, are similar to the ones discussed above.

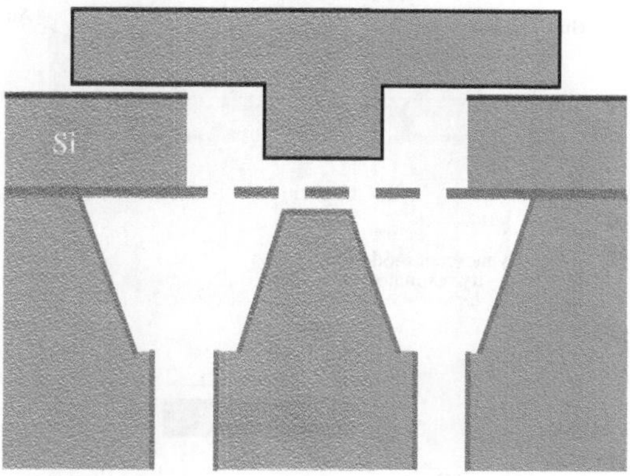


**Figure 4.9** Assembled silicon microfabricated evanescent-mode resonator for a BPF.

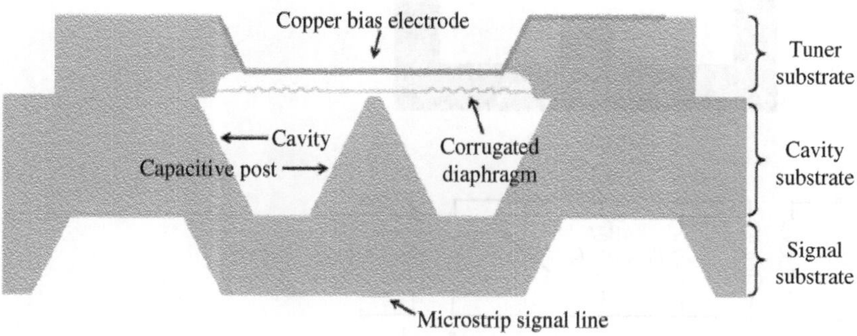


**Figure 4.10** Assembled silicon microfabricated evanescent-mode resonator for a BSF.

## 4.4 Mechanical Model and Power Handling

When a resonator is excited by high RF power, non linear behavior may emerge. This section focuses on RF self-actuation in evanescent-mode tunable resonators, which is one of the non-linearity concerns. To that end, an understanding of the mechanical behavior of the MEMS membrane needs to be developed. A simplified model for this is shown in Figure 4.11(a).

The mechanical behavior of the thin membrane tuner can be approximated by a simple 1-D spring-mass model shown in Figure 4.11(b). The diaphragm actuator is primarily subject to three forces:

1. The electrostatic force $F_{DC}$ from the bias electrode. Assuming an electric field only exists in the overlapping area between the bias electrode and the

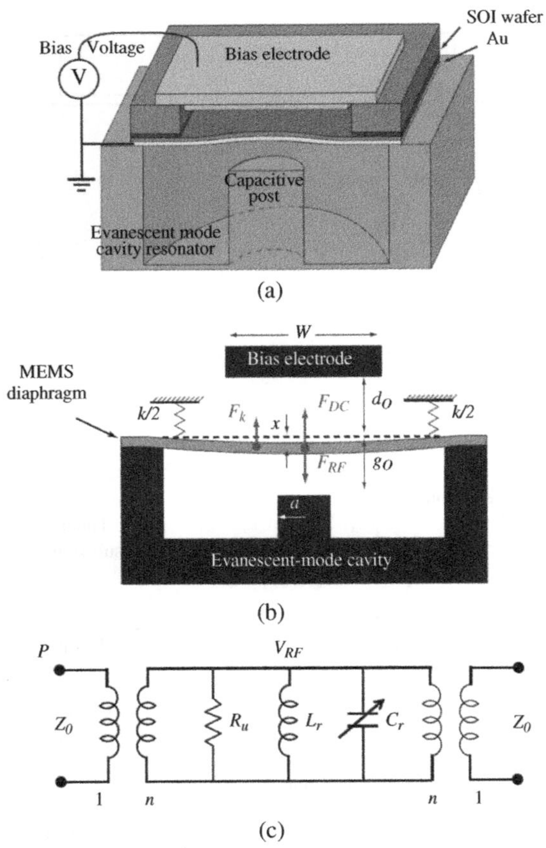


**Figure 4.11** (a) Concept drawing of MEMS evanescent-mode tunable resonator. (b) Spring-mass model of the MEMS diaphragm actuator. (c) Equivalent circuit of the MEMS evanescent-mode tunable resonator.

diaphragm actuator, $F_{DC}$ can be approximated by

$$F_{DC} = \frac{\epsilon_0 W^2 V_{DC}^2}{2(d_0 + x)^2} \quad (4.4)$$

where $W$ is the width of the bias electrode, $V_{DC}$ is the bias voltage, $d_0$ is the initial gap between the bias electrode and the membrane, and $x$ is the deflection of the diaphragm. This equation neglects the effect of the fringing field, which can be taken into account by the nonlinear circuit model.

2. The electrostatic force from the RF signal power $F_{RF}$ [43]. Using parallel-plate capacitance for $C_r$, the resonator capacitance, $F_{RF}$ is given by

$$F_{RF} = \frac{\epsilon_0 \pi a^2 |V_{RF}|^2}{4(g_0 - x)^2} \quad (4.5)$$

where $a$ is the post radius, $V_{RF}$ is the peak-peak RF voltage between the post and the diaphragm and $g_0$ is the initial gap between the post and the membrane. Again, the fringing field contribution to $F_{RF}$ is not taken into consideration here.

3. Mechanical restoring force $F_k$. Assuming linear deflection, $F_k$ is given by

$$F_k = kx \quad (4.6)$$

where $k$ is the spring constant of the diaphragm.

Within a resonator's tuning range, electromechanical equilibrium is achieved between these forces

$$F_{DC} + F_{RF} + F_k = 0 \quad (4.7)$$

At low input RF power, the deflection of the diaphragm actuator is dominated by the electrostatic force between the diaphragm and the DC biasing electrode. When the input power is increased, the RF-induced electrostatic force ($F_{RF}$) starts to affect the deflection of the diaphragm. Because the DC actuation electrode is separated from the RF signal path, the equivalent electrostatic force from RF power counteracts the DC electrostatic force, therefore pulling the diaphragm toward the capacitive post causing nonlinear responses. In a narrowband resonator/filter, this nonlinear response is "amplified" by the input and output transformers. Inserting equations (4.4), (4.5), and (4.6) into equation (4.7), we get

$$\frac{\epsilon_0 W^2 V_{DC}^2}{2(d_0 + x)^2} + \frac{\epsilon_0 \pi a^2 |V_{RF}|^2}{4(g_0 - x)^2} + kx = 0 \quad (4.8)$$

Note that the DC bias increases the capacitive gap and, therefore, reduces $F_{RF}$. In other words, when a DC bias is applied to increase the resonant frequency, the power handling capability will also increase. As a result, the power capability is the lowest when no DC bias is applied. In the following analysis, without loss of generality, we assume no DC bias and look at the nonlinear response of the evanescent-mode tunable resonators solely due to RF power. With no DC electrostatic force, (4.8) is simplified

$$\epsilon_0 \pi a^2 |V_{RF}|^2 = 4kx(g_0 - x)^2 \tag{4.9}$$

$V_{RF}$ can be calculated by linear circuit analysis of the model in Figure 4.11(c)

$$V_{RF} = \frac{2}{\left(j\omega C_r + \frac{1}{j\omega L_r} + \frac{1}{R_u}\right) + \frac{2}{n^2 Z_0}} \sqrt{\frac{2P_{RF}}{n^2 Z_0}} \tag{4.10}$$

where $n$ is the transformation ratio, $Z_0$ is the port impedance, and $P_{RF}$ is the RF power from the input port. Inserting equation (4.10) into equation (4.9), and using parallel-plate approximation for $C_r = \frac{\epsilon_0 \pi a^2}{g_0 - x}$, and rearranging both sides of the equation, we get

$$\frac{x}{g_0}\left[\left(\left(\frac{\omega}{\omega_0}\right)^2 - \frac{g_0 - x}{g_0}\right)^2 + \left(\frac{g_0 - x}{g_0}\right)^2 \left(\frac{\omega}{\omega_0}\right)^2 \frac{1}{Q^2}\right] = \left(\frac{\omega}{\omega_0}\right)^2 F \tag{4.11}$$

where

$$\omega_0^2 = \frac{1}{L_r C_r} = \frac{g_0}{\epsilon_0 \pi a^2 L_r}$$

is the small-signal resonant frequency of the resonator,

$$Q = \frac{1}{\omega_0 L_r \left(\frac{1}{R_u} + \frac{2}{n^2 Z_0}\right)}$$

is the doubly loaded quality factor of the resonator and

$$F = \frac{2 P_{RF} L_r}{k n^2 Z_0 g_0^2}$$

We now define a normalized gap

$$\hat{g} = \frac{g_0 - x}{g_0}$$

and normalized frequency

$$\hat{\omega} = \left(\frac{\omega}{\omega_0}\right)^2$$

Equation (4.11) can be now simplified as

$$(1 - \hat{g})\left[(\hat{\omega} - \hat{g})^2 + \frac{\hat{g}\hat{\omega}}{Q^2}\right] = \hat{\omega} F \tag{4.12}$$

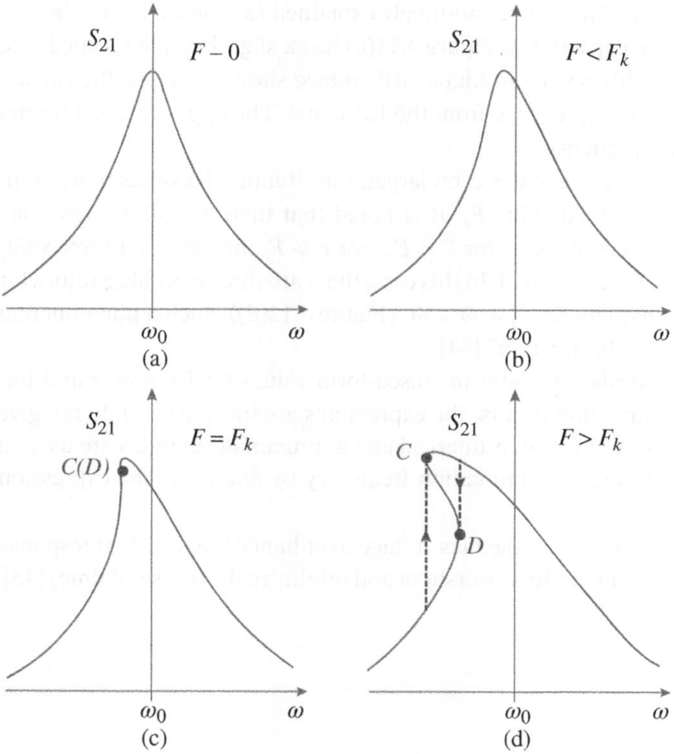


**Figure 4.12** Frequency responses of the nonlinear MEMS evanescent-mode resonator with different input power levels. (a) Symmetric response with very small input. (b) Frequency distortion with medium input power. (c) Onset of bifurcation. (d) Bifurcation.

Equation (4.12) is a nonlinear equation describing the relationship between the normalized deflection of the membrane and the RF power. It is a third-order equation in terms of $\hat{g}$ and has three solutions in the complex domain. Among the three solutions, the ones in the real domain give the amplitude of the normalized diaphragm deflection under certain external RF power.

For small input power, i.e., small $F$, only one solution is in the real domain. This corresponds to the case of small signal input (Figure 4.12(a)). In the limiting case of $F \to 0$, the frequency response of the resonator is symmetrical around the resonant frequency.

As $F$ increases, the resonant frequency shifts to a lower frequency, and the frequency response starts to "bend" toward lower frequency as well.

From this conceptual experiment, it is obvious that the evanescent-mode resonator reacts differently to input RF signals below and above the resonant frequency. The consequence of such different behaviors is the asymmetrical response shown in Figure 4.12. This can be intuitively explained as follows: below the resonant frequency, the capacitor in Figure 4.11(c) has a slightly higher impedance than at resonance. This results in higher RF voltage showing across the capacitor, causing higher pulling voltage from the RF signal. The opposite effect occurs above the resonant frequency.

When the input power becomes even larger, the situation becomes more complex as $F$ reaches a critical value $F_c$. It is noted that there is still a one-to-one correspondence between $\hat{g}$ and $\hat{\omega}$ for $F < F_c$. For $F > F_c$, however, all three solutions to equation (4.12) can be real. In this case, there are three possible $\hat{g}$ values for a certain range of frequencies $\hat{\omega}_1 < \hat{\omega} < \hat{\omega}_2$ (Figure 4.12(d)). Such a phenomenon is often referred to as "bifurcation" [44].

While further analysis can result in closed-form values for RF power and frequency at which bifurcation occurs, the expressions are involved and do not give any further insights. As a result, numerical and approximate analyses are used to find the power handling and bifurcation frequency by analyzing the expression $\partial\hat{g}/\partial\hat{\omega} \rightarrow \infty$.

It is also possible to engineer the bias voltage to enhance the transient response of the tunable membrane (reduce overshoot and minimize the transient time) [45].

# 5

# Injection Molding

## 5.1 Manufacturing Technology

In recent years, much interest has been shown to generate high-performance devices for wireless communications through new technologies such as additive manufacturing [46]. In [47], state-of-the-art performance is demonstrated with a measured $Q_u$ between 1,638 and 3,520 on a novel monolithic air-cavity bandpass filter using additive manufacturing. Simultaneously, 3D printing technologies continue to generate improved results as the industry delivers machines that can match higher standards in geometrical tolerances and expand into new materials with a broad portfolio of properties. Emphasis has been placed on their performance while maintaining low manufacturing cost, from design to end product.

Injection molding of thermoplastics is used widely on everyday-use products. In recent years, micro-injection molding technology has advanced to deliver parts with tight tolerances on the order of micrometers. These can be successfully metalized with finely finished surfaces while producing parts with complex 3D geometries, providing a significantly lower cost per part and lower weight per volume, with the same or better characteristics of a machined part. Moreover, micro-injection molding has the potential to generate components for higher-frequency bands as they can maintain the required micrometer-scale tolerances.

### 5.1.1 Device Concept

The conceptual structure of a tunable evanescent-mode air-filled cavity resonator is shown in Figure 5.1. Tunable evanescent-mode cavity resonators have two essential parts: (a) a static cavity with a post located in the center, and (b) a flexible

*Tunable Evanescent-Mode Filters: Principles, Implementation, and Applications*,
First Edition. Dimitrios Peroulis and Mohammad Abu Khater.

Companion website: www.wiley.com/go/Peroulis/TunableEvanescent-ModeFilters

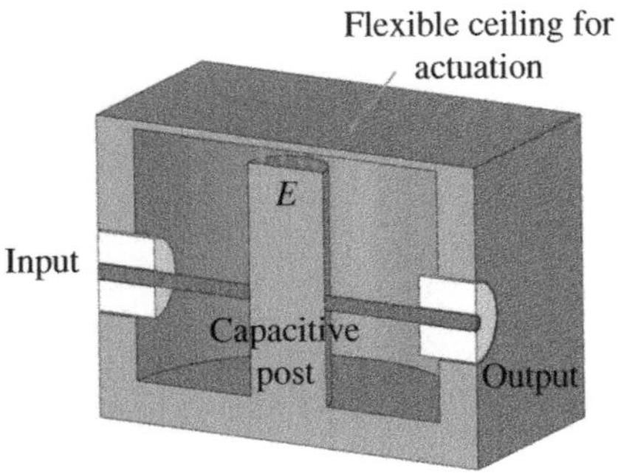


**Figure 5.1** Conceptual structure of an evanescent-mode air-filled cavity resonator.

ceiling that serves as a tuning element. The walls of the air-filled cavity must be formed using conductive metal. To realize such a cavity, we need to manufacture two parts and assemble them at a later step. Moreover, the tuning element requires a method of actuation that has been previously demonstrated using electrostatic [48], magneto-static [49], and piezo-static [50] technologies. Because our goal is to demonstrate the technology related to device performance, actuation is realized using a commercially available submicron linear positioner. The positioner's cost is not taken into consideration, given that it can be chosen in accordance with the desired application. Finally, one crucial structural parameter is the bonding of the two parts to create an air cavity. Different methods have been demonstrated using mechanical pressure, conductive epoxy [51], and thermo-compression bonding, as reported in [48]. In this chapter, we incorporate a mechanical bonding technique.

Technological limitations and capabilities are considered to define the best possible design for part manufacturing, part assembly, and device performance. One critical parameter is the flexible upper wall of the air-filled cavity, which serves as a tuning element. An external actuator facilitates its accurate movement to control the resonant frequency of the device. As such, the large diameter of the tuning membrane allows for higher deflection, which dictates the tuning range. The geometrical characteristics, such as the diameter of the post, the diameter of the cavity, the depth of the cavity, and the actuation length of the membrane, are considered in the RF design and are explained later in this chapter.

### 5.1.2 Injection Molding Technology

Injection molding technology is widely used to manufacture products we use in our everyday lives by injecting molten thermoplastic into a mold. It is the most commonly used mass manufacturing process, capable of providing competitive characteristics for the manufactured parts, such as surface finish, tolerance, and geometrical features. Moreover, it offers the capability of mass-producing parts at the lowest cost and fastest production time. The analysis conducted by Karania and Kazmer [52] clearly shows the advantage of injection molding compared to sister technologies when large-scale manufacturing is taken into consideration. Despite having a high initial setup cost, it is the most competitive technology when marginal cost and production time are considered, as shown in Figure 5.2 [52]. Their results for a simple experimental part showed that conventional injection molding is most suitable for quantities above ~10,000 parts. As shown in Figure 5.3, there is a threshold beyond which the injection molding cost per part is significantly lower than that of other competing technologies, such as Computer Numerical Control (CNC) machining. Moreover, injection molding can deliver parts that maintain precision in the order of micrometers [53], good surface finish, and excellent structural strength. Table 5.1 highlights the advantages of

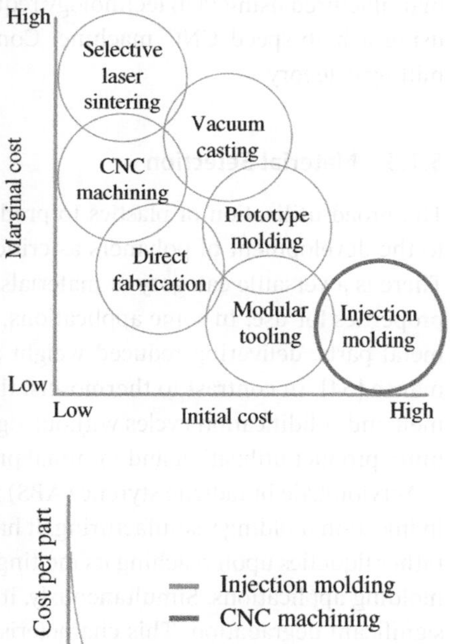


**Figure 5.2** Manufacturing cost comparison of plastic technologies [52].

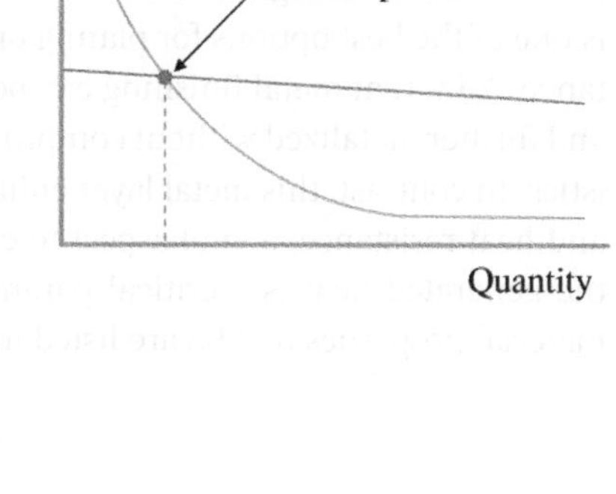


**Figure 5.3** Cost cut-off point in comparison between injection molding and CNC machining.

**Table 5.1** Additive manufacturing technology comparison [52].

| | Resin availability | Surface finish | Part integrity | Part complexity |
|---|---|---|---|---|
| Conventional injection molding | ↑↑↑ | ↑↑↑ | ↑↑↑ | ↑ |
| Fused deposition modeling (FDM) | ↓↓ | ↓↓ | ↓ | ↑↑↑ |
| Selective laser sintering (SLS) | ↓↓ | ↓↓ | ↓ | ↑↑↑ |
| CNC machining | ↑↑ | ↑↑ | ↑↑ | ↑ |

the injection molding technology compared to other manufacturing methods available for fast production. For this discussion we note that high-$Q$ resonators manufactured using PCB technology require drilling via holes, which are realized using a high-speed CNC machine. Consequently, we place them in the CNC milling category.

### 5.1.3 Material Selection

The broad utilization of plastics to produce end-use parts and products has led to the development of polymers as critical materials in modern manufacturing. There is a versatile category of materials with thousands of polymer options and properties for use. In some applications, thermoplastics are direct substitutes for metal parts, delivering reduced weight and cost without compromising performance [54]. In contrast to thermosets, thermoplastics can go through numerous melt and solidification cycles without significant degradation, allowing for maximum product utilization and minimal production cost.

Acrylonitrile butadiene styrene (ABS) is one of the most common plastics used in injection molding manufacturing. It has the valuable property of not burning. It rather liquefies upon reaching its melting point, which makes it ideal for injection molding applications. Simultaneously, it is possible to cool and reheat it without significant degradation. This characteristic provides a cost-effective option, as it reduces manufacturing waste. It is a tough, versatile, engineering thermoplastic that combines strength and durability with good heat resistance. Moreover, ABS is one of the best options for plating on plastic because of its high chemical resistance. Adherent metal finishing can be deposited through chemical pretreatment and further metalized without compromising the structural and surface characteristics. In contrast, this metal layer enhances material attributes related to strength and heat resistance, a vital aspect to consider in high-power applications, where the generated heat is a critical parameter impacting performance. The general material properties of ABS are listed in Table 5.2 [55].

**Table 5.2** ABS general material properties.

| Physical properties | ASTM* test method | Units | Nominal value |
|---|---|---|---|
| Density | D1505 | g/cc | 1.03 |
| Tensile strength @yield | D638 | psi | >6000 |
| Elongation @break | D638 | % | 40 |
| Flexural modulus | D790 | psi | 300,000 |
| Flexural yield strength | D790 | psi | 10,700 |
| Durometer | D785 | R scale | 102 |
| Izod impact | D256 | ft lbs/in$^2$ | 7.7 |
| Vicat softening temp. | D1525 | °F | 219 |
| Heat deflection temp. (66 psi) | D648 | °F | 201 |
| Flammability | UL94 | UL94 | HB |

*American Society for Testing and Materials.

### 5.1.4 Design for Moldability

There are three main principles that one needs to consider when designing a structure that will be injection molded: (1) undercuts, (2) uniformity, and (3) drafts [56]. The mold in our case is composed of two parts: a cavity and a core. When they come together, their enclosure forms the cavity to be filled by the molten thermoplastic, and their contact forms the parting line. They are machined with high precision in order to guarantee a tight-fit enclosure. Complex molds require slides or handloaded cores to form intricate shapes. This increases the complexity of the mold and, in rugged designs, the limit at which the parting line of the mold is located. Figure 5.4 shows the parting line located in the proposed design and how it can be modified to simplify the design of the mold and guarantee a successful injection-molded part.

Uniformity refers to the wall thickness of the injection-molded plastic part. It should be the same across it in order to maintain cosmetic and structural integrity. A sink occurs when a thicker area of a part cools at a different rate, pulling the faces toward its center. As depicted in Figure 5.5, a poor design results in sinking marks that can be avoided by modifying it to have a wall of uniform thickness.

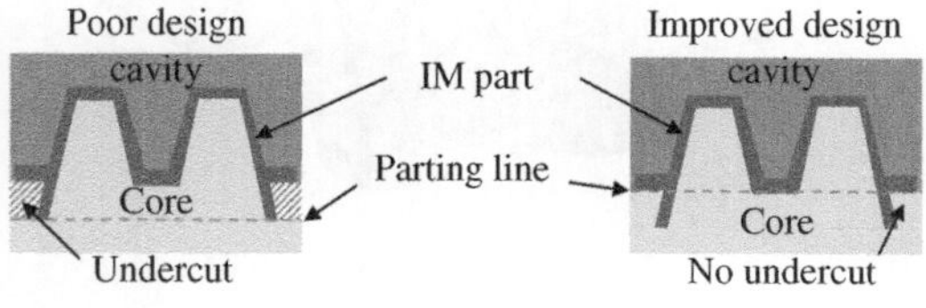


**Figure 5.4** Parting line location for the mold. Poor design versus improved design.

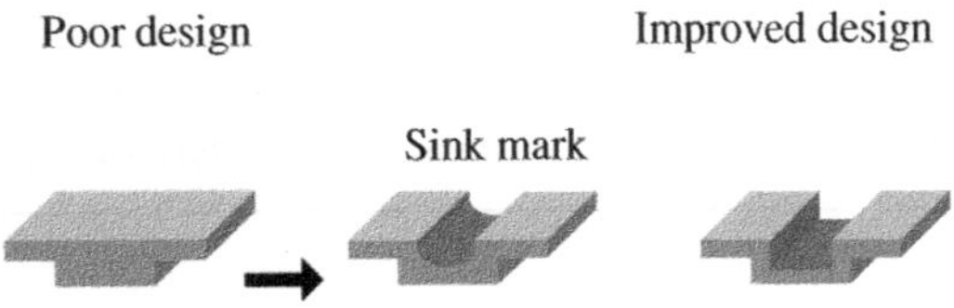


**Figure 5.5** Wall uniformity considerations for the mold. Poor design versus improved design.

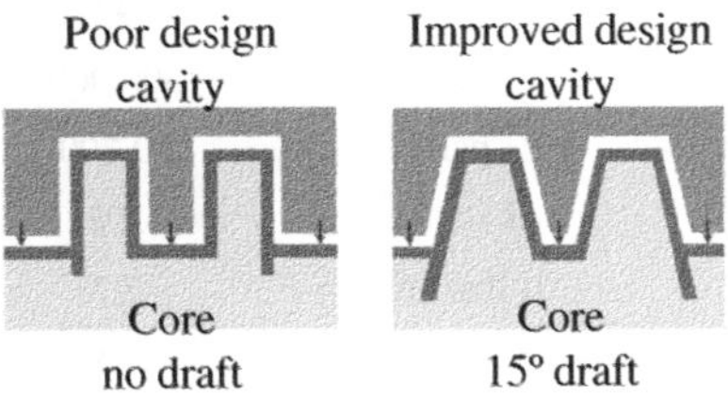


**Figure 5.6** Draft angle considerations for the mold. Poor design versus improved design.

The draft angles are necessary for the part to be released from the mold. A deep cavity increases the length of the pull, and a lack of draft causes drag marks or scratches on the sides of the part. In this design (Figure 5.6), a small draft of 15° is recommended to allow for a more accessible release of the part.

Geometrical characteristics are essential in the design of evanescent-mode devices and have been considered in the construction of resonators and second-order BPFs. Any possible undercuts are eliminated, and the part is designed to have a uniform thickness of 1 mm.

## 5.2 Resonator and Filter RF Design

This section is adopted from [167].

### 5.2.1 Resonator

Figure 5.7 presents a 3-D view of the cavity resonator and shows three critical dimensions of its geometry for design optimization: (1) the RF gap ($h_g$) between

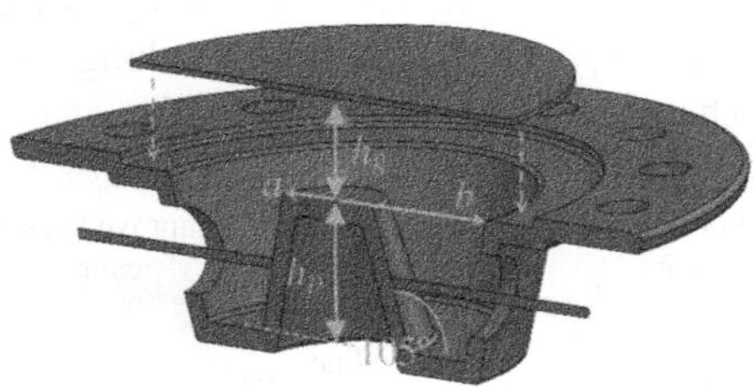

**Figure 5.7** Injection-molded cavity resonator.

**Table 5.3** Designed evanescent-mode cavity resonator parameters.

| Parameter | Value |
|---|---|
| Post radius, $a$ | 3.0 mm |
| Gap (min–max), $h_g$ | 45.7–540 μm |
| Post height, $h_p$ | 6.8 mm |
| Cavity radius, $b$ | 10.0 mm |

the flexible membrane and the center post; (2) the height of the post ($h_p$); and (3) the 105° angled walls, which are critical for maintaining smooth surface roughness during the injection molding process. The tuning element on top of the post-loaded cavity resonator is implemented using a liquid crystal polymer (LCP) membrane metalized with copper on one side (Ultralam 3850 by Rogers Corp.). A commercial linear actuator (M3L-S by New Scale Technologies) controls the membrane movement by changing the gap ($h_g$) vertically. Based on the dimensions of the evanescent-mode air-filled cavity summarized in Table 5.3, its resonance frequency ($f_0$) can be calculated using the analysis derived in Section 2.2.1.

The analytical equation of $Q_u$ for a traditional evanescent-mode cavity resonator is derived in [57]. As discussed earlier, resonators suffer losses due to conductor, dielectric, and radiation losses. With an air-filled cavity, the injection molding resonator has the lowest possible dielectric loss. Radiation loss is also low because of the close contact between the flexible ceiling and the cavity. To minimize conductor loss, the injection molding technology implementation requires a high surface finish for the part, which is maintained during metalization by utilizing a reverse power plating (RPP) method ($R_a$ = 0.31 μm) [58].

### 5.2.2 Filter

A two-pole reconfigurable BPF utilizing the evanescent-mode cavity resonator is designed using standard coupling synthesis techniques to demonstrate the proposed manufacturing method and validate its performance. The design procedure for tunable evanescent-mode cavity filters has been extensively reported in the literature [59–61], thus, only a brief discussion is included here. A 3-D sketch of the filter is shown in Figure 5.8 with final design dimensions summarized in Table 5.4. The flexible membrane that encloses the cavity serves as the tuning element. A section cut of the device shows the internal structure of the cavity and its design parameters.

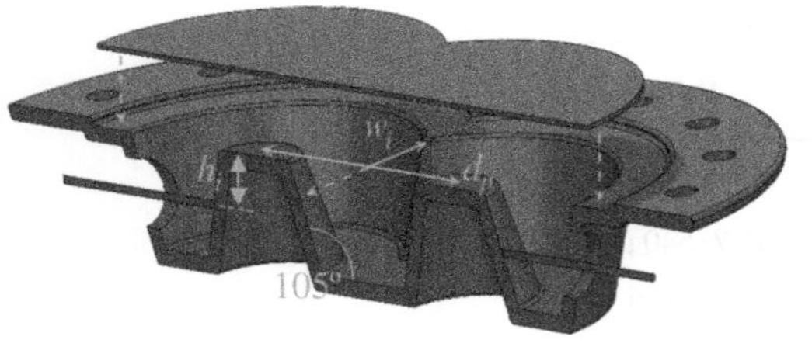

**Figure 5.8** Injection-molded cavity filter.

**Table 5.4** Designed evanescent-mode cavity filter parameters.

| Parameter | Value |
|---|---|
| Post radius, $a$ | 3.0 mm |
| Gap (min–max), $h_g$ | 101–588 μm |
| Post height, $h_p$ | 6.8 mm |
| RF input height, $h_i$ | 3.5 mm |
| Cavity radius, $b$ | 10.0 mm |
| Post distance, $d_p$ | 16.0 mm |
| Iris width, $w_i$ | 13.0 mm |

The filter response type and fractional bandwidth ($FBW$) depend on the cavities' external input/output and inter-resonator couplings. The external couplings are evaluated by the external quality factor $Q_e$, and the inter-resonator coupling $k_{12}$ is evaluated by the even-mode ($f_m$) and odd-mode ($f_e$) resonance frequencies of a coupled-resonator system [62]. For this particular design, a Chebyshev BPF response at 2.8 GHz ($h_g$ = 101 μm) with 7.6% $FBW$ and 30 dB minimum return loss is initially designed, with calculated coupling values $Q_e$ = 4.76 and $k_{12}$ = 0.22.

The required external coupling is realized by an RF connector inserted through the sidewall of the cavity and short-circuited on the metalized sidewall of the capacitive post. The coupling strength is controlled by the distance of the pin from the top of the post, as shown by dimension $h_i$ in Figure 5.8. The closer the pin is to the top of the post (smaller $h_i$), the stronger the external coupling becomes. Ensuring that the tip of the pin makes good contact with the sidewall of the post is also important. If the contact between the pin and the post is poor, the coupling is reduced. To determine the dimension $h_i$, a single evanescent-mode cavity resonator fed by an RF connector is simulated, and the required $Q_e$ for a given $h_i$ is extracted from $S_{11}$ using the group delay method, where $Q_e = 2\pi f_0 \tau(f_0)/4$, and $\tau(f_0)$ is the reflection group delay at resonance [63]. For this design, the pin is located 3.5 mm below the top of the post to achieve the desired $Q_e$.

The iris located between the two resonators implements the necessary inter-resonator coupling, where the dimensions $W_i$ and $d_p$, shown in Figure 5.8, along with the overall shape of the iris, determine the coupling strength [59]. The closer the two posts are together (smaller $d_p$), or the wider the iris opening becomes (larger $W_i$), the stronger the coupling is. To determine the dimensions $W_i$ and $d_p$, two iris-coupled evanescent-mode cavity resonators without RF connectors are simulated. For a given set of dimensions $W_i$ and $d_p$, the even- and odd-mode resonance frequencies are extracted and the coupling coefficient is calculated using $k_{12} = (f_e^2 - f_m^2)/(f_e^2 + f_m^2)$ [62]. The iris dimensions providing the desired $k_{12}$ are determined to be $W_i = 13$ mm and $d_p = 16$ mm. A more comprehensive design procedure for tunable evanescent-mode cavity BPFs has been reported in [59], with considerations for a constant bandwidth or constant *FBW* across the entire tuning range. Although not implemented here, the proposed structure allows for such improved performance features.

Finally, a flexible membrane is used to enclose the cavity top to provide vertical movement and control the air gap ($h_g$) to tune the filter. An LCP layer is mechanically pressed against the sidewalls to achieve good contact, thus maintaining a high $Q_u$. Two external microactuators individually control the two separate tuners, changing $h_g$ from 101 to 588 μm, thus tuning the filter from 2.8 to 5.2 GHz.

## 5.3 Fabrication and Measurements

### 5.3.1 Resonator

A cavity core is manufactured in aluminum through precision CNC machining and used in industrial-level injection molding equipment by Proto Labs Inc. During the manufacturing process, molten ABS material is injected through the mold, and the part is released at the end of the cycle. The produced cavity parts are shown in Figures 5.9 and 5.10, for the resonator and BPF, respectively. The manufactured parts are solvent-cleaned, and the surface is prepared for metalization. Initially, the surface is treated with an argon plasma for 60 seconds in a desktop reactive ion etcher (RIE) to increase the adhesion of the seed metal. Next, the samples are inserted in a magnetron sputtering equipment to deposit a total of 0.5 μm seed layer of Ti/Au (20 nm of Ti) to electroplate a thick copper layer. The samples are

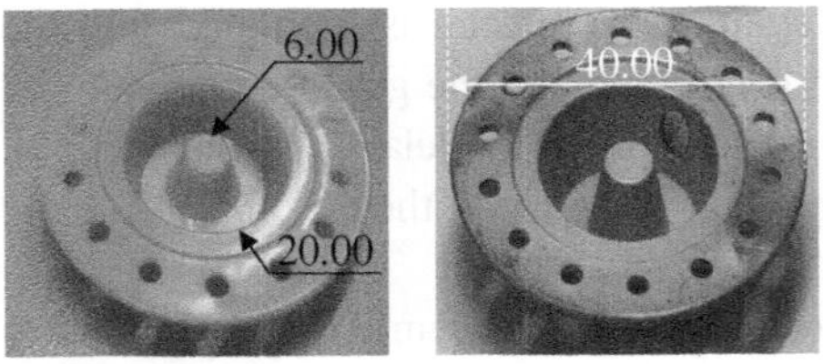


* all dimensions in mm
(a) (b)

Figure 5.9 ABS injection molded cavity resonator. (a) Molded cavity, (b) copper electroplated cavity.

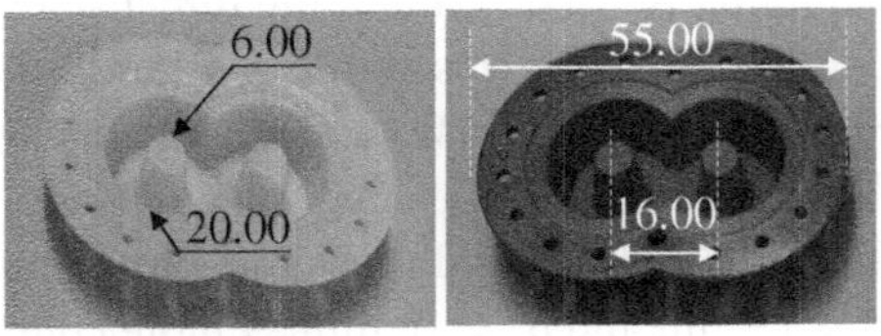


* all dimensions in mm
(a) (b)

Figure 5.10 ABS injection molded cavity filter. (a) Molded cavity, (b) copper electroplated cavity.

placed in a copper sulfate bath, and an RPP method is used to obtain a high density and fine surface finish [58]. The LCP membrane used as a tuning element is 25 μm thick and is metalized with a 9 μm copper layer.

The final assembly also has a connector launch pad to facilitate the transition of the SMA connector inside the cavity. It has been CNC machined from bulk copper and serves as a pressure plate for the cavity and membrane seal. In addition, a CNC-machined fixture is produced to align and hold the microactuators in position. The microactuator pillars are fixed on top of the tuning membrane using a conventional fast-drying adhesive layer. These fixtures can be eliminated from the final product by integrating into one piece with the injection molding cavity. Images of the assembled devices are shown in Figure 5.11(a) and (b).

A weakly coupled evanescent-mode cavity is also manufactured to evaluate the $Q_u$ of the tunable resonator. Referring to Figure 5.7, the feeding pins are disconnected from the post to achieve weak coupling. The resonator is tuned to different gap heights by adjusting the position of the membrane above the post using a microactuator. Figure 5.12(a) shows the measured results compared to simulation. The measured $Q_u$ is extracted from the measured $S_{21}$ for different tuning gap heights using the well-known equation $Q_u = Q_L/(1 - S_{21}(f_0))$ [64], where $Q_L$ is the loaded quality factor, and $S_{21}(f_0)$ is the measured insertion loss at resonance frequency $f_0$, as shown in Figure 5.12(b). The simulated $Q_u$ of the resonator is evaluated without including the effect of the connectors, likely accounting for the discrepancy between simulation and measurement, which includes SMA connector loss. Furthermore, the ideal bulk conductivity of copper ($\sigma = 5.8 \times 10^7$ S/m)

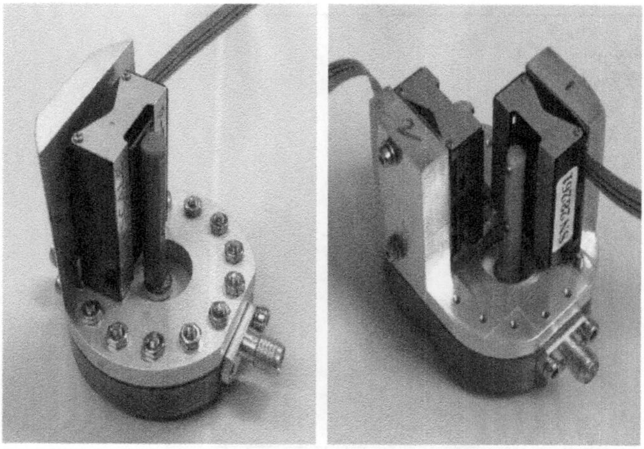

(a) (b)

**Figure 5.11** Assembled manufactured devices. (a) Resonator and (b) bandpass filter.

is used for all simulations; in comparison, the measured cavity suffers further losses due to copper surface roughness and other fabrication tolerances or assembly errors. The actual conductivity of the electroplated copper was not measured. The resonator is tuned from 2.0–5.0 GHz using gap heights ($h_g$) of 45.7–520 μm, respectively, and achieves a measured $Q_u$ of 1,548–2,573.

### 5.3.2 Filter Performance Measurements

In Figure 5.13, the measured performance of the fabricated two-pole injection molding evanescent-mode cavity bandpass filter is compared with the simulation results, with good agreement. Adjusting the simulated gap height from 101–588 μm, the filter's center frequency is tuned from 2.8–5.2 GHz, for a tuning range of 1.9:1. The return loss ($|S_{11}|$) of the fabricated device is shown in Figure 5.13(a), with each tuning state achieving better than 20 dB matching. In Figure 5.13(b), the insertion loss ($|S_{21}|$) of the filter is shown, with two insets giving a closer look at the in-band response for the lowest and highest bandpass frequencies. A copper conductivity of $\sigma = 5.8 \times 10^7$ S/m has been used in the simulation, along with the Teflon of the SMA connector with dielectric constant $\epsilon_r = 2.08$ and loss tangent $\tan\delta = 0.001$. The measured results include SMA connector loss. At a center frequency of 2.8 GHz, the measured filter has a minimum in-band insertion loss of 0.07 dB and a 3 dB *FBW* of 7.6%, while at a center frequency of 5.2 GHz, the measured filter has a minimum in-band insertion loss of 0.07 dB and an *FBW* of 8.4%. The minimum insertion loss across the entire tuning range varies between 0.06–0.1 dB. A smoothing function on the measured data has been used to remove

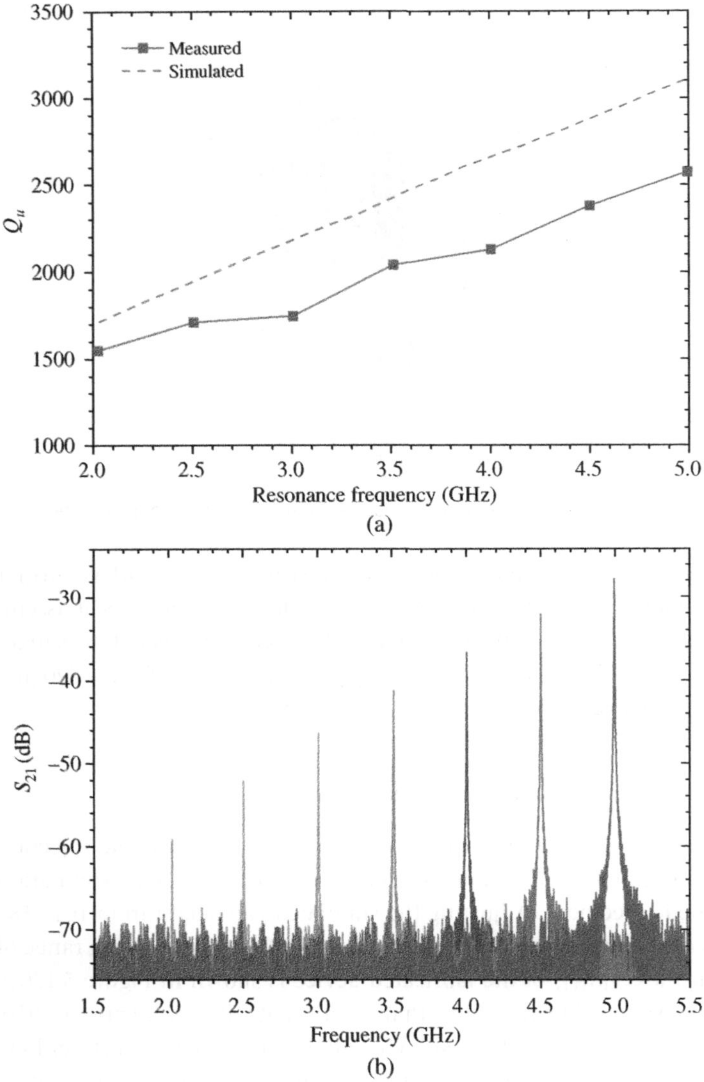


**Figure 5.12** (a) Measured and simulated extracted $Q_u$ of the weakly coupled single cavity resonator, and (b) measured insertion loss ($|S_{21}|$) of the weakly coupled single cavity resonator, with the colors representing different tuning gap heights ($h_g$) used to extract the measured $Q_u$ at several resonance frequencies across the resonator's tuning range.

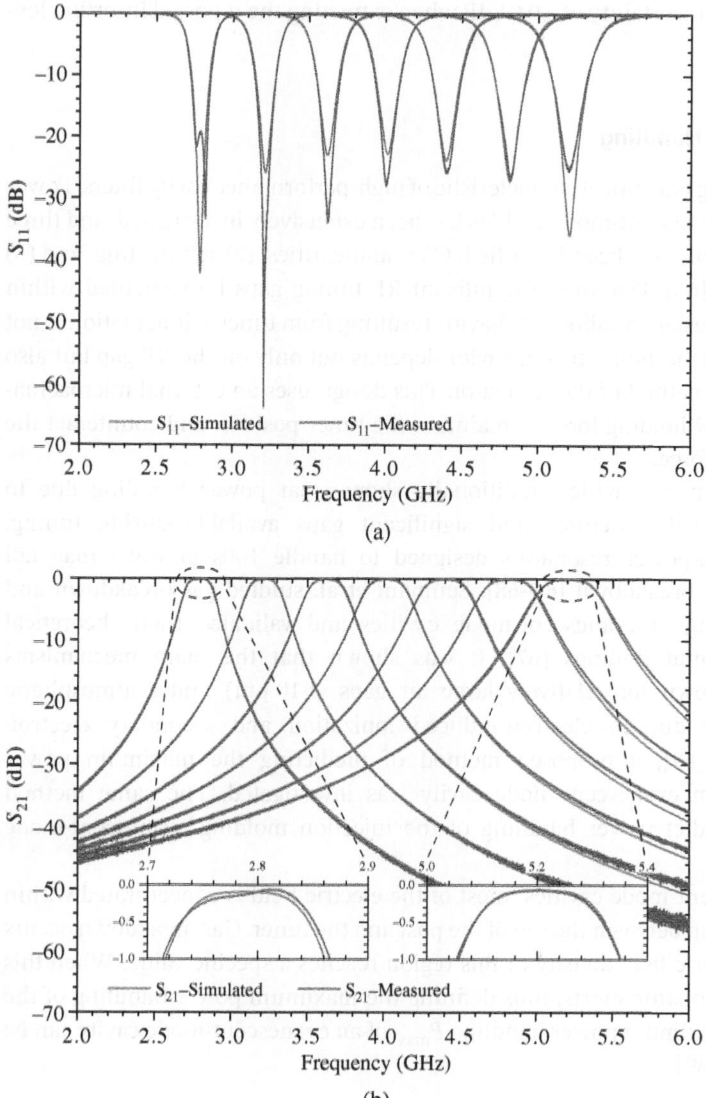


**Figure 5.13** (a) Measured and simulated return loss ($|S_{11}|$) of the tunable cavity filter, and (b) measured and simulated insertion loss ($|S_{21}|$) of the filter, with insets showing a closer view of the in-band insertion loss for the lowest and highest bandpass filter responses (results include SMA connector loss).

measurement uncertainty of ±0.01 dB when extracting the reported insertion loss values above.

### 5.3.3 Power Handling

Power handling is a critical characteristic of high-performance cavity filters. Power handling of evanescent-mode cavities has been extensively investigated, and three main limitations have been identified: (1) nonlinearities, (2) self-heating, and (3) gas discharge [65]. Due to the significant RF tuning gaps implemented within the proposed design, nonlinear behavior resulting from tuner self-actuation is not significant. Furthermore, this behavior depends not only on the RF gap but also on the tuning method of the resonator. This design uses an external microactuator with a 0.5 N holding force to maintain the tuner position and counteract the RF-generated force.

Evanescent-mode cavities traditionally show great power handling due to their geometrical structures and significant gaps available during tuning. However, high-power resonators designed to handle 100s of watts may fail because of gas breakdown [66–68]. Semnani et al. studied gas breakdown and power handling of evanescent-mode cavities and validated their theoretical and experimental findings [67]. It was shown that the main mechanisms of gas breakdown for relatively large air gaps (>10 μm) under atmospheric pressure were due to electron-induced ionization and secondary electron emission. In [69], a proposed method of predicting the maximum power handling of an evanescent-mode cavity was investigated. The same method is used to predict power handling of the injection molding evanescent-mode cavities.

For evanescent-mode cavities, most of the electric field is concentrated within the small air gap between the top of the post and the tuner. Gas breakdown occurs when the electric field density in this region reaches a specific value. When this occurs, the resonator shorts, thus defining the maximum power handling of the device. The maximum power handling $P_{\max}$ of an evanescent-mode cavity can be calculated as [69]

$$P_{max} = \left(\frac{E_{breakdown}}{E_{max\text{-}norm}}\right)^2 \times \frac{1}{W_{stored\text{-}max}} \tag{5.1}$$

where $E_{breakdown}$ is the electric field breakdown threshold of air, $E_{max\text{-}norm}$ is the maximum magnitude of the simulated electric field inside the cavity using the eigenmode solver of HFSS normalized to 1 nJ, and $W_{stored\text{-}max}$ is the maximum stored energy inside the cavity in nanojoules. For air gaps on the order of a few 10s of micrometers, $E_{\mathrm{breakdown}}$ has been reported to be 7.9 V/μm [66].

The maximum stored energy inside the evanescent-mode cavity can be calculated from

$$W_{stored\text{-}max} = \frac{1}{2} C_{eqv} V^2 \tag{5.2}$$

where $C_{eqv}$ is the post-equivalent capacitance. By exciting the simulated cavity with 1 W of input power, the nodal voltage $V$ across $C_{eqv}$ can be extracted by multiplying the maximum magnitude of the electric field by the gap height.

In Figure 5.14, the simulated electric field distribution on the post for an air gap $h_g = 43.1$ μm is shown, along with the calculations of $P_{\max}$ for three different air gaps. A close-up of the top of the post is also shown in Figure 5.14(b), where it is clear that the majority of the field is located within this region. The calculated values for $P_{\max}$ were 41.9, 55.9, and 260.4 W for air gaps $h_g = 43.1$, 48.1, and 92.8 μm, and cavity resonance frequencies 2.0, 2.1, and 2.8 GHz, respectively.

The power handling of the manufactured resonator is investigated to compare the predicted power handling with measurements. A weakly coupled resonator reflects too much power and does not provide a good indication of the resonator's power handling. Thus, the amount of input/output coupling of the resonator was increased by adjusting the feed pin location to make contact with the post and

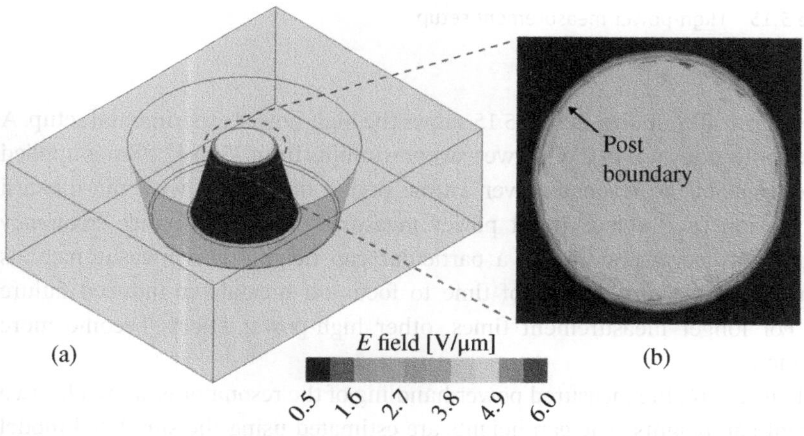


| Air gap $h_g$ (μm) | $E_{max\text{-}norm}$ (V/μm) | $W_{stored/max}$ (nJ) | $P_{max}$ (W/dBm) |
|---|---|---|---|
| 43.1 | 0.65 | 3.51 | 41.9/46.2 |
| 48.1 | 0.60 | 3.10 | 55.9/47.5 |
| 92.8 | 0.35 | 1.96 | 260.4/54.2 |

**Figure 5.14** (a) Simulated electric field distribution for air gap $h_g = 43$ μm on the post inside the cavity (tuner is not shown for better display) and (b) a closer view of the field strength on the top of the post, along with the calculated maximum power handling $P_{max}$ for three different air gap sizes.

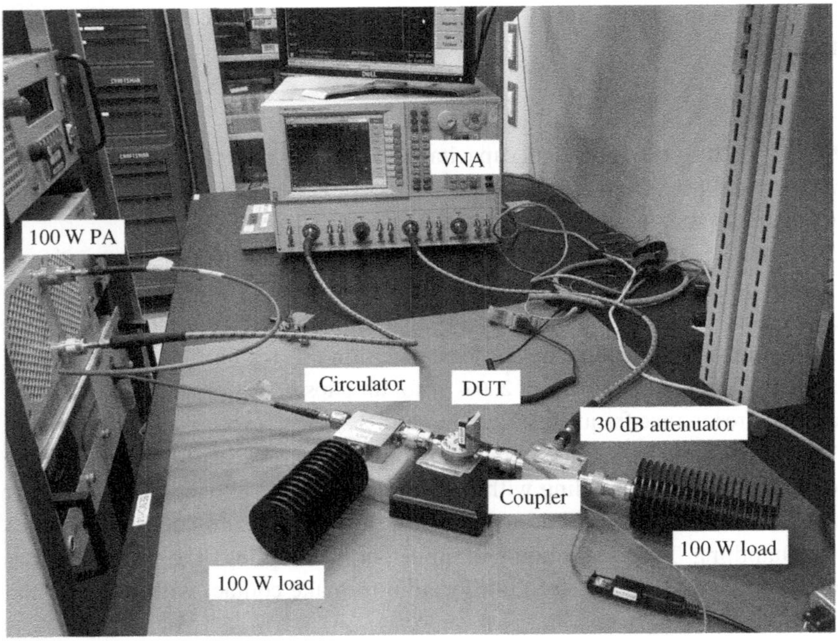


**Figure 5.15** High-power measurement setup.

achieve critical coupling. Figure 5.15 shows the high-power experimental setup. A single continuous wave (CW) power-swept stimulus from 33 to 47 dBm is applied at the input of the resonator over a time period of 185 ms, with the forward transmission ($S_{21}$) versus input power measured at the resonance frequency of the cavity corresponding to a particular gap height. This measurement is conducted over short periods of time to focus on breakdown-induced failure [66]. For longer measurement times, other high-power effects become more prominent.

In Figure 5.16, the measured power handling of the resonator is shown for two different gap heights. The gap heights are estimated using the simulated model by correlating the measured frequency with the equivalent simulated gap. At a resonance frequency of 2.004 GHz, with an estimated gap height of 43 μm, no breakdown is observed at a power-sweep time of 185 seconds up to 47 dBm. This indicates that the simulated power handling shown in Figure 5.14 is a conservative estimate. Increasing the power-sweep time to 20 seconds, breakdown occurs at 43.7 dBm, and at 60 seconds, breakdown occurs at 42.2 dBm. At a resonance frequency of 2.104 GHz, with an estimated gap height of 48 μm, breakdown did not occur at any measured sweep time.

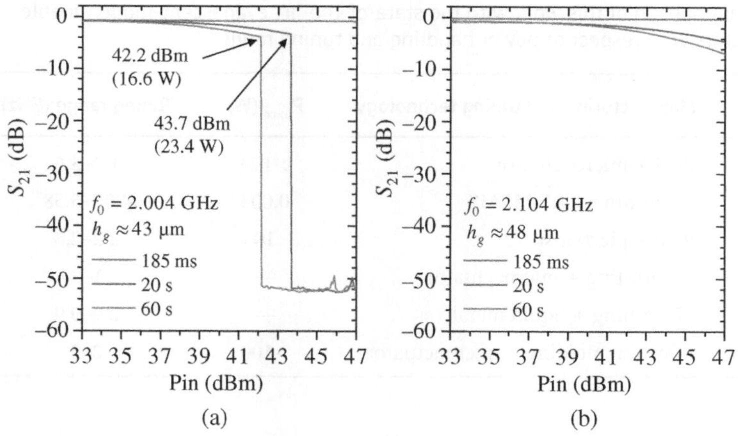


**Figure 5.16** High-power measurement results for the critically coupled cavity at (a) 2.004 GHz and (b) 2.104 GHz.

The method in [69] is also used to predict the power handling of the evanescent-mode cavity BPF. The filter response at 2.8 GHz with a gap of 101 μm is evaluated because it has the narrowest *FBW* and smallest gap height, resulting in the worst-case scenario for power handling with the longest group delay and thus the largest amount of stored energy [70]. First, the maximum electric field magnitude for each individual resonator of the filter is extracted from the simulation model, noting that the peak field occurs at the edges of the passband and has a different magnitude for each resonator. By exciting the model with an input power of 1 W, the first resonator is found to have a greater field magnitude. From this value, the maximum stored energy, $W_{stored\text{-}max}$, is calculated using (5.2) and is found to be 2.32 nJ. Next, the filter is simulated normalized to 1 nJ, and the maximum normalized electric field $E_{max\text{-}norm}$ is extracted with a value of 0.33 V/μm. Using (5.1), the estimated power handling $P_{max}$ of the evanescent-mode cavity BPF is determined to be 244 W, or 53.9 dBm.

## 5.4 Discussion

Tables 5.5 and 5.6 relate this work's manufactured resonator and filter, to state-of-the-art evanescent-mode resonators and filters in the literature. The implementation of injection molding technology provides a high surface finish cavity, which is essential for a high-*Q* performance. In conjunction with applying a soft adhesive to ensure optimal contact with the flexible membrane, mechanical pressure bonding allows for superior electrical connections by eliminating

**Table 5.5** Comparison of this work with the state-of-the-art evanescent-mode tunable cavity resonators with respect to power handling and tuning range.

| References | Manufacturing and tuning technology | $P_{max}$ (W) | Tuning range (GHz) |
|---|---|---|---|
| [60] | PCB + microactuator | >100 | 1.3–3.6 |
| [71] | Machining + RF MEMS | 0.004 | 4.07–5.58 |
| [72] | PCB + piezo disk | 10 | 2.2–2.8 |
| [73] | 3D printing + microactuator | — | 1–3 |
| [74] | Machining + liquid metal | — | 2.9–9.9 |
| This work | Injection molding + microactuator | >100 | 2–5 |

**Table 5.6** Comparison of this work with the state-of-the-art evanescent-mode cavity filters with respect to insertion loss and quality factor.

| References | FBW (%) | IL (dB) | Tuning range (GHz) | Quality factor |
|---|---|---|---|---|
| [47] | 13 | 0.35 | 5.2 | 3257 |
| [75] | 8.8 | 0.08 | 3.53 | 1833 |
| [76] | 8.4–9.1 | 0.77–0.26 | 2.2–4.2 | 500–700 |
| [48] | — | 3.14–0.78 | 18.9–39.6 | 265–510 |
| [73] | 4.1–8.5 | <1.3 | 1.3–2.2 | >500 |
| [74] | — | <1 | 3.4–7.5 | 120–625 |
| [77] | 5.1 | 2.1 | 10.26 | 214 |
| [78] | 1.1 | 3.57–1.56 | 0.98–3.48 | 300–650 |
| This work | 7.6–8.4 | 0.06–0.1 | 2.8–5.2 | 1548–2573 |

the air gaps and radiation leakage, resulting in the best possible performance. The simplicity of the assembly minimizes manufacturing errors and produces measured results that are in good agreement with the simulated results.

Figure 5.14 presents the simulated breakdown of the evanescent-mode resonator based on different tuning gaps. These results are in good agreement with the measured results presented in Figure 5.16(a) and (b). In addition, Figure 5.14 shows the expected breakdown power of the resonator at 2.8 GHz (92.8 μm gap height), the lowest measured frequency response of the filter. The simulated breakdown of the evanescent-mode BPF at 2.8 GHz is also determined, with only a slight decrease in power handling compared with the resonator. Thus, based on the designed and manufactured tunable filter's minimum gap of 101 μm, along with the good correlation between the single resonator's simulation and experimental results conducted at smaller gap heights, we can reliably extract the filter's high-power handling (>100 W) capabilities within its tuning frequency range.

In summary, both manufactured devices show outstanding performance. The resonator and filter exhibit simultaneously low insertion loss, wide tuning range, high-power handling, and high unloaded quality factors. In Table 5.5, one can observe these tradeoff characteristics. In [60], insertion loss was sacrificed for higher power handling, and in [74], the quality factor was sacrificed for a wider tuning range. In [75], a comparable insertion loss was shown but with a fixed-frequency device. Table 5.6 also summarizes a comparison with the state-of-the-art taking into consideration the insertion loss and the quality factor.

## 5.5 Conclusion

This chapter investigates the potential use of injection molding technology as a large-scale low-cost manufacturing method to realize high-*Q* high-power tunable filters. For demonstration, an injection-molded tunable evanescent-mode cavity resonator and a bandpass filter were manufactured and tested. The measured filter achieved an insertion loss between 0.06–0.1 dB within a tuning range from 2.8–5.2 GHz, and an unloaded quality factor of 1,548–2,573, which is comparable to existing state-of-the-art devices for a fraction of the cost per part produced. To demonstrate the power handling of the proposed technology, a strongly coupled resonator was shown to handle high-power loads at small gap heights. The presented results confirm the potential of this technology for low-cost mass production of high-performance tunable filters in the sub-6 GHz range.

## 5.6 How to Choose the Right Manufacturing Technology

As already discussed in the previous three chapters, there are several manufacturing technologies available for implementing evanescent-mode tunable filters. We recommend the designer consider the following key questions when choosing the appropriate technology:

- **Operating frequency:** High-frequency filters require smaller dimensions and tighter manufacturing tolerances. Consequently, silicon micromachining and micro-injection molding may be best for filter cavities in the range of 100s μm or less, which will likely be required for frequencies above 40 GHz.

- **Production cost and volume:** While CNC machining or 3D printing may be attractive for prototyping purposes, injection molding or micro-injection molding may be optimal for high-volume applications (10s of millions to billions of units). PCB manufacturing is best suited for medium-volume applications (thousands to a few million units).
- **Tuning mechanism:** Mechanical tuners will, in general, result in the best possible RF performance in terms of loss, linearity, and power handling. Consequently, they can be a good choice when highest system performance is desired. However, tuning speed and manufacturing costs will be sub-optimal. Conversely, electronic tuning (using switches or varactors) offers lower performance, but it can be acceptable, particularly when we consider the increased tuning speed, lower cost, and improved integration.

In practice, there will be several additional questions to consider such as temperature range, manufacturing repeatability, and vibration/shock requirements to name a few. The optimal choice will often be a compromise of various requirements, performance, and fabrication technologies.

## 5.7 How to Choose the Right Actuator for the Filter

Similar to selecting manufacturing technology, a few parameters need to be considered when selecting tuning actuators. In general, the designer will select actuators based on

- **Size:** Linear mechanical actuators can have sub-micrometer resolution. Their volume, however, might prohibit their use in size-constraint applications.
- **Repeatability:** Mechanical actuators with built-in motion closed-loop control are typically preferred.
- **Mechanical rigidity:** A filter may need to tolerate substantial mechanical vibrations, shocks, and/or temperature variations. Its tuning actuators will need to handle such constraints.
- **Power consumption:** While some mechanical actuators can achieve all design parameters mentioned so far, this might come at the cost of DC power consumption. As a result, the designer has to keep in mind the acceptable amount of power that can be consumed by an actuation device.

Chapter 12 discusses some closed-loop control methods that are specifically designed for evanescent-mode resonators. Although these are not an exhaustive list, they do present a strong potential for tuning evanescent-mode filters with minimal overhead on the wireless system.

# Part III

# Applications

# 6

# Absorptive Bandstop Filters

## 6.1 Design Principles of Absorptive Filters

### 6.1.1 Analysis of a Two-Pole Absorptive Bandstop Filter

A schematic representation of a two-pole absorptive bandstop filter is shown in Figure 6.1(a). This circuit was first disclosed in [79, 80].

The filter consists of two resonators coupled to a source-to-load transmission line of length $\theta$ with denormalized coupling coefficients $k_{E1,2}$,[1] and coupled to each other with coupling coefficient $k_{12}$. Though represented as shunt-parallel RLC resonators in Figure 6.1(b), the resonators can be implemented as any resonators, which have parallel RLC equivalent circuits near resonance. The coupling elements are implemented as admittance inverters scaled by the resonator and system characteristic impedances as defined in [24], and the source-to-load coupling is assumed to be an ideal TEM transmission line. With the sign convention used in this analysis, positive coupling provides a +90° insertion phase, whereas negative coupling yields a −90° insertion phase. The source and load impedances are assumed in this analysis to be identical to the characteristic impedance of the transmission line. The expressions in Figure 6.1 are defined as follows

$$Y_{res1,2} = \frac{1}{Z_R}\left(\frac{1}{Q_u} + p \pm jB\right) \tag{6.1}$$

$$p = j\left(\frac{\omega}{\omega_0} - \frac{\omega_0}{\omega}\right) \tag{6.2}$$

$$J_{12} = k_{12}/Z_R \tag{6.3}$$

1 The $k$ coupling coefficients are sometimes used instead of $M$ from Chapter 1. For external coupling $k_{ext} = FBW \times M_{01}$ and for inter-resonator coupling $k_{int} = \sqrt{FBW} \times M_{12}$.

*Tunable Evanescent-Mode Filters: Principles, Implementation, and Applications*,
First Edition. Dimitrios Peroulis and Mohammad Abu Khater.

Companion website: www.wiley.com/go/Peroulis/TunableEvanescent-ModeFilters

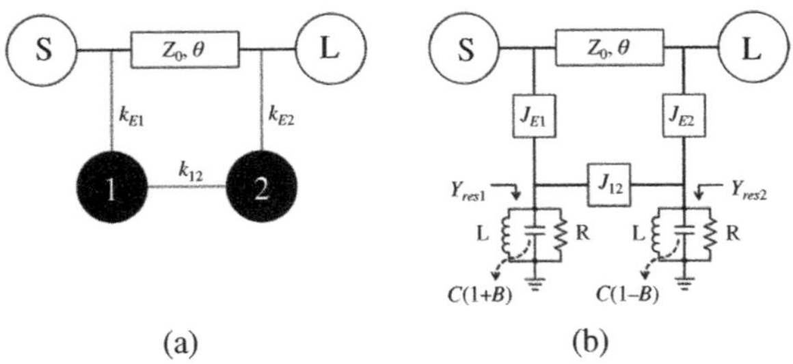


**Figure 6.1** (a) Schematic representation of a two-pole absorptive bandstop filter. (b) Equivalent circuit of (a). Source-to-load coupling is implemented by a transmission line of characteristic impedance $Z_0$ and electrical length $\theta$, resonator coupling elements are implemented by admittance inverters, and resonators are represented as parallel RLC resonators.

$$J_{E1,2} = k_{E1,2}/\sqrt{Z_0 Z_R} \tag{6.4}$$

$$Z_R = \sqrt{L/C} \tag{6.5}$$

$$\omega = 1/\sqrt{LC} \tag{6.6}$$

$$Q_u = \omega_0 RC \tag{6.7}$$

$Y_{res1,2}$ represents the admittance of each resonator and is simply the parallel combination of the inductor, capacitor, and resistor, which model each resonator, slightly rearranged and reduced using the definitions for the frequency variable (6.2), the resonators' impedances (6.5), and the resonators' unloaded quality factors (6.7). The capacitors are differentially tuned by a factor of $1 \pm B$, which allows for asynchronous tuning of the resonators if $B$ is chosen to be nonzero. The frequency-invariant reactance $B$ in equation (6.1), which appears as a result of this differential capacitance tuning is only approximate – in reality, the reactance would have frequency dependence, but in the narrowband case it can be approximated as constant.

To obtain the transmission and reflection coefficients ($S_{21}$ and $S_{11}$) of the circuit in Figure 6.1(b), all components are first converted into their representative ABCD matrices (Figure 6.1(a)) using the expressions in Figure 6.2.

The elements in the bottom branch of the circuit (the resonators and coupling elements) are cascaded together by multiplying their ABCD matrices (Figure 6.3(b)). The resulting matrix is converted into its equivalent Y-parameter matrix (Figure 6.3(c)), which is then added to the Y-parameter matrix of the transmission line due to their parallel configuration (Figure 6.3(d)). The resulting Y-parameter matrix is converted into its equivalent S-Parameter matrix (Figure 6.3(e)). The resulting transmission and reflection coefficients are extracted using symbolic math tools and are shown in Figure 6.4. Inspection of

**Figure 6.2** ABCD parameters for each block in Figure 6.3.

| | |
|---|---|
| $A_1 = \cos(\theta)$ | $A_{3,5} = 1$ |
| $B_1 = jZ_0\sin(\theta)$ | $B_{3,5} = 0$ |
| $C_1 = j/Z_0\sin(\theta)$ | $C_{3,5} = Y_{res1,2}$ |
| $D_1 = \cos(\theta)$ | $D_{3,5} = 1$ |
| $A_{2,6} = 0$ | $A_4 = 0$ |
| $B_{2,6} = -j/J_{E1,2}$ | $B_4 = -j/J_{12}$ |
| $C_{2,6} = -j \cdot J_{E1,2}$ | $C_4 = -j \cdot J_{12}$ |
| $D_{2,6} = 0$ | $D_4 = 0$ |

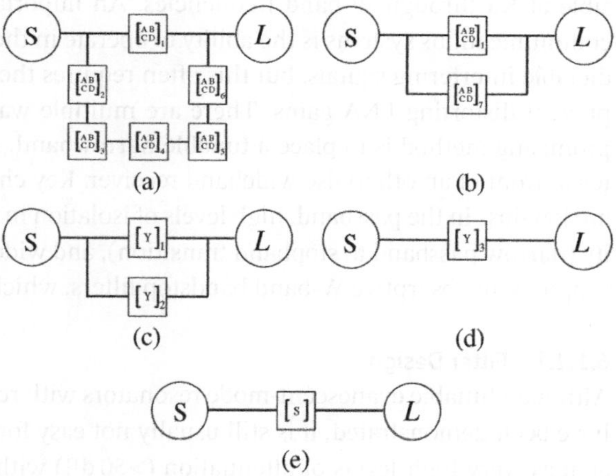


**Figure 6.3** Step-by-step process for calculating the S-parameters of the circuit in Figure 6.1.

$$S_{21} = \frac{e^{-j\theta}\left(p^2 + \frac{2}{Q_U}p + B^2 + k_{12}^2 + \frac{1}{Q_U^2} + k_{12}k_{E1}k_{E2}\sin\theta\right)}{p^2 + \frac{4+(k_{E1}^2+k_{E2}^2)Q_U}{2Q_U}p + je^{-j\theta}k_{E1}k_{E2}\left(k_{12} + \frac{1}{2}k_{E1}k_{E2}\sin(\theta)\right) + \frac{1}{Q_U^2} + \frac{1}{2Q_U}\left(k_{E1}^2 + k_{E2}^2\right) + B^2 + k_{12}^2 + j\frac{B}{2}\left(k_{E1}^2 - k_{E2}^2\right)}$$

$$S_{11} = \frac{-\frac{1}{2}\left(k_{E1}^2 + e^{-2j\theta}k_{E2}^2\right)p + \mu}{p^2 + \frac{4+(k_{E1}^2+k_{E2}^2)Q_U}{2Q_U}p + je^{-j\theta}k_{E1}k_{E2}\left(k_{12} + \frac{1}{2}k_{E1}k_{E2}\sin(\theta)\right) + \frac{1}{Q_U^2} + \frac{1}{2Q_U}\left(k_{E1}^2 + k_{E2}^2\right) + B^2 + k_{12}^2 + j\frac{B}{2}\left(k_{E1}^2 - k_{E2}^2\right)}$$

$$\mu = -je^{-j\theta}k_{12}k_{E1}k_{E2} + \frac{e^{-2j\theta}k_{E2}^2\left(-2 + 2jBQ_U + k_{E1}^2Q_U\right)}{4Q_U} - \frac{k_{E1}^2\left(2 + 2jBQ_U + k_{E2}^2Q_U\right)}{4Q_U}$$

**Figure 6.4** $S_{11}$ and $S_{21}$ of the circuit in Figure 6.3.

the expression of $S_{21} = 0$ at the filter's center frequency ($\omega - \omega_0$, or alternatively $p = 0$) yields

$$\frac{1}{Q_u^2} + B^2 + k_{12}^2 + k_{12}k_{E1}k_{E2}\sin(\theta) = 0 \tag{6.8}$$

If this equation is satisfied, the filter has theoretically infinite attenuation even with finite $Q_u$ resonators, and thus it is the basic governing equation for absorptive bandstop filters.

Many combinations of $k_{E1,2}$, $k_{12}$, $B$, $Q_u$, and $\omega$ can provide valid solutions to (6.8), and thus it is instructive to examine the bounds placed on each variable and to see how the choice of each variable affects the filter's transfer function. Those bounds are studied in detail in [81].

### 6.1.2 W-Band Absorptive Bandstop Filter

Recent advances in millimeter-wave components such as antennas, LNAs, and power amplifiers have made functional radar and communication systems possible at Ka through W-band frequencies. An important characteristic of robust communications systems is the ability to operate in the presence of strong, unpredictable interfering signals, but this often requires the use of dynamic filtering to prevent distorting LNA gains. There are multiple ways to achieve this, but one promising method is to place a tunable narrowband, high-isolation bandstop filter in front of an otherwise wideband receiver. Key characteristics of such a filter are low-loss in the passband, high levels of isolation in the stopband, high selectivity (narrow passband to stopband transition), and wide tuning range. This section proposes an absorptive W-band bandstop filters, which exhibits these qualities.

#### 6.1.2.1 Filter Design

Although tunable evanescent-mode resonators with relatively high quality factors have been demonstrated, it is still usually not easy for tunable bandstop filters to achieve very high levels of attenuation (>50 dB) with narrow bandwidths (<3% fractional bandwidth). One technique to overcome the limitations of limited resonator quality factor is to add a small amount of interresonator coupling between the two resonators, resulting in the absorptive response.

It is interesting to note that with this topology, the maximum attenuation of the filter is not limited by the quality factor of the resonators, but the minimum bandwidth is. The external coupling must be at least $\sqrt{2/Q_u}$, which limits the minimum possible 3-dB fractional bandwidth to $2/Q_u$ [96].

The absorptive evanescent-mode filter for W-band requires a fine fabrication process. As a result, it is designed and fabricated using the microfabrication process, as discussed in Chapter 4. A conceptual drawing of that is shown in Figure 6.5.

The evanescent-mode resonators are realized with gold-plated wet-etched silicon cavities each containing a conical post in the center. This structure approximates a short length of coaxial transmission line loaded with a capacitance. The source-to-load coupling is realized with a 270° microstrip transmission line on a high-resistivity silicon substrate, with apertures in the ground plane, which implement the external coupling by allowing a portion of the transmission line's magnetic field to couple with the magnetic field of the cavity at resonance.

**Figure 6.5** Conceptual drawing of a W-band tunable bandstop filter. The top element is a MEMS electrostatic actuator, the middle element is the cavity substrate, and the bottom element is the signal substrate.

The inter-resonator coupling is created by introducing an iris between the two resonators, coupling the magnetic fields of the two resonators.

The strength of the external coupling is determined by both the width and length of the coupling aperture and can be extracted from the S-parameters obtained from fullwave EM simulations, as discussed in Section 2.3. The size of the coupling apertures was chosen to be 0.18 mm × 0.34 mm, yielding a coupling coefficient $k_{ext}$ = 0.13 at 95 GHz in order to realize a 1.5% 3 dB fractional bandwidth notch.

The apertures in the ground plane present series inductance to the through transmission line, which can seriously degrade the passband performance of the filter. This effect can be mitigated by adding capacitive patches to the transmission line directly over the ground plane apertures [82]. However, this combination of series inductance and shunt capacitance adds a significant phase shift to the through-line, which must be taken into account when designing the through transmission line. Using 3D EM simulations to determine the actual phase of the through-line, including the phase shift from the coupling apertures, the distance between the two coupling apertures is chosen to be 0.71 mm so that the total phase shift between coupling slots (center to center) is 270° at 95 GHz.

It was observed through 3D EM simulations that above 100 GHz, a significant amount of power is to parasitic propagating surface modes. To mitigate this problem and reduce passband insertion loss, the substrate on either side of the microstrip line was etched away, preventing the propagation of these spurious modes.

The inter-resonator coupling is realized with an inductive iris, which is essentially a section of below-cutoff waveguide, which allows the magnetic fields of the resonators to couple with each other at resonance. Increasing the width of the coupling iris increases the strength of the coupling, as does reducing the spacing between the resonators. Because the resonator spacing was fixed after choosing the length of the through transmission line, the only free variable was the width of the coupling iris. The resonator quality factor is estimated to be 400 from simulations, and thus the desired inter-resonator coupling value is $1/Q_u = 0.0025$. The coupling iris width is chosen to be 0.5 mm in order to attain this coupling value. Those dimensions are shown in Figures 6.6 and 6.7.

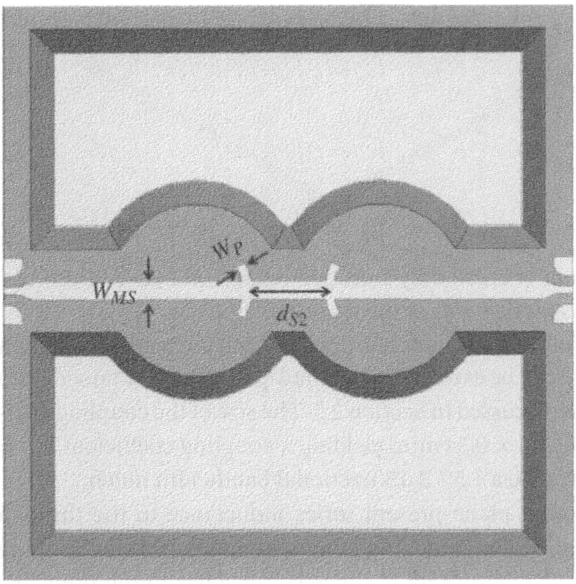


**Figure 6.6** Signal-side of cavity substrate. $W_{MS} = 0.155$ mm, $W_P = 0.060$ mm, $d_{s2} = 0.710$ mm.

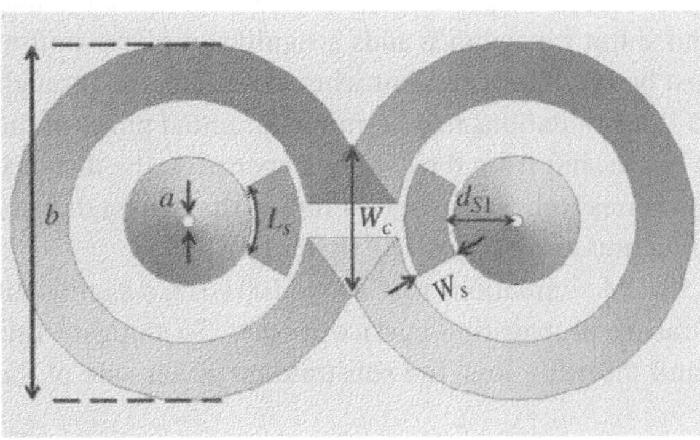


**Figure 6.7** Cavity-side of cavity assembly. $b = 1.68$ mm, $a = 0.060$ mm, $L_S = 0.340$ mm, $W_C = 0.500$ mm, $W_S = 0.180$ mm, $d_{s1} = 0.330$ mm.

#### 6.1.2.2 Fabrication and Measurements

The 200-μm high-resistivity silicon substrate used for the signal substrate is bonded to a 300-μm silicon substrate using a gold intermediate layer, and the cavities are wet-etched using a TMAH and Triton X-100 solution [83]. The cavity and transmission lines are metalized and patterned with a 1 μm layer of sputtered gold.

The tuner's bias electrodes are created by wet-etching cavities in the backside of the tuner substrate, which are then electroplated with a thick layer of copper. The corrugated diaphragm is created by etching circular corrugations in the silicon substrate, metalizing the corrugations, then etching the silicon from under the diaphragm using the $XeF_2$ dry-release process to leave a flexible, free-standing diaphragm.

After release, the tuner is aligned and bonded to the cavity structure using gold-to-gold thermocompression bonding. Those steps are summarized in Figure 6.8. SEM images of the fabricated device are shown in Figure 6.9.

Figure 6.10 shows the measured response of a filter tuned to 96 GHz. Its notch depth is greater than 70 dB, and the passband insertion loss varies from 1.7 dB at 71 GHz to 3.25 dB at 109 GHz. The relatively high passband return loss (8 dB) is due to parasitic reactances caused by the CPW to microstrip transition required to measure the device with RF probes. From simulations, the return loss of the filter itself would be better than 15 dB across the passband if the effects of the transition were removed.

The measured responses of two assembled filters are shown in Figure 6.11. The two filters are identical in all respects except for the initial gap between the tuning diaphragm and the posttip. The first filter has a capacitive gap, which ranges from 3 to 10 μm, and the second filter has a gap, which ranges from 6 to 13 μm. The first filter has an analog tuning range of 75–103 GHz with an applied bias of 0–90 V. The second filter tunes from 96 to 108 GHz with 0–80 V applied bias. Below 90 GHz,

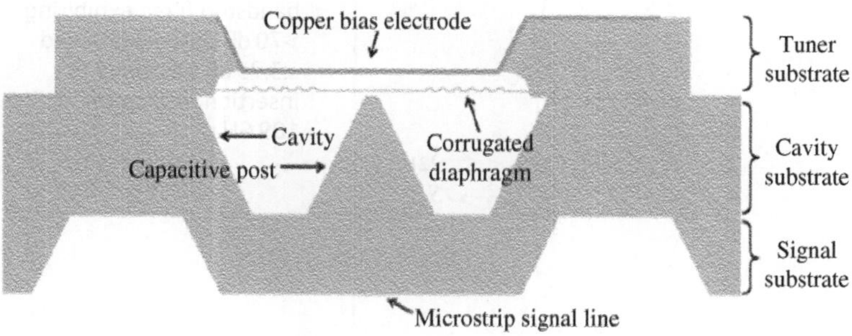


**Figure 6.8** A cross section of the final filter structure showing the tuner substrate with the bias and corrugated diaphragm (top), cavity substrate (middle), and the signal substrate (bottom).

Figure 6.9 SEM images of fabricated device. (top left) Corrugated tuner diaphragm. (top right) Cavities with capacitive posts and coupling apertures. (bottom) Photograph of assembled filter.

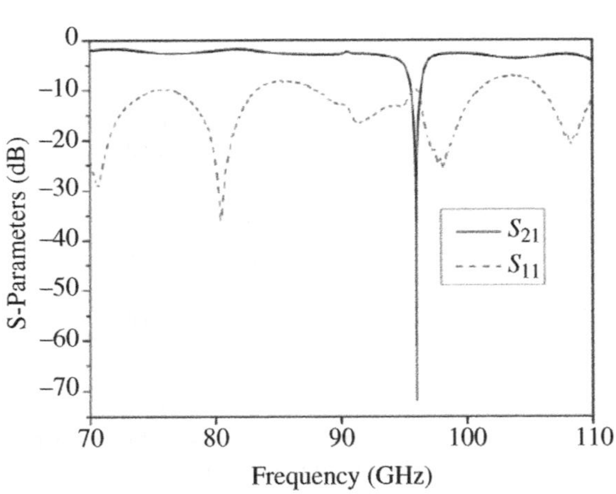


Figure 6.10 Measured response of the W-band bandstop filter, exhibiting >70 dB notch depth and <3.25 dB passband insertion loss up to 109 GHz.

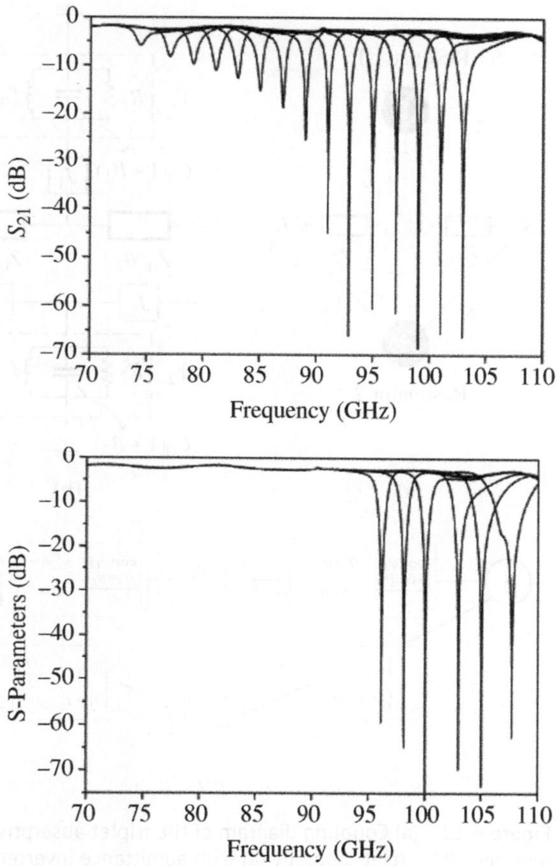


**Figure 6.11** Performance of the measured filters. (a) Filter with 3–10 μm tuning gap covering 75–103 GHz. (b) Filter with 6–13 μm tuning gap covering 96–108 GHz.

the filters are not able to obtain deep notches because the through-line is no longer 270° and the phase relationship, which is required for absorptive operation is no longer valid. The unloaded quality factor of the resonators are extracted to be 290, and the filters have a 1.5% 3-dB fractional bandwidth (calculated after deembedding the passband insertion loss) at 95 GHz.

## 6.2 Triplet Quasi-absorptive Topology

The quasi-absorptive architecture is similar to the absorptive one in the sense that it can achieve a theoretically infinite rejection with a limited quality factor. The tuning range of the absorptive filter, however, is limited by the 90/270° transmission line between the employed resonators. The triplet quasi-absorptive architecture does not suffer from this limitation.

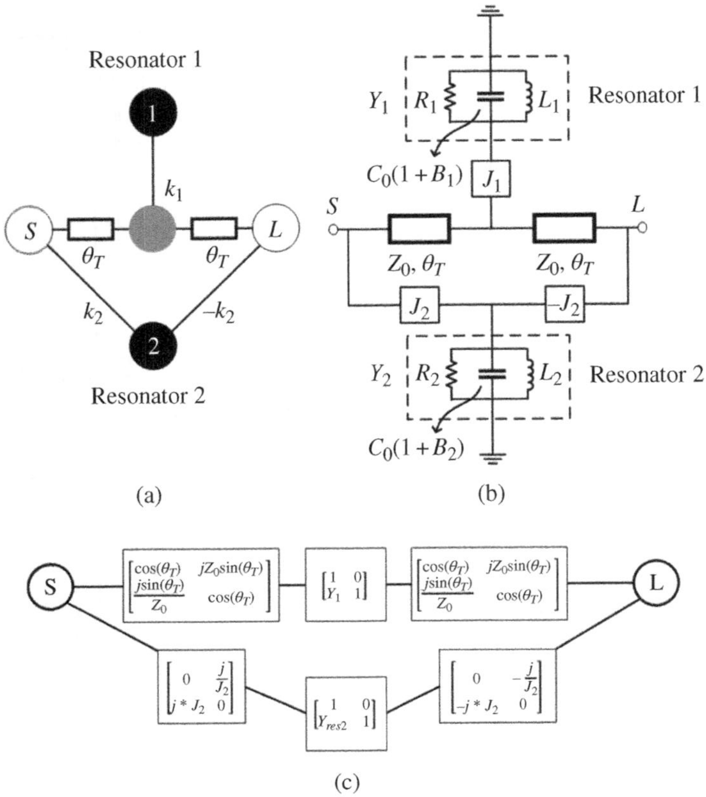


**Figure 6.12** (a) Coupling diagram of the triplet absorptive topology with admittance inverters. (b) Circuit topology (a) with admittance inverters. (c) ABCD matrix representation of (b).

Mixed-parameter analysis can be used to describe the response of this filter as shown in Figure 6.12, and also discussed earlier in this chapter. Two independent conditions, however, can be satisfied. Setting $S_{21}$ to 1 yields an all-pass filter. On the other hand, setting $S_{21}$ to 0 yields a quasi-absorptive condition with a theoretically infinite rejection even with a finite quality factor. The difference between the two is merely the de-tuning of the two resonators [84].

In the case of a continuously tuned resonator, such as the evanescent-mode one, the transition between the two functions is smooth. As a result, it is theoretically possible to get any rejection using this architecture.

The quality factor primarily affects the insertion loss in the all-pass as shown in Figure 6.13. For the quasi-absorptive response, it slightly affects the bandwidth as shown in Figure 6.14.

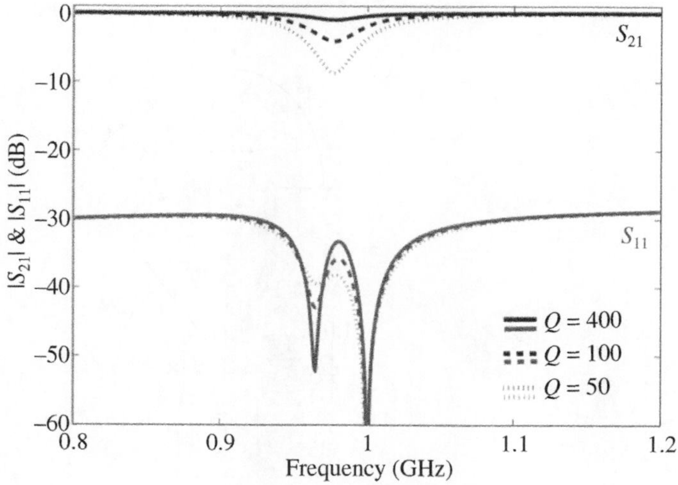


**Figure 6.13** All-pass state response with three different resonator $Q_u$: 50, 100, and 400. The coupling $k_1$ is 0.2.

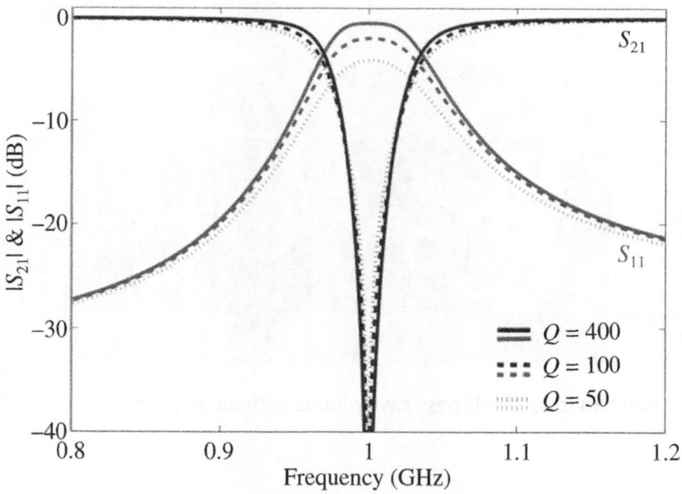


**Figure 6.14** Simulated absorptive bandstop state response with three resonator $Q_u$: 50, 100, and 400. The coupling $k_1$ is 0.2.

### 6.2.1 Quasi-absorptive Filter Design and Implementation

The evanescent-mode filter is designed with a frequency range from 1 to 2 GHz. The resonators are tuned with piezoelectric actuators. The gap is designed to tune from 10 to 50 μm. The exploded view of the model for the filter is shown in

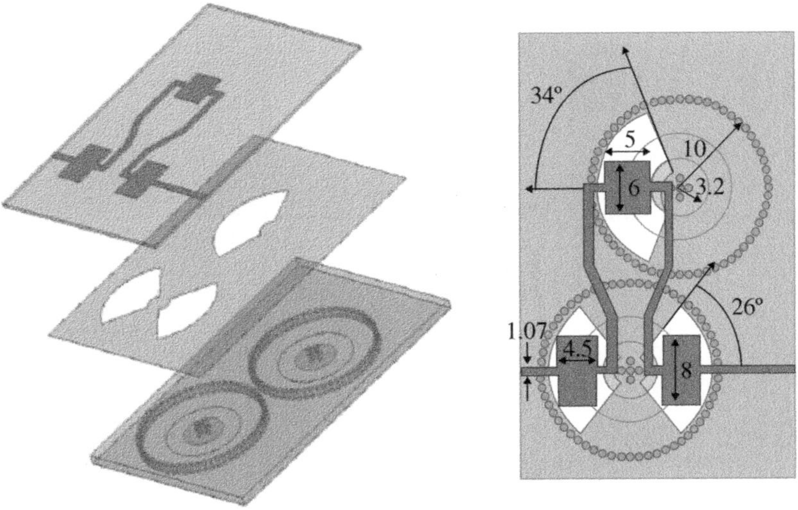


**Figure 6.15** Exploded view of the model and the dimensions in mm for the evanescent-mode quasi-absorptive cavity filter.

**Figure 6.16** Fabricated bandstop-to-all-pass cavity filters without and with piezoelectric actuators.

Figure 6.15. The top layer with a 50 Ω-microstrip line source-to-load coupling is designed on a 0.508-mm thick Rogers 4003 substrate. The cavities are designed on a 1.524-mm thick Rogers 4003 substrate. The filter couplings are realized through slots at the ground plane between the microstrip lines and the cavities. The pads on top of the coupling slots add shunt capacitance to the microstrip line to compensate the series inductance introduced by coupling slots. The simulated unloaded resonator $Q$-factor is approximately 440. The fabricated filter is shown in Figure 6.16.

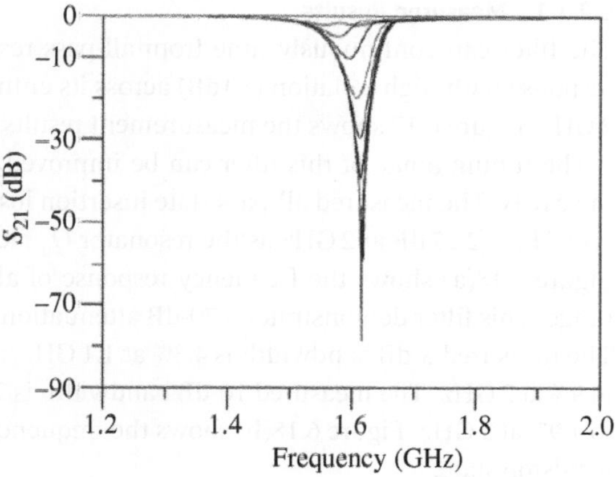


**Figure 6.17** Measured bandstop response with tunable attenuation levels from 80 to 2.16 dB.

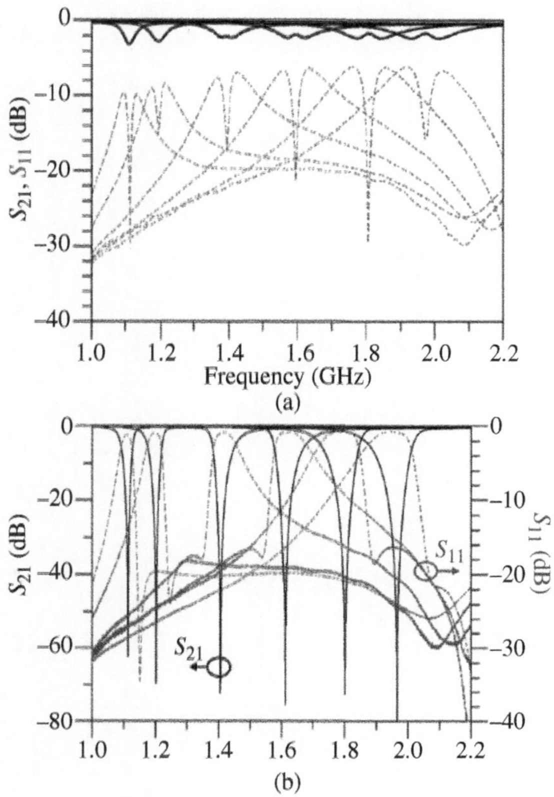


**Figure 6.18** Measured frequency response of (a) all-pass state across its entire tuning range and (b) absorptive bandstop state.

#### 6.2.1.1 Measured Results

The filter can continuously tune from all-pass response to absorptive bandstop response with high isolation (80 dB) across its entire frequency range from 1.1 to 2 GHz. Figure 6.17 shows the measurement results for various attenuation levels.

The tuning range of this filter can be improved by reducing the initial gap of the cavity. The measured all-pass state insertion loss is decreasing from 3.14 dB at 1.1 GHz to 2.27 dB at 2 GHz as the resonator $Q_u$ factor increases along frequency. Figure 6.18(a) shows the frequency response of all-pass states across its tuning range. This filter demonstrates a 70-dB attenuation across its entire tuning range. The measured 3-dB bandwidth is 4.3% at 1.1 GHz, and it monotonically increases to 9% at 2 GHz. The measured 10-dB bandwidth is 2.3% at 1.1 GHz and increasing to 4.9% at 2 GHz. Figure 6.18(b) shows the frequency response of quasi-absorptive bandstop state.

# 7

# Bandwidth and Passband Control

## 7.1 Bandwidth Control for Bandpass Filters

Controlling a bandpass filter's bandwidth in addition to its center frequency is often crucial for many applications. This is due to the fact that filter bandwidth can vary greatly over a large tuning range since it is proportional to its center frequency [85]. Varactor diodes [86, 87] or micro-electromechanical systems (MEMS) cantilevers [88] are examples of tunable components that can be added to dynamically control a filter's interresonator coupling as a function of its center frequency. Unfortunately, this is not always possible as the performance of a high-$Q$ filter can be severely deteriorated by low-$Q$ tunable components, which is typically the case for most lumped components. This section presents a different strategy that is especially impactful for high-$Q$ narrowband filters where insertion loss is particularly sensitive to loading of low $Q$ tuning components.

The key idea is that the bandwidth of a narrowband filter can be changed significantly with only a small variation in its inter-resonator coupling. Consequently, low-$Q$ tuning varactors do not have to dominate inter-resonator coupling. Instead, tuning varactors can be incorporated in parallel with a static low-loss inter-resonator coupling, which sets the initial bandwidth of the filter. To achieve a good impedance match to this bandwidth control scheme, a tunable external coupling design using similar low-$Q$ varactors also needs to be implemented. This design method has the additional advantage of allowing control over the shape/group delay of the filter. The tunable resonators that control center frequency tuning are shielded from these low-$Q$ varactors through impedance transformers. In summary, bandwidth and external coupling dynamic control can co-exist with high-$Q$ resonators and excellent filter performance.

*Tunable Evanescent-Mode Filters: Principles, Implementation, and Applications*,
First Edition. Dimitrios Peroulis and Mohammad Abu Khater.

Companion website: www.wiley.com/go/Peroulis/TunableEvanescent-ModeFilters

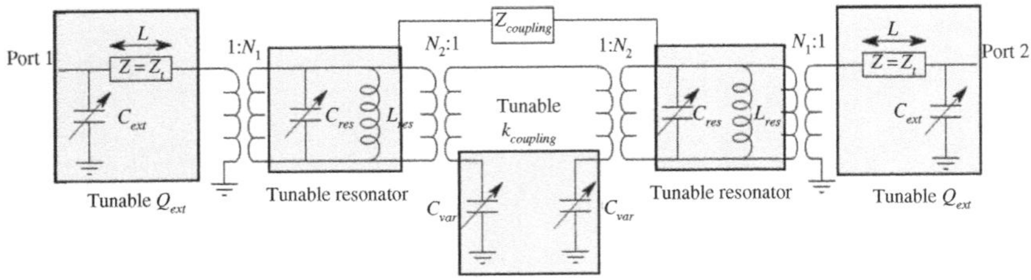


**Figure 7.1** Basic design of a second-order filter with center frequency and bandwidth control. Note that the external couplings are also adjustable.

To better illustrate this, we will focus on a basic second-order reconfigurable filter design as shown in Figure 7.1. The displayed design allows for both center frequency and bandwidth tuning. At its center frequency ($\omega_0$), the filter primarily consists of two high-$Q$ tunable resonators that are approximated as lumped resonators with a tunable capacitor $C_{res}$ and a static inductance $L_{res}$. The filter's initial bandwidth is set by the impedance $Z_{coupling}$, which indicates the low-loss static inter-resonator coupling between the two resonators. For a narrowband filter, we need $Z_{coupling} \gg \omega_0 L_{res}$ or $Z_{coupling} \gg 1/(\omega_0 C_{res})$ for inductive or capacitive coupling, respectively [24]. Lumped varactors $C_{var}$ are employed in parallel with $Z_{coupling}$ to adjust the filter bandwidth. These varactors are coupled to each resonator via transformers with $N_2$ turns. Without loss of generality, a transformer simulates a coupling that excludes a −90° phase shift.

At each port of the filter, we have included a transmission line with impedance $Z_t$ and length $L$ to set the primary external coupling values. Moreover, we integrate lumped varactors $C_{text}$ to regulate the external couplings over the filter tuning range. Similar to the bandwidth control capacitors, these capacitances are isolated from the high-$Q$ resonators by transformers of $N_1$ turns. It should be clear now that, for sufficiently large values of $N_1$ and $N_2$, neither bandwidth control nor impedance control varactors severely degrade the performance of the filter as they do not dominate its high-$Q$ resonators.

### 7.1.1 Filter Design

#### 7.1.1.1 Bandwidth Variation

Let us now quantify these ideas with simple design equations. The even and odd modes of the filter, shown in Figure 7.2(a) and (b), respectively, provide insight into how $C_{var}$ affects the filter bandwidth. Since $C_{var}$ only modifies the filter's odd mode, it may be used to alter the filter bandwidth. Equations (7.1) and (7.2) describe the odd modes of the filter for inductive (i.e., $Z_{coupling} = \omega_0 L_{coupling}$)

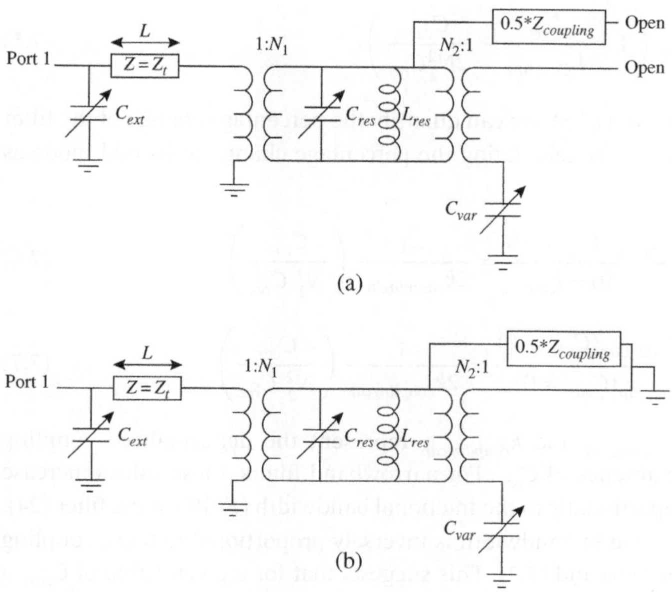


**Figure 7.2** Equivalent circuit displaying the filter resonant modes for (a) the even mode and (b) the odd mode.

and capacitive coupling (i.e., $Z_{coupling} = 1/(\omega_0 C_{coupling})$) networks, whereas equation (7.3) describes the even mode.

$$f_{Odd\ Cap} = \frac{1}{2\pi} \frac{1}{\sqrt{L_{res}\left(C_{res} + \frac{C_{var}}{N_2^2} + 2C_{Coupling}\right)}} \tag{7.1}$$

$$f_{Odd\ Ind} = \frac{1}{2\pi} \frac{1}{\sqrt{\left(L_{res} \parallel \frac{L_{Coupling}}{2}\right)\left(C_{res} + \frac{C_{var}}{N_2^2}\right)}} \tag{7.2}$$

$$f_{Even} = f_{res} = \frac{1}{2\pi} \frac{1}{\sqrt{L_{res} C_{res}}} \tag{7.3}$$

When we select small coupling values (i.e., large values for $N_1$ and $N_2$), we will yield a narrowband filter such that $\frac{C_{var}}{N_2^2} + 2C_{Coupling} \ll C_{res}$ or $L_{Coupling} \gg L_{res}$. Consequently, (7.1) and (7.2) can be approximated as

$$f_{Odd\ Cap} \simeq f_{res}\left(1 - \frac{C_{var}}{2N_2^2 C_{res}} - \frac{C_{Coupling}}{C_{res}}\right) \tag{7.4}$$

$$f_{Odd\ Ind} \simeq f_{res}\left(1+\frac{L_{res}}{L_{Coupling}}-\frac{C_{var}}{2N_2^2C_{res}}\right) \tag{7.5}$$

Using (7.3), (7.4), and (7.5), we can quantify the percentage change of the filter bandwidth due to $C_{var}$ by calculating the percentage change in its odd mode as follows:

$$\frac{f_{Odd\ Ind}-f_{Odd\ Ind}(C_{var}=0)}{f_{Odd\ Ind}(C_{var}=0)-f_{even}}=\frac{-1}{2k_{couplingind}}\left(\frac{C_{var}}{N_2^2\ C_{res}}\right) \tag{7.6}$$

$$\frac{f_{Odd\ Cap}-f_{Odd\ Cap}(C_{var}=0)}{f_{even}-f_{Odd\ Cap}(C_{var}=0)}=\frac{1}{2k_{couplingcap}}\left(\frac{C_{var}}{N_2^2\ C_{res}}\right) \tag{7.7}$$

The symbols $k_{couplingind}$ and $k_{couplingcap}$ represent the denormalized coupling coefficients in the absence of $C_{var}$. For narrowband filters, these values increase approximately proportionally to the fractional bandwidth (*FBW*) of the filter [24]. The percentage change in bandwidth is inversely proportional to these coupling coefficients, as per (7.6) and (7.7). This suggests that for a given value of $C_{var}$, a greater percentage change would be possible with a narrower starting bandwidth, which corresponds to a smaller coupling coefficient. Using (7.6) and (7.7) as a guide, the minimum value of the varactor capacitance is determined by the least desirable percentage bandwidth change from the original bandwidth, represented by $\Delta_{min}$.

$$C_{var}(min) = 2\Delta_{min}k_{coupling}N_2^2C_{res} \tag{7.8}$$

#### 7.1.1.2 Quality Factor Impact

Next, we need to estimate the impact of the lumped-element varactor $Q$ on the original filter $Q$, denoted by $Q_{initial}$. This is done as follows:

$$Q_{new} \approx \omega_0\left(R_{res} \parallel N_2^2R_{var}\right)C_{res} \tag{7.9}$$

The resistance $R_{var}$ represents the shunt equivalent resistance of $C_{res}$. We can develop additional intuition here by simplifying equation (7.9) as follows:

$$Q_{new} \approx \frac{1}{\dfrac{1}{Q_{initial}}+\dfrac{C_{var}}{N_2^2Q_{var}C_{res}}} \tag{7.10}$$

As expected, the tuning varactor adversely affects the unloaded quality factors of the filter resonators. By setting the minimum desirable $Q_{new}$, we can decide the maximum capacitive load $C_{var}(max)$ we can tolerate.

#### 7.1.1.3 Impedance Matching

To enable impedance matching to this adjustable bandwidth across wide tuning ranges, external coupling control must be achieved. The external quality factor $Q_e = 1/k_e^2$ for a bandpass filter can be written as [24]

$$Q_e = \frac{1}{FBW \times M_{01}^2} \tag{7.11}$$

According to (7.11), a higher $Q_e$ is needed to achieve a reduced fluctuation of bandwidth in a wide range of settings. In order to achieve excellent impedance matching, it is necessary to reduce the external coupling at the upper end of the tuning range. This is accomplished by tailoring the external coupling as a function of the tuning range using a short transmission line (see Figures 7.1 and 7.3(a)) of length $L < \lambda/4$ and impedance $Z_t > Z_0$. Additionally, dynamic control over external coupling is achieved by using the varactor $C_{ext}$. By comparing the real components of admittances $Y_{in\ new}$ and $Y_{in\ old}$, as illustrated in Figure 7.3(a) and (b), we can estimate the impact of this external coupling section. The real part of $Y_{in}$ may be connected to $Q_e$ using

$$Q_e = \frac{N_1^2 \omega_0 C_{res}}{Re(Y_{in})} \tag{7.12}$$

Assuming that $N_1$ remains constant, we have

$$\frac{Q_{e\ new}}{Q_{e\ old}} = \frac{Re(Y_{in\ old})}{Re(Y_{in\ new})} \tag{7.13}$$

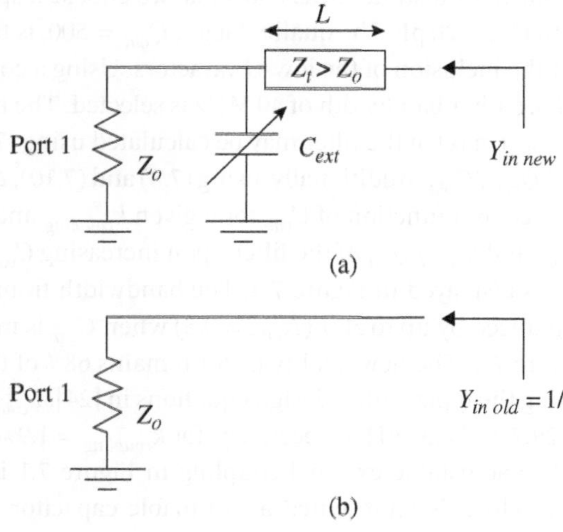


**Figure 7.3** External coupling section model (a) with and (b) without an additional external coupling network.

By standard transmission line theory, we can rewrite this as

$$\frac{Q_{e\ new}}{Q_{e\ old}} = (1 - Z_t\beta L\omega C_{ext})^2 + \left(\frac{Z_t\beta L}{Z_0}\right)^2 \tag{7.14}$$

The expression above demonstrates that an increase in the value of $C_{ext}$ results in a reduction of $Q_e$. To assess the impact of the transmission line independently, lets us examine the specific scenario where $C_{ext} = 0$. In this case, equation (7.14) simplifies to

$$\frac{Q_{e\ new}}{Q_{e\ old}} = 1 + \left(\frac{Z_t\beta L}{Z_0}\right)^2 \tag{7.15}$$

Consequently, a short high-impedance line can be used to increase $Q_e$, as desired. Additionally, the linear dependency of $\beta$ on $\omega$ will result in a significantly greater reduction of the external coupling at the upper end of the tuning range. As a result, a good impedance match to the decreased bandwidth fluctuation will be possible. Although the transmission line alone helps, the external coupling fluctuation as a function of frequency cannot be easily altered. Therefore, we seek the solution of dynamically controlling the external coupling by adding the varactor $C_{ext}$ before the transmission line. Over wide tuning ranges, the filter response may be adjusted to produce the required out-of-band attenuation/group delay using the two variable controls (T-line and $C_{ext}$) of the external coupling. Another benefit of this design is that $C_{ext}$ can be a low-$Q$ varactor since the input and output coupling transformers with $N_1$ turns protect it from the high-$Q$ tank resonators.

#### 7.1.1.4 Simulated Results

To further illustrate these concepts, we choose a specific high-$Q$ filter at 1.5 GHz with $C_{res} = 20$ pF. The quality factor, $Q_{old} = 500$, is the initial $Q$ for this filter without the inclusion of the low-$Q$ varactors. Using a coupling coefficient of $k_{coupling} =$ 1.9%, a filter bandwidth of 40 MHz is selected. The maximum coupling coefficient $k_{coupling}$(max) of the filter may be calculated using (7.8) and (7.10) for a given $\Delta_{min}$ and $Q_{new}/Q_{old}$. Additionally, using (7.8) and (7.10), $\Delta_{min}$ and $Q_{new}/Q_{old}$ may be calculated as a function of $C_{var}$ for a given $k_{coupling}$ and $N_2$. The estimated change in $\Delta_{min}$ and $Q_{new}/Q_{old}$ of the filter upon increasing $C_{var}$ ($Q_{var} = 20$ at 1.5 GHz) for $N_2$ = 3 is displayed in Figure 7.4. The bandwidth from the original value of 40 MHz is reduced by up to 50% ($\Delta_{min} = 0.5$) when $C_{var}$ is increased to 3.4 pF, as shown in Figure 7.4. The new quality factor remains 68% of the old one ($Q_{new}/Q_{old} = 0.68$). Using the typical filter design equations in [24], $L_{coupling}$ and $L_{res}$ may be calculated as 29.7 and 0.57 nH, respectively, for $k_{coupling} = 1.9\%$ and $C_{res} = 20$ pF.

The adjustable external coupling in Figure 7.1 is achieved using the varactor $C_{ext}$, which is represented as a tunable capacitor with a quality factor of 20 at

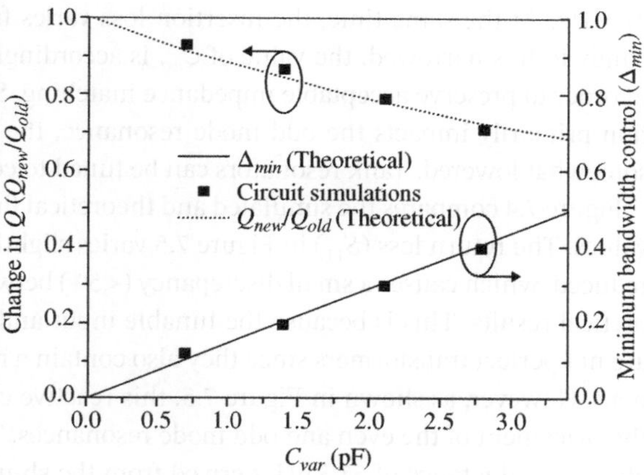


**Figure 7.4** Simulated changes in quality factor and bandwidth as a function of $C_{var}$ for a specific numerical example.

1.5 GHz and a maximum capacitance of 1 pF, which is needed to provide adequate impedance matching to this tunable bandwidth. The transmission line parameters are set to $Z_t = 100\ \Omega$ and an electrical length of 30° ($L = \lambda/12$) at 1.5 GHz. As the bandwidth decreases with rising $C_{var}$, the external coupling should be decreased from 1 to 0.1 pF because a bigger value of $C_{ext}$ leads to an increased external coupling. When $C_{var} = 0$ and $C_{ext} = 1$ pF, the number of turns $N_1$ may be found using (7.11) and (7.12) for the starting bandwidth of 40 MHz. A value of 2.5 is found for $N_1$ in the case of a second-order maximally flat filter.

Now that every parameter in Figure 7.1 has been established, linear simulators may be used to acquire the filter response. Figure 7.5 plots the simulated results. By increasing $C_{var}$ from 0 to 3.4 pF, the bandwidth can be decreased from 40 to

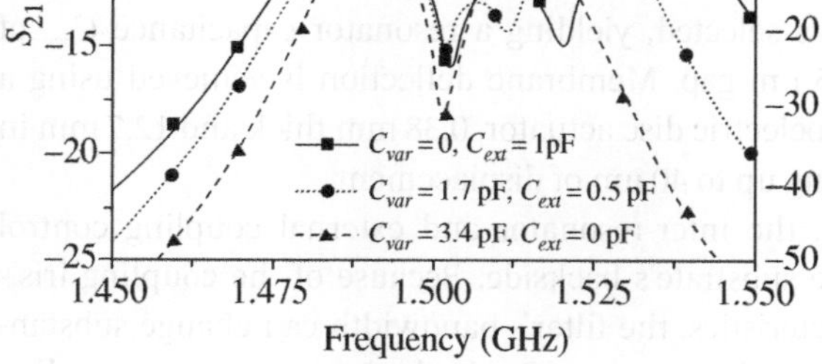


**Figure 7.5** Simulated results as a function of important control parameters.

21 MHz. At the same time, the insertion loss varies from 1.3 and 2.3 dB. As the bandwidth is narrowed, the value of $C_{ext}$ is accordingly decreased from 1 to 0 pF in order to preserve acceptable impedance matching. Since the bandwidth reduction primarily impacts the odd mode resonance, the filter's center frequency is somewhat lowered. Tank resonators can be tuned to compensate for this.

Figure 7.4 compares the simulated and theoretical findings for four distinct $C_{var}$ values. The return loss ($S_{11}$) in Figure 7.5 varies slightly as the filter bandwidth is reduced, which causes a small discrepancy (<5%) between the simulated and theoretical results. This is because the tunable input and output coupling networks are not perfect transformers since they also contain a reactive admittance component. However, as shown in Figure 7.5, this reactive component has no effect on the placement of the even and odd mode resonances. The reason is that since the effective inductance of 55.3 nH, derived from the shunt reactance of the external coupling network, is much greater than the resonator inductance of 0.57 nH, it is effectively absorbed into the resonator circuit. Consequently, a good impedance match across a controllable bandwidth can be achieved through the use of a tunable external coupling network based on $C_{ext}$.

### 7.1.2 Filter Implementation

This section applies the previously discussed design principles to develop a high-$Q$ evanescent-mode cavity filter with tunable bandwidth. By varying the capacitance introduced by the tuning post, the filter can be adjusted across a wide frequency range. Combining this broad tunability and high-$Q$ performance with precise control of inter-resonator and external coupling enables the filter response to be tailored to specific design requirements.

Figure 7.6 illustrates the layout of a high-$Q$ evanescent-mode cavity filter that uses low-$Q$ varactors to control both inter-resonator and external couplings. The filter is implemented on a 3.175-mm thick Rogers TMM substrate with a dielectric constant of 3.2 and a loss tangent of 0.002. Two resonators, each with a central capacitive post, are formed by conductive vias and are connected via a coupling iris (Figure 7.6(a)).

A post-radius of 3.5 mm is selected, yielding a resonator capacitance $C_{res}$ of approximately 50 pF for a 5 μm gap. Membrane deflection is achieved using a commercially available piezoelectric disc actuator, 0.38 mm thick and 12.7 mm in diameter, capable of providing up to 40 μm of displacement.

As seen in Figure 7.6(b), the inter-resonator and external coupling control networks are located on the substrate's backside. Because of the coupling iris's frequency-dependent characteristics, the filter's bandwidth can change substantially over wide tuning ranges. The varactors $C_{var}$ in the inter-resonator coupling control network (Figure 7.6(b)) can compensate for this bandwidth variation.

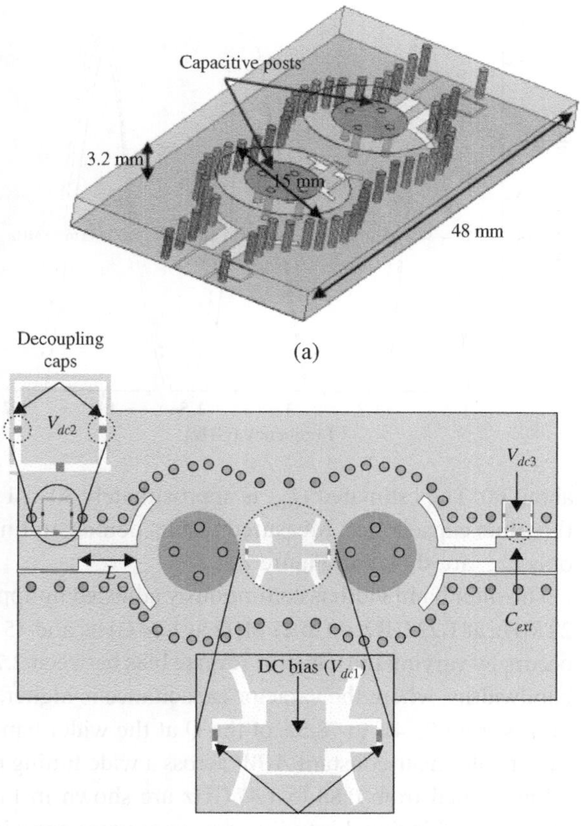


**Figure 7.6** Two-pole filter formed by two resonators, each with a central capacitive post. (a) Top side. (b) Bottom side. Notice the feed lines and the sections implementing the tunable bandwidth and external coupling.

Proper design can achieve a constant absolute bandwidth (*ABW*) across the entire tuning range. The same voltage, $V_{dc1}$, is used to bias both varactors in Figure 7.6(b). The slots in the bandwidth control network can be viewed and modeled as transformers ($N_2$ turns).

To minimize bandwidth variation across a wide tuning range, external coupling is also controlled using varactors corresponding to $C_{ext}$ in Figures 7.1 and 7.6(b). Independent bias voltages, $V_{dc2}$ and $V_{dc3}$, are applied to the input and output varactors to shape the filter response and control group delay. A high-impedance ($Z_t = 100\ \Omega$) transmission line segment of length $L = 5.3$ mm is used to moderate external coupling across the tuning range. Commercially available BB857 varactors from Infineon Technologies are used for both $C_{var}$ and $C_{ext}$, enabling control of inter-resonator and external couplings. These varactors offer a capacitance range of 0.5–6.5 pF at 1 MHz as the reverse bias $V_R$ varies from 28 to 1 V. At 470 MHz and $V_R = 5$ V, we have $C_{var} \simeq 2.5$ pF with a series resistance of 1.5 Ω, yielding a $Q$ of

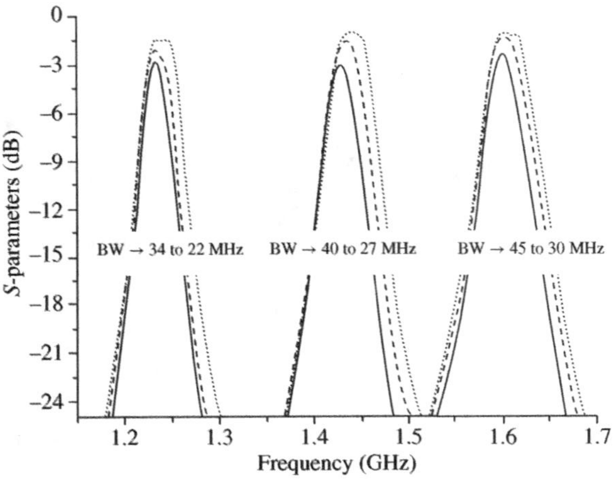


**Figure 7.7** Measured filter results showing the effect of bandwidth control.

about 60. The estimated $Q_{var}$ is approximately 30 at 1 GHz and 20 at 1.5 GHz for the same capacitance. Measured results demonstrating bandwidth control using only $C_{var}$ are shown in Figure 7.7.

The filter bandwidth is continuously adjusted by approximately 33%: from 34 to 22 MHz at 1.23 GHz, 40 to 27 MHz at 1.44 GHz, and 45 to 30 MHz at 1.6 GHz. This occurs by varying the varactor reverse bias between 2.7 and 25 V. At the narrower bandwidths, where the varactor capacitance is higher, the simulated resonator $Q$ ranges from 71.4% to 78.5% of the $Q$ at the wider bandwidths. This level of control enables near-constant *ABW* across a wide tuning range. Measured results for a filter tuned from 0.88 to 1.47 GHz are shown in Figure 7.8. Simulations estimate that achieving this tuning range requires approximately 25 μm of membrane deflection. Over this range, the varactor bias is swept from 25 V at 0.88 GHz to

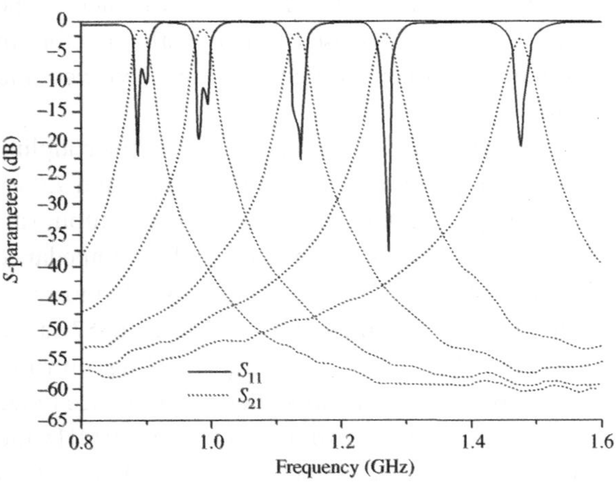


**Figure 7.8** Measured filter scattering parameters. Notice the near-constant bandwidth.

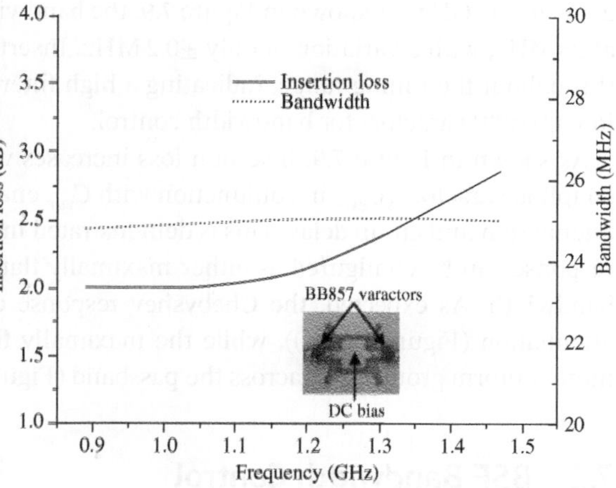


**Figure 7.9** Measured filter insertion loss and bandwidth across its tuning range. Source: With permission of Joshi et al. [114]/IEEE.

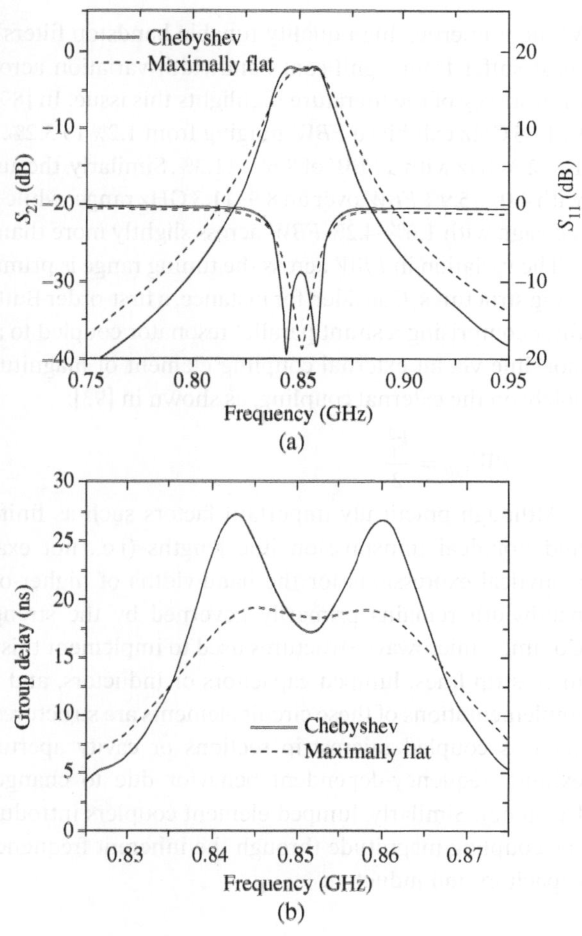


**Figure 7.10** Measured (a) scattering parameters and (b) group delay as a function of frequency.

2.9 V at 1.47 GHz. As shown in Figure 7.9, the bandwidth remains nearly constant at 25 MHz, with a variation of only ±0.2 MHz. Insertion loss remains below 3 dB throughout the tuning range, indicating a high filter $Q$ (>250) despite the use of low-$Q$ (<30) varactors for bandwidth control.

As shown in Figure 7.9, insertion loss increases with frequency. The external coupling varactors ($C_{ext}$) in conjunction with $C_{var}$ enable control over out-of-band attenuation and group delay. This is demonstrated in Figure 7.10, where the filter response can be configured as either maximally flat or Chebyshev for the same bandwidth. As expected, the Chebyshev response exhibits greater out-of-band attenuation (Figure 7.10(a)), while the maximally flat configuration provides a more uniform group delay across the passband (Figure 7.10(b)).

## 7.2 BSF Bandwidth Control

While numerous high-quality tunable bandstop filters have been demonstrated, most suffer from significant bandwidth variation across wide tuning ranges. A brief survey of the literature highlights this issue. In [89], a filter tuning from 0.65 to 1.65 GHz exhibits a *FBW* ranging from 1.2% to 3.2%. The design in [90] covers 1.3–2.3 GHz with a *FBW* of 8.6%–11.3%. Similarly, the authors of [91] report a filter with 4.0%–5.9% *FBW* over an 8.9–11.3 GHz range, while the authors of [92] present a design with 1.6%–4.2% *FBW* across slightly more than an octave of tuning.

The variation in *FBW* across the tuning range is primarily attributed to the coupling structures. Consider, for instance, a first-order Butterworth lossless bandstop filter comprising a shunt parallel resonator coupled to a source-to-load transmission line via an external coupling element of magnitude $k_0$. The *FBW* depends solely on the external coupling, as shown in [93]:

$$BW_{3dB} = \frac{k_0^2}{2} \tag{7.16}$$

Although practically important factors such as finite unloaded quality factor and nonideal transmission line lengths (i.e., not exactly 90°) complicate the analytical expressions for the bandwidths of higher-order bandstop filters, the bandwidth remains primarily governed by the strength of external coupling. Common microwave structures used to implement this coupling include coupled microstrip lines, lumped capacitors or inductors, and cavity apertures. Physical implementations of these circuit elements are structures with specific geometries, such as coupled microstrip sections or cavity apertures (Figure 2.13), which exhibit frequency-dependent behavior due to changes in electrical size with frequency. Similarly, lumped element couplers introduce frequency variation on the coupling magnitude through the inherent frequency-dependent reactance of capacitors and inductors.

Maintaining a constant *ABW* across a wide tuning range is often even harder, as *ABW* is defined by the product of *FBW* and the center frequency $f_0$. As a result, even if *FBW* remains constant, *ABW* will double over an octave tuning range. To achieve constant *ABW*, *FBW* must decrease linearly with frequency. The external coupling coefficient must be inversely proportional to the square root of frequency to yield this.

Several approaches have been proposed to address bandwidth variation. The authors of [94] and [95], for example, address this question for absorptive bandstop filters with about an octave tuning range. Inductive admittance inverters combined with capacitively tuned resonators yield relatively constant *ABW* in this case. However, due to the reliance on lumped inductors and capacitors, this technique is limited to low frequencies (e.g., below approximately 2 GHz).

Another approach involves breaking the dominance of either electric or magnetic coupling in the external coupling sections. For example, this can be done using an electrically long section of a coupled transmission line. By optimizing the length of this coupling line, electric and magnetic couplings can be blended, and their opposing frequency dependencies can partially cancel out, resulting in a relatively stable bandwidth. This technique is demonstrated in [96] and [97], with [96] achieving a 92% center frequency tuning range and only 24% variation in its 3-dB bandwidth. Nevertheless, this method is limited to microstrip or other electrically long resonators capable of supporting both electric and magnetic field couplings.

Using adjustable coupling elements to lower coupling at higher frequencies and preserve constant *ABW* is a third method for creating constant-bandwidth filters. For example, microstrip resonators loaded with varactors at both ends are used in [93] and [98]. The resonator's voltage and current distributions can be altered by differentially tuning the varactors, which also modifies the filter's bandwidth and external coupling coefficients. The capacity to tune the 3-dB bandwidth from 70 to 140 MHz and sustain a steady 100 MHz 3-dB bandwidth over a center frequency tuning range of 1.2–1.6 GHz is demonstrated in [98]. In [93], the presented filter can adjust its 3-dB bandwidth between 26 and 143 MHz and keep it steady between 0.67 and 1.0 GHz. A through-line with varactors is connected to substrate-integrated-waveguide cavity resonators in [99], which enables the filter to sustain a steady 83 MHz 3-dB bandwidth over a tuning range of 0.77–1.25 GHz. Despite these demonstrations, it is also important to consider the associated limitations. More specifically, although tunable coupling components may allow filters to have a fixed bandwidth, they may also reduce filter linearity, increase loss, and add control complexity.

The bandwidth fluctuation of high-*Q* evanescent-mode cavity resonators could not be passively controlled until recently. This kind of resonator is incompatible with the lumped element coupling mechanism of [94], especially at high frequencies where lumped-element components can become prohibitively lossy. Moreover, it is challenging to simultaneously realize both electric and magnetic couplings. The approach taken by [97] is inapplicable because the magnetic field is approximately evenly distributed throughout the resonator's volume, whereas the electric field is concentrated into a very small portion of it (see Figure 2.3). Although tunable coupling components can be employed, as in [99], in practice it is often advisable to keep things simple and avoid complexities arising from such extra tuning elements. This makes passive compensation techniques particularly interesting.

In [100], a passively compensated coupling technique for evanescent-mode bandstop filters was introduced, enabling bandwidth control for evanescent-mode bandstop filters. In contrast to the usual coupling method for these filters, this paper demonstrated a constant-*ABW* filter with only 27% variation in its 3-dB *ABW* over an octave center-frequency tuning range. It also introduced a method for passively compensating for the frequency variation of the filter's coupling coefficients. Building on this work, this chapter examines this new bandwidth control method in greater detail, explains how it works, and assesses design considerations for the method, including the relationship between bandwidth variation and tuning range, as well as the effects of transmission line length and coupling sign. We demonstrate that this coupling technique not only significantly minimizes bandwidth fluctuation but also lessens the phase variation of the transmission line between a two-pole bandstop filter's resonators. This lessens the variance in the transfer function's shape. Four filters with tuning ranges centered around 4.5 GHz are designed using the coupling method: a two-pole constant *FBW* filter, which has an octave tuning range and a 3-dB bandwidth of 1.16%–1.3%; a two-pole constant *ABW* filter, which has an octave tuning range and a 3-dB bandwidth of 50.3–56.5 MHz; a two-pole constant *ABW*, which has a 50% tuning range and a 52–54 MHz 3-dB bandwidth; and a constant *ABW* octave-tunable four-pole filter, which can maintain a constant (to within measurement limits) 50 MHz 10-dB bandwidth.

### 7.2.1 Constant Bandwidth Coupling Concept

This chapter's coupling topology is depicted in Figure 7.11. Originally proposed in [100] for constant-bandwidth filters, this circuit was initially shown in [93] to realize intrinsically-switchable bandstop filters.

The circuit of Figure 7.11(a), which consists of a resonator coupled twice to a through line of length $\theta_0$ with coupling elements $k_{E1}$ and $k_{E2}$, can be demonstrated to be equivalent to the circuit of Figure 7.11(b), which consists of a resonator coupled to a through-line with only one coupling element $k_E$, followed and preceded by transmission lines of lengths $\theta_1$ and $\theta_2$. Since we are utilizing widely tunable

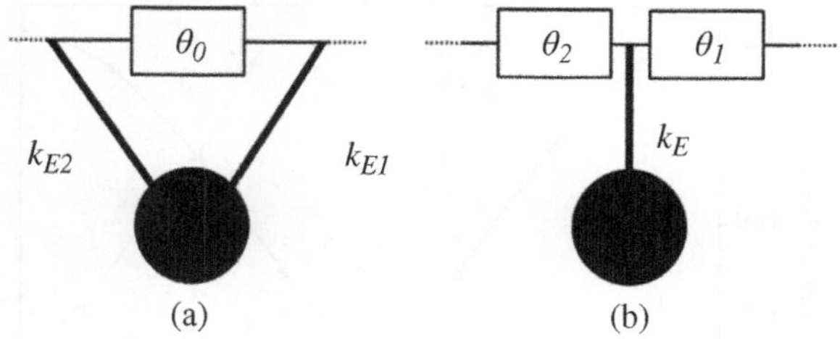


**Figure 7.11** (a) Twice-coupled resonator that is key for realizing constant bandwidth filters. (b) Equivalent circuit of (a).

resonators that can correct for small frequency offsets, we may ignore the slight resonant frequency offset between the resonators in Figure 7.11(a) and (b). The following equations connect the expressions in the original circuit and the equivalent circuit:

$$k_E = \sqrt{k_{E1}^2 + k_{E2}^2 + 2k_{E1}k_{E2}\cos(\theta_0)} \tag{7.17}$$

$$\theta_1 = \frac{1}{2}\left(\pi - \arctan\left(\frac{\dfrac{k_{E1}}{k_{E2}} + e^{-j\theta_0}}{\dfrac{k_{E1}}{k_{E2}} + e^{j\theta_0}}\right)\right) \tag{7.18}$$

$$\theta_2 = \theta_0 - \theta_1 \tag{7.19}$$

If the frequency dependence of the two coupling elements $k_{E1}$ and $k_{E2}$ is approximately the same, but one is a fraction of the other, for example,

$$k_{E1} \approx rk_{E2} \tag{7.20}$$

where $r$ is a constant, then (7.17) can be approximated as

$$k_E \approx Fk_{E1} \tag{7.21}$$

$$F = \sqrt{1 + r^2 + 2r\cos(\theta_0)} \tag{7.22}$$

The total equivalent coupling coefficient is now equal to one of the initial coupling coefficients times a shaping factor $F$. This has a sinusoidal-like frequency dependence since the electrical length of the transmission line $\theta_0$ is proportional to frequency. Figure 7.12 shows a plot of the shaping factor $F$'s frequency dependence. It is evident that $F$ has a negative-sloped frequency dependency for $0° < \theta_0 < 180°$ and a positive frequency dependence for $180° < \theta_0 < 360°$ when the two coupling coefficients have the same sign, that is $r$ is positive. On the other hand, the opposite tendency is seen when the two coupling coefficients have the opposite sign, that is $r$ is negative: $F$ has a negative frequency dependency for $180° < \theta_0 < 360°$ and a positive frequency dependence for $0° < \theta_0 < 180°$. The key idea is that it is possible to at least partially offset the positive frequency dependence present

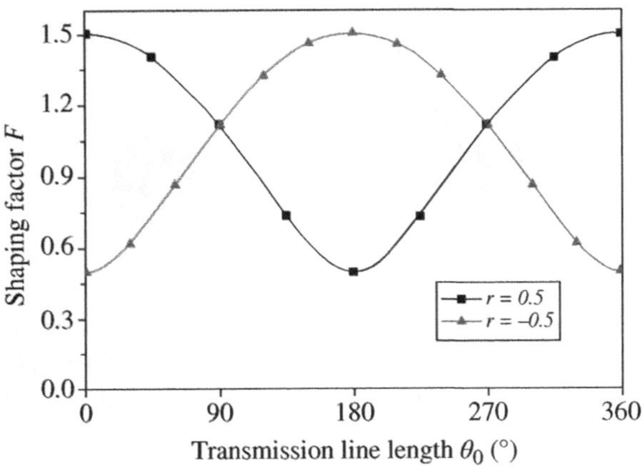


**Figure 7.12** The sinusoidal-like frequency dependence of the shaping factor $F$. Notice that the positive and negative slopes depend on the sign of $r$.

in the original coupling structure by using the regions where $F$ has negative frequency dependence ($0° < \theta_0 < 180°$ when the coupling coefficients have the same sign, and $180° < \theta_0 < 360°$ when they have opposite signs).

We will now apply this method to a frequency-dependent coupling coefficient and examine how the different design parameters impact the frequency variation of the composite coupling coefficient. This will help us understand how we can realize constant coupling coefficients for constant-*FBW* filters and coupling coefficients that decrease with frequency for constant-*ABW* filters. To simplify the analysis that follows, we define $k_{E1}$ so that it has a nominal value of 1 at the lowest tuned frequency of the resonator and rises by 50% over an octave tuning range:

$$k_{E1} = 1 + 0.5\left(\frac{f_0}{f_{min}} - 1\right) \tag{7.23}$$

This frequency dependence is a realistic approximation of the practical coupling coefficients found in many filters considered later. Figure 7.13 illustrates how altering $r$, the ratio of the two coupling elements, affects the coupling coefficient fluctuation. As we can observe, the composite coupling coefficient $k_E$ increases by 50% throughout an octave tuning range and is equal to that of a single coupling element when $r$ is zero. The coupling increases at lower frequencies and decreases at higher frequencies when $r$ is increased. Notice that we can obtain a wide range of frequency dependencies. The coupling coefficient reaches zero at the frequency where $\theta_0$ equals 180° when $r = 1$. This makes intuitive sense because a net zero coupling coefficient should arise from the cancellation of two coupling structures of equal magnitude that are separated by a 180° transmission line.

Figure 7.14 plots the coupling variation as a function of the length of the transmission line $\theta_0$ that separates the coupling elements. The specified $\theta_0$ values are defined at 2 Hz. It is evident that the coupling coefficient has a concave-down form for lengths of $\theta_0$ less than 170° and a concave-up shape for lengths of $\theta_0$ more

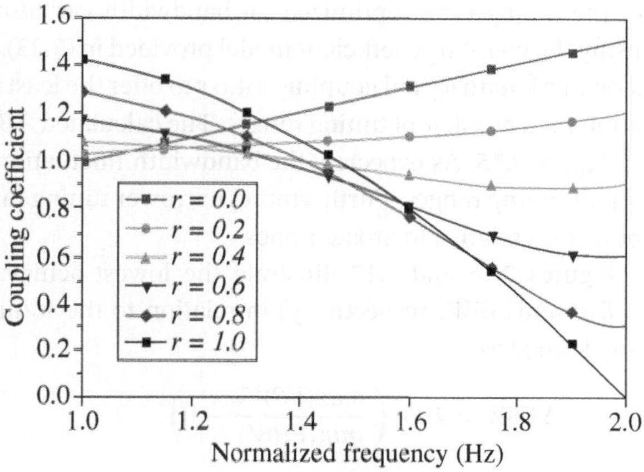


**Figure 7.13** The equivalent coupling coefficient frequency dependence as a function of the ratio $r$. The transmission line length $\theta_0 = 180°$ at 2 Hz for the calculations in this figure.

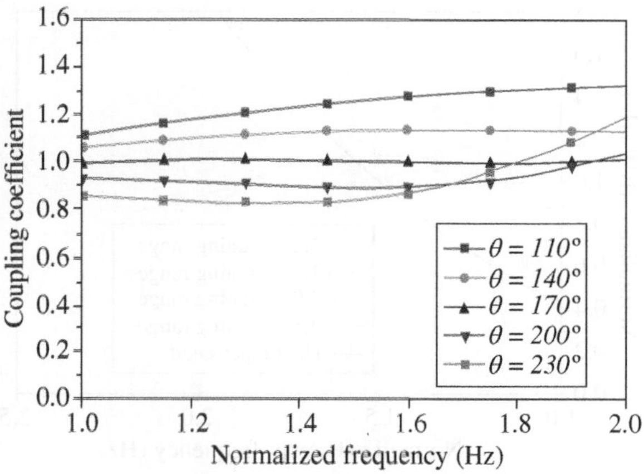


**Figure 7.14** The equivalent coupling coefficient frequency dependence as a function of $\theta_0$ specified at 2 Hz. The ratio $r = 0.3$ for the calculations in this figure.

than 170°. A transmission line of length 170° at 2 Hz and coupling ratio $r = 0.28$ produces a nearly-constant coupling coefficient as required for constant *FBW* for the particular frequency dependence considered in this example. Moreover, a transmission line of length 180° at 2 Hz and coupling ratio $r = 0.6$ causes the coupling coefficient to decrease with frequency as required for constant *ABW*.

#### 7.2.1.1 BW Variation Versus T-Line Length and Tuning Range

The shaping factor $F$ can lessen the amount of coupling fluctuation over large tuning ranges since it has a negative slope with respect to frequency. However, over narrow ranges of $\theta_0$, the slope of $F$ is much more linear (in the neighborhood of $\theta_0 = 90°$ when $r$ is positive and $\theta_0 = 270°$ when $r$ is negative). As a result, it is expected that this method will be even more effective when used over narrow tuning ranges. The reduction in bandwidth variation is generally a nonlinear function of the tuning range optimized for bandwidth variation. We demonstrate this by using the coupling coefficient model provided in (7.23). We optimize the transmission line length $\theta_0$ and coupling ratio $r$ to offer the least amount of bandwidth variation for a number of tuning ranges. The calculated *ABW* variations are displayed in Figure 7.15. As expected, the bandwidth fluctuation is substantially improved for all tuning ranges. Furthermore, narrower tuning ranges demonstrate superior outcomes relative to broader ones.

Figures 7.16 and 7.17 illustrate the lowest achievable bandwidth variations (*FBW* and *ABW*, respectively) in relation to the tuning range. These variations are defined as:

$$\Delta FBW = 100\left(\frac{max(FBW)}{min(FBW)} - 1\right) \tag{7.24}$$

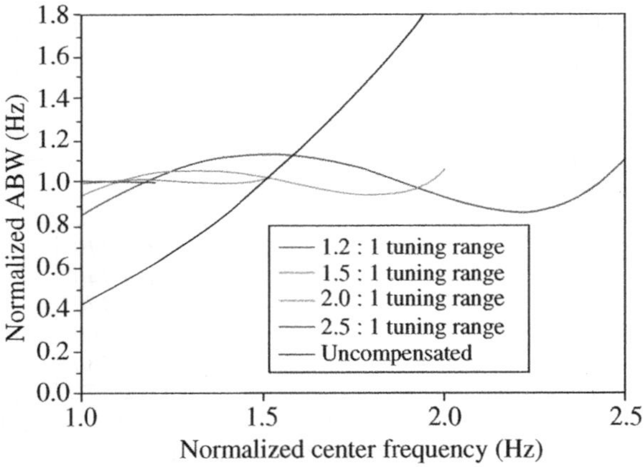


**Figure 7.15** Calculated frequency dependence of the improved and uncompensated absolute bandwidths for different tuning ranges. Notice that while improvement is possible for all tuning ranges, narrower tuning ranges demonstrate superior outcomes relative to broader ones.

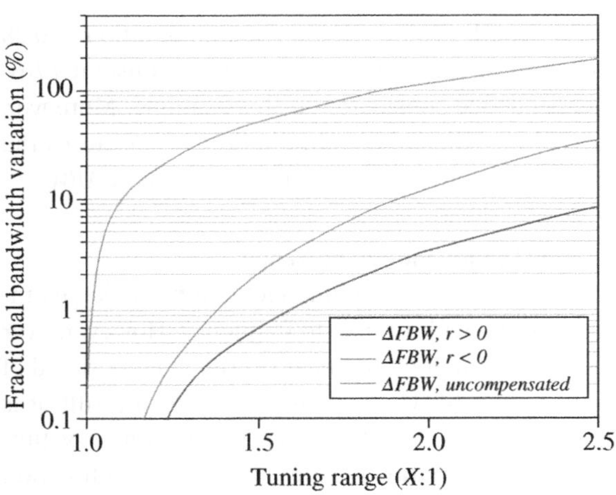


**Figure 7.16** Minimum achievable *FBW* variation as a function of the center frequency tuning range.

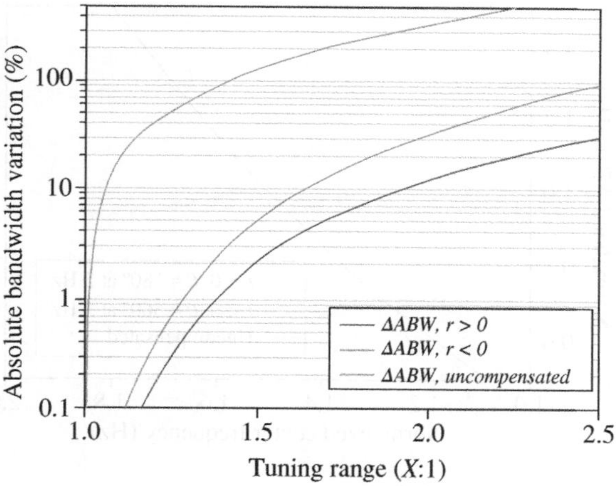


**Figure 7.17** Minimum achievable *ABW* variation as a function of the center frequency tuning range.

$$\Delta ABW = 100\left(\frac{max(ABW)}{min(ABW)} - 1\right) \tag{7.25}$$

Figures 7.16 and 7.17 illustrate the extent to which the suggested strategy mitigates bandwidth variance. The proposed approach can diminish the *FBW* variance from 150% to 3.5% and the *ABW* variation from 350% to 12% over an octave (2:1) tuning range. Halving the tuning range to 1.5:1 significantly diminishes bandwidth variation. The *FBW* fluctuation can be diminished to around 0.7%, while the *ABW* variation can be lowered to about 2%.

The presented data also illustrate the difference between employing coupling elements of identical polarity, necessitating a transmission line of approximately 180° at $f_{max}$, and utilizing coupling elements of opposing polarity, which requires a transmission line of approximately 360° at $f_{max}$. It is obvious that considerably lower bandwidth fluctuation can be achieved when employing coupling elements of identical sign. In contrast to a nominally 180° transmission line, the nominally 360° transmission line undergoes double the phase variation across a specified frequency tuning range, resulting in a more nonlinear shaping factor $F$ that is less effective in compensating for the linear frequency dependence of the coupling element. Consequently, it is preferable, when feasible, to employ coupling elements of identical polarity to facilitate the utilization of a <180° transmission line. This may not always be feasible, particularly at higher frequencies or when employing high-permittivity substrates (as referenced in [101] and [102]), which could render it physically impracticable to achieve a transmission line with less than 180° between the coupling elements. Figure 7.18 illustrates the frequency dependence of the coupling coefficient for a 360° transmission line in comparison to the 180° realization.

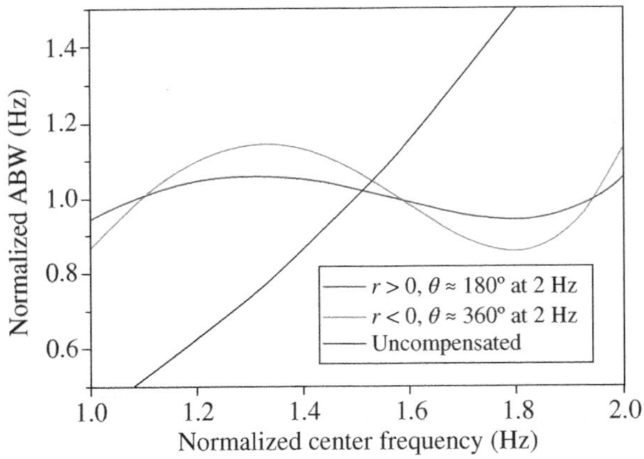


**Figure 7.18** Coupling coefficient as a function of frequency for 180° and 360° transmission lines. Notice also the uncompensated case.

#### 7.2.1.2 Phase Variation

A two-pole absorptive bandstop filter based on the coupling topology of Figure 7.11(a) can be realized based on the topology illustrated in Figure 7.19(a). A transmission line of length $\theta_3$ is required between the two resonators to achieve a 90° phase difference, essential for a symmetric bandstop response [103]. Substituting the twice-coupled resonators in Figure 7.19(a) with their equivalent circuits Figure 7.11(b) results in the circuit depicted in Figure 7.19(b). From this equivalent circuit, it is clear that the phase contributed by the constant coupling structure must be considered when determining the length of the transmission line $\theta_3$. Substituting equation (7.20) into equation (7.18) allows for calculating $\theta_1$ as follows:

$$\theta_1 = \frac{1}{2}\left(\pi - arg\left(-\frac{r + e^{-j\theta_0}}{r + e^{j\theta_0}}\right)\right) \tag{7.26}$$

Equation (7.26) remains the same for $\theta_2$.

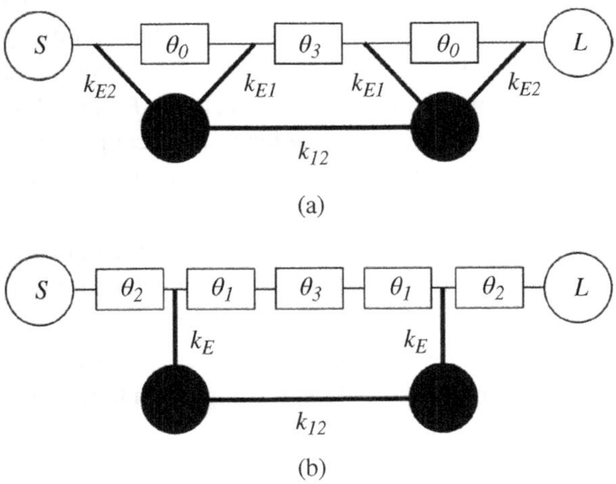


**Figure 7.19** (a) Two-pole bandstop filter based on the constant-bandwidth coupling structure of Figure 7.11(a). (b) Two-pole filter equivalent circuit based on the equivalent circuit of the coupling structure from Figure 7.11(b).

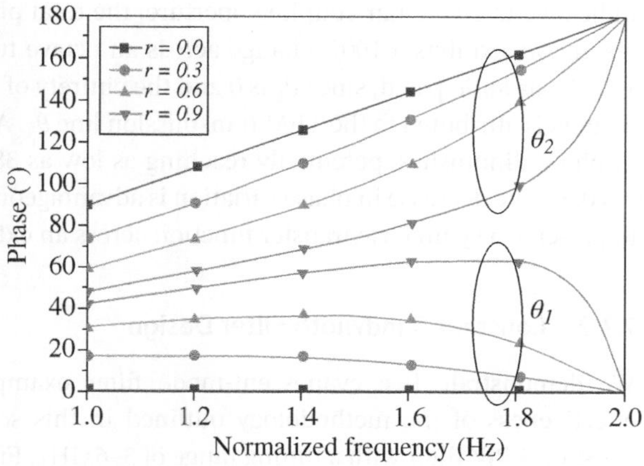


**Figure 7.20** Frequency variations of phase lengths $\theta_1$ and $\theta_2$ from Figure 7.19. Notice the negative slopes for $\theta_1$ for portions of the frequency range.

The equivalent transmission line lengths that establish the coupling reference plane are influenced by both the length of the transmission line in the coupling section and the ratio of the two coupling values. Upon analyzing the frequency variation of $\theta_1$ and $\theta_2$ across different values of $r$ (illustrated in Figure 7.20), we observe that for $0° < \theta_0 < 180°$, the effective phase length nearest to the larger coupling aperture ($\theta_1$, as depicted in Figure 7.19(b) when $|r| < 1$) exhibits a negative frequency slope at certain frequencies. It is also interesting to note that $\theta_1$ consistently equals zero when $\theta_0$ is 180°.

The negative phase-versus-frequency slope is advantageous, as it diminishes the frequency variation of the phase length between the two resonators. The total phase between the two resonators, comprising the additional transmission line length $\theta_3$ and twice the length $\theta_1$, exhibits reduced phase variation within a specified tuning range compared to a single Transverse Electro-Magnetic (TEM) transmission line. This is illustrated in Figure 7.21. For a coupling ratio of $r = 0$,

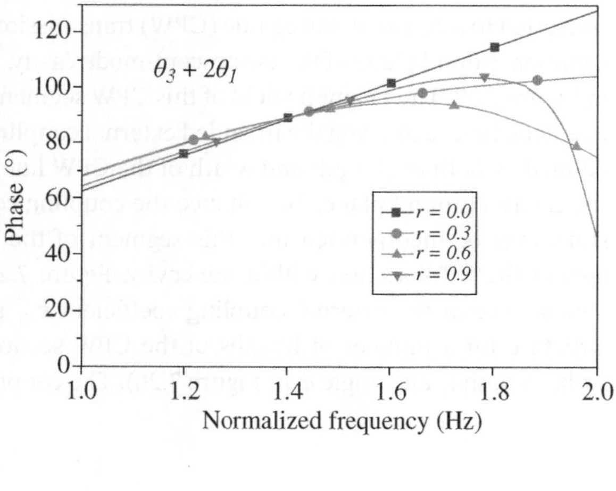


**Figure 7.21** Frequency variation of the total phase between the resonators of the two-pole filter. This phase is equal to $\theta_3$ because of the physical transmission line between the resonators plus $2\theta_1$, which is the phase contributed by the coupling structure.

indicative of a singular coupling aperture, the total phase variation between the resonators exhibits a 100% change across an octave tuning range (from 63.6° to 127.3°), as anticipated, since $\theta_1$ is 0 and the entirety of the phase between the resonators is attributed to the TEM transmission line $\theta_3$. As $r$ increases, the variation in phase diminishes, potentially reaching as low as 38% (from 68.5° to 94.7° for $r = 0.6$). The decrease in phase variation is advantageous, since it enables the filter to preserve a symmetric transfer function across an extensive frequency range.

### 7.2.2 Constant Bandwidth Filter Design

We demonstrate five evanescent-mode filter examples to clearly assess the effectiveness of the methodology outlined in this section. Filter A is a 1.25% constant *FBW* filter with a tuning range of 3–6 GHz. Filter B is a 53 MHz constant *ABW* filter with a tuning range of 3–6 GHz. Filter C is a 53 MHz constant *ABW* filter with a tuning range of 3.5–5.5 GHz. Filter C is designed to illustrate that significantly reduced bandwidth variation can be achieved within a narrower tuning range. Filter D is a four-pole filter formed by a back-to-back Filter C units. Filter E is an uncompensated filter that employs a single coupling element rather than the constant-bandwidth coupling approach presented above. This filter facilitates a clear assessment of the performance improvements achieved by the constant-bandwidth coupling scheme. An absorptive bandstop filter design is employed to enhance stopband rejection. By selecting an external coupling coefficient above $\sqrt{2/Q_u}$ and an interresonator coupling coefficient of $1/Q_u$, where $Q_u$ denotes the unloaded quality factor of the individual resonators, the filter can theoretically attain infinite stopband attenuation, even with finite-$Q_u$ resonators. Chapter 6 presents a comprehensive analysis and design of absorptive filters.

#### 7.2.2.1 External Coupling

This section discusses a coupling structure where the microstrip feeding line is converted to a coplanar waveguide (CPW) transmission line integrated within the common ground plane of the evanescent-mode cavity. This structure is illustrated in Figure 7.22. The magnetic field of this CPW segment interacts with the cavity's magnetic field, achieving the intended external coupling. The coupling strength is defined by both the length and width of the CPW line segment integrated within the cavity's ground plane. To enhance the coupling coefficient, radially oriented stubs may be incorporated into this segment of the CPW line to augment the area of the CPW section within the cavity. Figure 7.23 illustrates the frequency dependence of the external coupling coefficient $k_{E1}$ generated by this coupling structure for a number of lengths of the CPW section and angles of the radial stubs (length $l_1$ and angle $\phi$ in Figure 7.26). The coupling coefficient has a nearly

**Figure 7.22** Detailed view of the different layers needed to realize the two-pole constant-bandwidth filters.

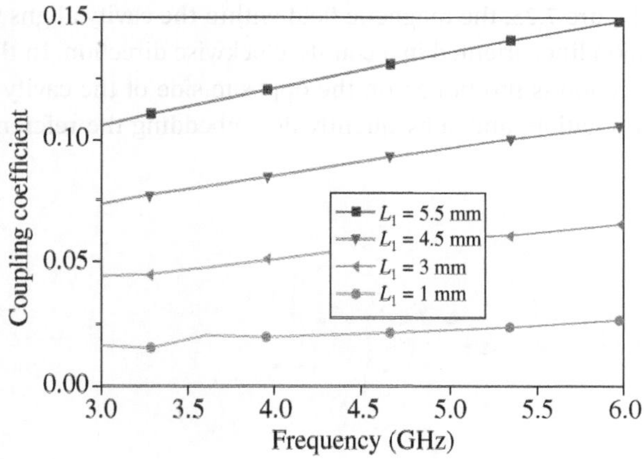


**Figure 7.23** Simulated frequency dependence of the coupling coefficient $k_{E1}$ for various $l_1$ lengths.

linear relationship with frequency, increasing by approximately 50% across an octave tuning range, so validating the model for the coupling coefficients outlined in (7.23).

The needed value of $k_{E2}$ is significantly smaller than $k_{E1}$, necessitating the utilization of a thinner and shorter CPW coupling line. This is realized by controlling the size $l_2$ in Figure 7.26. The frequency dependence of this coupling element is illustrated in Figure 7.24 for various CPW lengths.

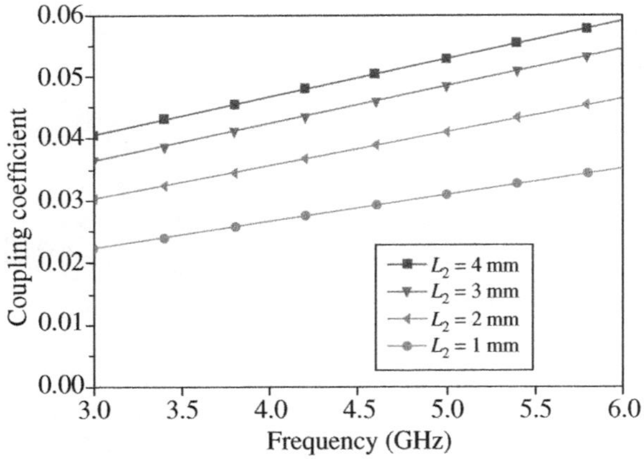


**Figure 7.24** Simulated frequency dependence of the coupling coefficient $k_{E2}$ for various $l_2$ lengths.

#### 7.2.2.2 External Coupling Structures: Polarity Design

Figure 7.12 illustrates that knowledge of the relative sign of the external coupling elements is essential for the design of a constant-bandwidth coupling structure. When a cavity utilizing the previously described coupling structure is excited by a signal from the left side, and the excitation is de-embedded to establish the reference plane at the midpoint of the coupling section, as illustrated in Figure 7.25, the magnetic field within the cavity aligns with that of the transmission line, oriented in a counterclockwise direction. In the event that the coupling section is positioned on the opposite side of the cavity while applying the same excitation, and subsequently de-embedding the reference plane to the center of

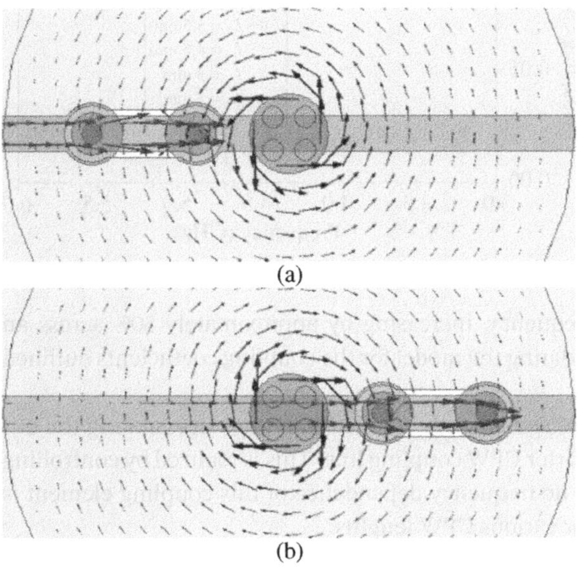


**Figure 7.25** Current density on the microstrip line and the magnetic field within the cavity when the incident signal propagates (a) from the exterior of the cavity to the interior, and (b) from the interior of the cavity to the outside. The magnetic fields' opposing directions in the two cases result in opposite signs for the two couplings.

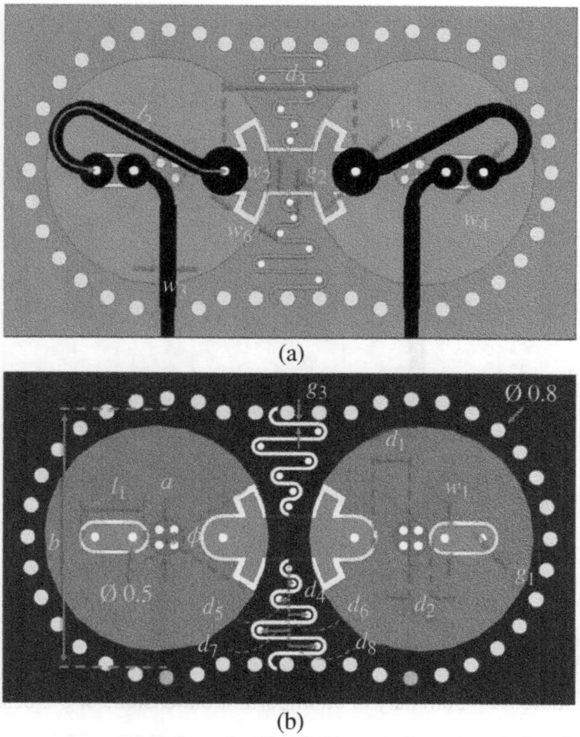


**Figure 7.26** Detailed layouts of Filters A, B, and C. The following dimensions are in millimeters: $a = 1.9$, $a = 13.8$, $d_4 = 0.2$, $d_5 = 0.5$, $d_6 = 1$, $d_7 = 1.5$, $d_8 = 1.5$, $g_3 = 0.15$, $w_3 = 0.86$.

the coupling element, the magnetic field within the cavity exhibits an orientation that is opposite to the previous configuration. Consequently, the two coupling structures, identical in shape, can produce opposite coupling signs based on the direction of signal propagation through the coupling section relative to the cavity's orientation (Figure 7.27).

It is clear that to utilize a nominally 180° transmission line within the coupling structure for the purpose of minimizing bandwidth variation, both coupling elements must exhibit the same polarity of coupling. As a result, the orientation of the coupling sections must ensure that the direction of signal propagation through each element is consistent, either from the interior of the cavity to the exterior or from the exterior to the interior of the cavity. To achieve this, the transmission line connecting the coupling elements is configured in a loop around the smaller coupling section. This arrangement ensures that a signal initiated from the input will propagate from the interior of the cavity to the exterior for both coupling elements.

Utilizing the frequency-dependent coupling values $k_{E1}$ and $k_{E2}$, along with the length of the transmission line connecting them, equation (7.17) was applied to optimize the structure. The objective was to minimize the variation in the coupling coefficient for Filter A and to ensure the *ABW* for Filters B-D remained consistent throughout the tuning ranges of the filters.

**Figure 7.27** Detailed layout of the four-pole filter. This layout is a back-to-back cascade of two Filter Cs.

**Table 7.1** Summary of all important dimensions of the designed filters in millimeters.

| Filter | $d_1$ | $d_2$ | $d_3$ | $g_1$ | $g_2$ | $l_1$ | $l_2$ |
|---|---|---|---|---|---|---|---|
| A | 0.71 | 2.0 | 10.2 | 0.21 | 0.24 | 2.35 | 14.0 |
| B | 1.8 | 1.11 | 6.9 | 0.19 | 0.23 | 3.3 | 13.7 |
| C | 1.8 | 1.29 | 7.0 | 0.19 | 0.23 | 3.13 | 14.3 |
| | $\varphi$ | $W_1$ | $W_2$ | $W_4$ | $W_5$ | $W_6$ | |
| A | 0° | 1.0 | 1.9 | 1.4 | 2.0 | — | |
| B | 63° | 1.33 | 2.2 | 1.83 | 2.61 | 1.5 | |
| C | 63 | 1.33 | 2.2 | 1.83 | 2.61 | 1.5 | |

#### 7.2.2.3 Inter-Resonator Coupling

The inter-resonator coupling between evanescent-mode cavity resonators is often achieved through a coupling iris, which comprises a below-cutoff section of a substrate-integrated waveguide, as illustrated in Figure 2.17. To implement an absorptive bandstop filter, a transmission line measuring 270° in length is required between the resonators. This configuration facilitates destructive signal interference, resulting in significant levels of stopband attenuation [104, 105]. On the other hand, utilizing a 90° length of transmission line is advantageous, as it leads to reduced passband insertion loss. More significantly, it provides an extended tuning range for the filter to maintain high stopband rejection. This is due to the ability to preserve the necessary phase relationship between the inter-resonator coupling and the transmission line across a broader tuning range. The inter-resonator coupling must assume the opposite sign for this effect.

The approach implemented in this filter is based on the techniques for obtaining negative inter-resonator coupling as outlined in [106, 107]. An array of vias connects the top and bottom conductors of the coupling iris section, while a slot is cut into the copper of the upper conducting layer. The slot obstructs the current on the upper conductor and redirects it to the lower conductor through the vias. This action reverses the current direction in the coupling section, resulting in a coupling value that is opposite to that of the original coupling iris. The inter-resonator coupling dimensions, such as the spacing between the vias and the width of the coupling iris, were determined through full-wave electromagnetic simulations in order to yield the coupling required for the absorptive bandstop filter ($k_{12} = 1/Q_u$). The final dimensions are presented in Figure 7.26.

### 7.2.3 Fabrication and Measurements

The filters were produced utilizing a commercial printed circuit board (PCB) milling, laminating, and plating system. The dimensions for the filters are summarized in Table 7.1. The signal and cavity substrates were constructed from Rogers 5880 and were laminated together utilizing Rogers 2929 bondply material. A 12.7 mm diameter piezoelectric disk (Piezo Systems T216-A4NO-273X) was utilized as the tuning element. The disks were coated with copper and affixed to the cavities utilizing low-temperature solder paste. Figure 7.28 illustrates the fabricated filters. A voltage source of ±200 V was utilized to bias the piezoelectric discs.

**Figure 7.28** Pictures of the manufactured filters. Notice the piezoelectric disks for their tuning. Source: With permission of Hickle et al. [135]/IEEE.

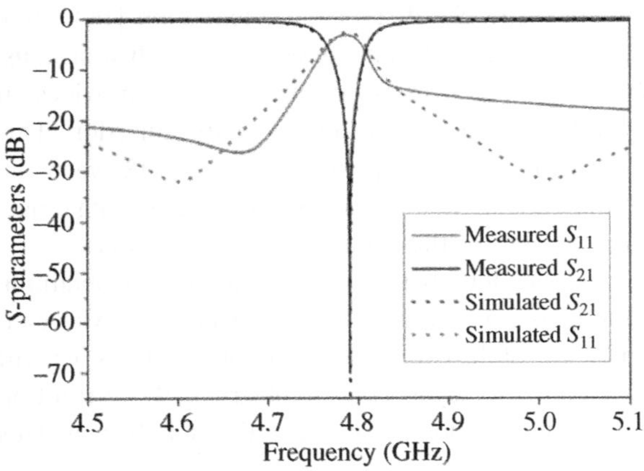


**Figure 7.29** Measured scattering parameters of Filter A at 4.8 GHz.

#### 7.2.3.1 Constant *FBW* Filter

Figure 7.29 illustrates the measured response of Filter A at a tuning frequency of 4.8 GHz. The filter exhibits a stopband rejection exceeding 70 dB, attributable to its absorptive design. The passband demonstrates low-loss of less than 0.5 dB up to 7 GHz. It also has good impedance matching with a passband return loss greater than 15 dB for the same frequency range. The full scattering parameters of Filter A throughout its tuning range of 3.2–6.4 GHz are plotted in Figure 7.30. The filter achieves a stopband rejection of over 60 dB across the tuning range.

To assess the improvement achieved through the constant-bandwidth coupling method, measurements of the 3-dB and 10-dB bandwidths for both the uncompensated filter (Filter E) and the constant *FBW* filter (Filter A) were conducted across their tuning ranges. The results are presented in Figure 7.31. The uncompensated filter exhibits 3-dB and 10-dB bandwidths that range from 1.25% to 2.3%

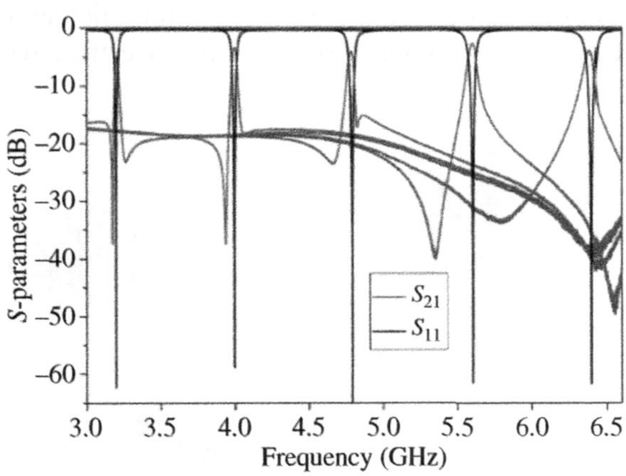


**Figure 7.30** Measured scattering parameters of Filter A across its octave tuning range.

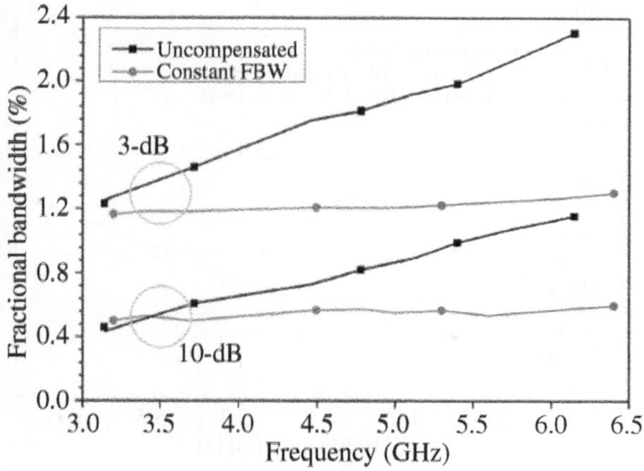


**Figure 7.31** Measured 3-dB and 10-dB fractional bandwidths for two filters: Filter A (constant *FBW*) and Filter E (uncompensated design).

(an 84% variation) and from 0.43% to 1.16% (a 170% variation), respectively. The constant *FBW* filter has significantly minimized bandwidth variation, achieving a 3-dB bandwidth range of 1.16%–1.3% (representing a 12% variation) and a 10-dB bandwidth range of 0.5%–0.6% (indicating a 20% variation). The constant *FBW* filter demonstrates an 86% reduction in 3-dB *FBW* variation and an 88% reduction in 10-dB *FBW* variation when compared to the uncompensated filter.

#### 7.2.3.2 Constant *ABW* Filter

Measurements were conducted on two filters optimized for constant *ABW*: Filter B, which operates over a 2:1 tuning range, and Filter C, which operates over a 1.5:1 tuning range. The S-parameters are illustrated in Figure 7.32.

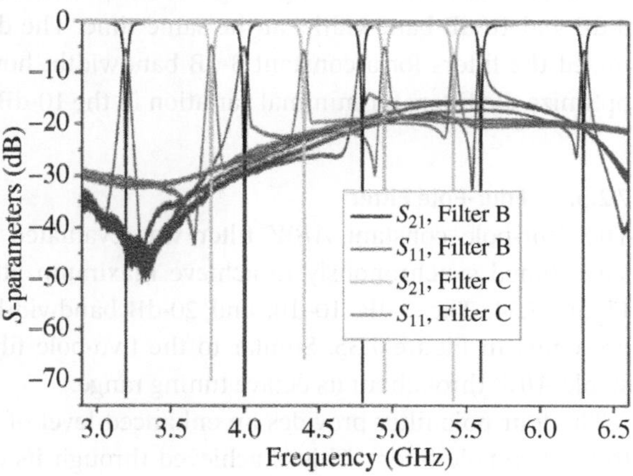


**Figure 7.32** Measured scattering parameters of two constant absolute bandwidth filters: Filter B (optimized for 2:1 tuning range) and Filter C (optimized for 1.5:1 tuning range).

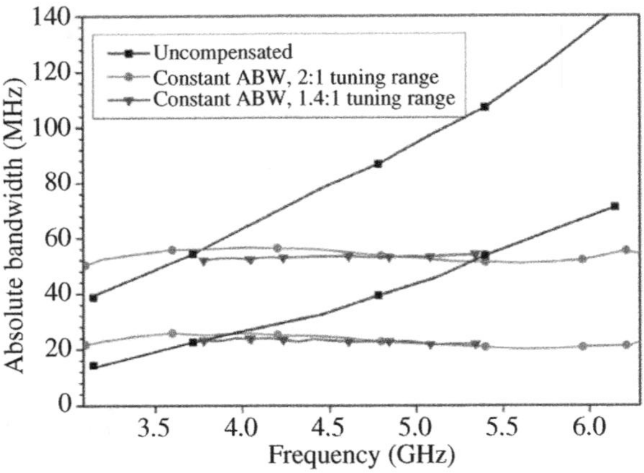


**Figure 7.33** Measured 3- and 10-dB bandwidths of Filters B and C. The results are compared to Filter E that represents the uncompensated bandwidth case.

The measured *ABW* values for both filters, as well as for Filter E, are presented in Figure 7.33. The 3-dB and 10-dB *ABWs* of the uncompensated Filter E are measured at 39–142 MHz, indicating a variation of 264%, and 14.4–71 MHz, indicating a variation of 393%, respectively. The filter designed for constant bandwidth across a 2:1 tuning range (Filter B) exhibits significantly reduced *ABW* variation, with a 3-dB bandwidth ranging from 50.3 to 56.5 MHz (resulting in a 12.3% variation) and a 10-dB bandwidth spanning from 20 to 25.8 MHz (yielding a 29% variation). Filter B achieves a 95% reduction in 3-dB bandwidth variation and a 93% reduction in 10-dB bandwidth variation when compared to the uncompensated filter.

The filter, optimized for a 1.5:1 tuning range, exhibits minimal bandwidth variation, with a 3-dB bandwidth ranging from 52 to 54 MHz (a variation of 3.8%) and a 10-dB bandwidth ranging from 21.8 to 24 MHz (a variation of 10%).

It is important to note that the unloaded quality factors of the resonators vary throughout the filter's tuning range, making it unfeasible to minimize both the 3-dB and 10-dB bandwidths at the same time. The designs discussed here optimized the filters for a constant 3-dB bandwidth; however, it is also possible to optimize the filters for minimal variation in the 10-dB bandwidth.

#### 7.2.3.3 Four-Pole Filter

The four-pole constant *ABW* filter was evaluated with the notches of each stage tuned synchronously to achieve maximum attenuation, as illustrated in Figure 7.34. The 3-dB, 10-dB, and 20-dB bandwidths were measured and are presented in Figure 7.35. Similar to the two-pole filters, it exhibits a relatively stable *ABW* throughout its octave tuning range.

The four-pole filter provides an enhanced level of reconfigurability compared to the two-pole filters. This is achieved through its design, which includes two

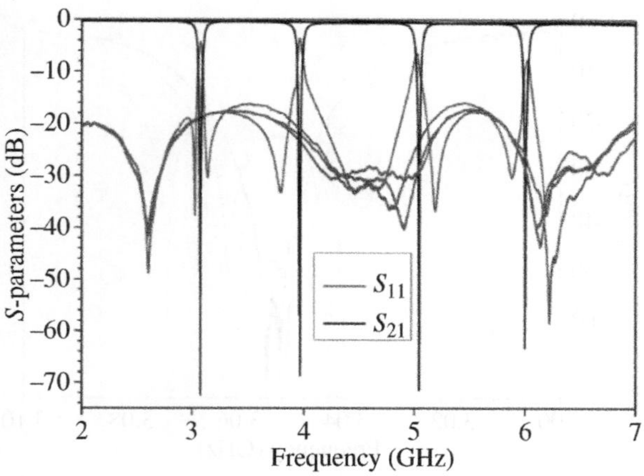


**Figure 7.34** Measured scattering parameters of the four-pole constant absolute bandwidth filter. The filter notches are synchronously tuned to maximize stopband attenuation.

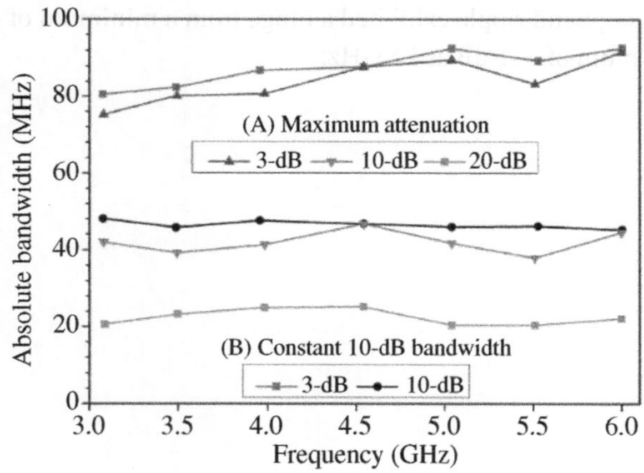


**Figure 7.35** Bandwidth measurements of the four-pole filter presented under two conditions: (A) both notches are tuned synchronously to achieve maximum attenuation, and (B) the notches are tuned asynchronously to ensure a consistent 10-dB bandwidth.

cascaded absorptive notch sections, allowing for independent control of the center frequency for each notch. The filter is capable of generating multiple transfer functions, as illustrated in Figure 7.36. The notches may be adjusted to the same frequency to achieve optimal attenuation, or they can be tuned asynchronously to deliver a Chebyshev frequency response. Asynchronous tuning of the notches allows for slight adjustments to the filter's bandwidth. This tunability can be utilized to address any remaining bandwidth variations that may occur following the implementation of the constant-bandwidth method discussed in this chapter. The filter was tuned from 3 to 6 GHz using this method, while the notches were asynchronously adjusted to ensure a consistent 50 MHz 10-dB bandwidth was maintained. The 10-dB bandwidth remains consistently stable, with limitations arising solely from the precision of the tuning of the two notches. Figure 7.35 illustrates

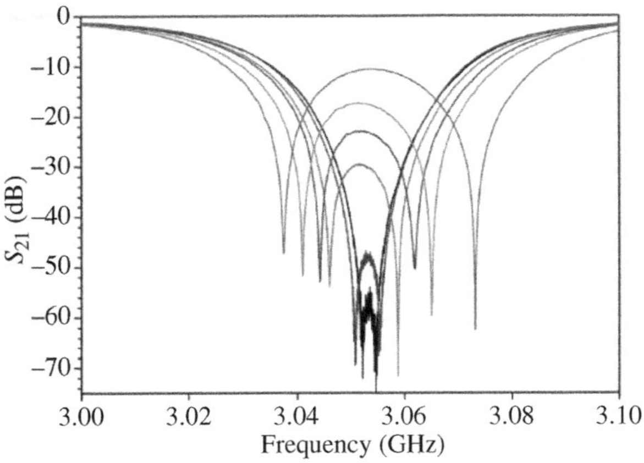


**Figure 7.36** Response measurement of four-pole filter when adjusted to various levels of stopband ripple and expanded bandwidth.

the 3-dB and 10-dB bandwidths. To achieve a consistent 10-dB bandwidth, the stopband ripple exhibited a range from a minimum of −55 dB at 4 GHz to a maximum of −15 dB at 5.5 GHz.

# 8

# High-Order and Fractional-Order Evanescent-Mode Filters

## 8.1 High-Order Evanescent-Mode Filters

Dual-mode resonators are briefly discussed in Section 2.5.1. In this section, we will discuss their applications to filters.

### 8.1.1 Independently Tunable Dual-Mode Evanescent-Mode Filter

Tunable dual-mode resonators are suitable for realizing a wide variety of multi-band reconfigurable filter responses while maintaining a compact size [37]. More specifically, the ability to control the resonant frequency of each mode independently is quite attractive for adaptive systems.

This section discusses a dual-mode evanescent-mode resonator with a high-$Q$. Two vertically movable tuners above the resonator are used to regulate it, allowing each mode to be independently tuned. Over a 59% tuning range from 1.7 to 2.7 GHz, the even resonant mode maintains a $Q$ of more than 200. Over a 94% tuning range from 1.7 to 3.3 GHz, it remains higher than 150. With a tuning range of more than 27% and a $Q$ higher than 260, the odd mode's frequency can be adjusted to be lower or higher than the even mode's frequency.

A reconfigurable second-order bandpass filter, based on a doublet structure, is implemented using the designed resonator. With this setup, the filter can be placed on the upper or lower side of the passband, giving it a configurable finite transmission-zero (TZ). Over the whole tuning range, the filter's insertion loss (IL) is better than 1.1 dB, and its center frequency can be adjusted between 2.3 to 2.2 GHz.

*Tunable Evanescent-Mode Filters: Principles, Implementation, and Applications*,
First Edition. Dimitrios Peroulis and Mohammad Abu Khater.

Companion website: www.wiley.com/go/Peroulis/TunableEvanescent-ModeFilters

#### 8.1.1.1 Resonator Design

The general coupling diagram in Figure 8.1(a) uses a dual-mode resonator. In this illustration, there is no inter-resonator interaction between the resonator's two parallel modes. In Figure 8.1(b) and (c), the proposed resonator's side view and 3D view are displayed together with its primary physical dimensions. Two center patches are electrically isolated from the top of a traditional Substrate Integrated Waveguide (SIW) cavity resonator and connected to the cavity bottom via two shunt vias to create the resonator. According to [108], this resonator can be represented as two TEM-mode resonators with dispersed coupling. However, since it is sufficiently precise for this work, we shall assume a basic LC resonator model. Therefore, the modes' resonant frequencies can be described by

$$f_e = \frac{1}{2\pi\sqrt{L_e C_1}}, \qquad f_o = \frac{1}{2\pi\sqrt{L_o(C_1 + 2C_2)}} \tag{8.1}$$

The inductance values for the odd and even modes are denoted by $L_o$ and $L_e$, respectively. According to Figure 8.1(c)), $C_1$ represents the capacitance between the two center patches and the resonator top layer, and $C_2$ represents the capacitance between the two center patches themselves. It is discovered that the

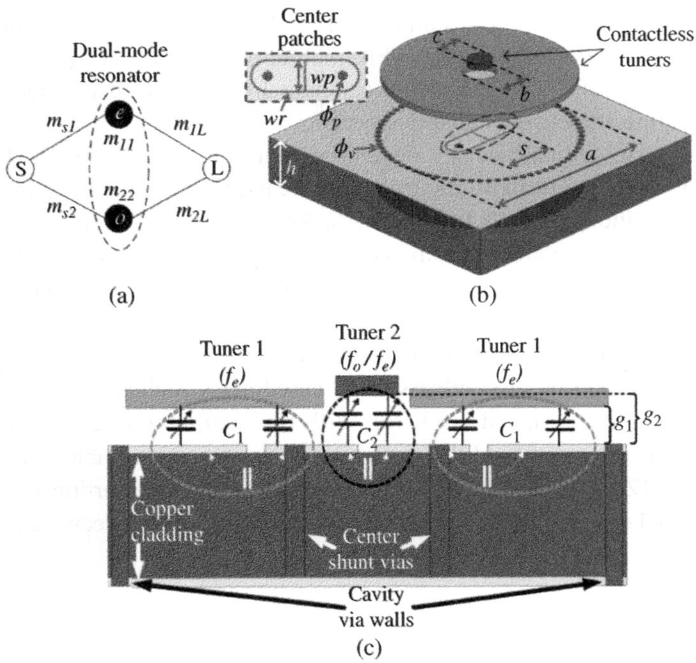


**Figure 8.1** (a) The coupling diagram of a dual-mode resonator. (b) Dual-mode SIW resonator implementation. (c) A cross section of the implementation in (b).

inductance and capacitance ratios are directly proportional to the ratio between the modes' resonant frequencies, as demonstrated in

$$\frac{f_o}{f_e} = \sqrt{\frac{L_e}{L_o} \frac{C_1}{C_1 + 2C_2}} \tag{8.2}$$

The distance between the center vias ($s$) determines the inductance ratio ($L_e/L_o$). Additionally, $C_2$ primarily depends on the patch width ($w_p$) prior to the tuners being added. Figure 8.2 shows the frequency ratio dependence that results from simulating the resonator without the tuners in order to illustrate how these factors affect the frequency ratio.

A convenient method to separately tune $f_e$ and $f_o$ is provided by the frequency dependency on capacitance displayed in (8.1) and (8.2). Two flat tuners with conductive surfaces are placed on top of the resonator to allow for this tuning, resulting in two distinct adjustable gaps ($g_1, g_2$). Both capacitances, $C_1$ and $C_2$, are altered by the tuners' presence (Figure 8.1). Changing $g_1$ in the resultant structure modifies $C_1$, which adjusts $f_e$ and $f_o$. The frequency ratio is tuned by changing $g_2$, which only modifies $C_2$, which only modifies $f_e$. The tuning capacity is demonstrated by varying $g_1$ from 20 to 1000 μm and sweeping $g_2$ from 10 to 1000 μm. The results of the tuning simulations are displayed in Figure 8.2(b). Clearly, $g_2$ can tune $f_o$ to a lower or a higher value than $f_e$ at any value for $g_1$.

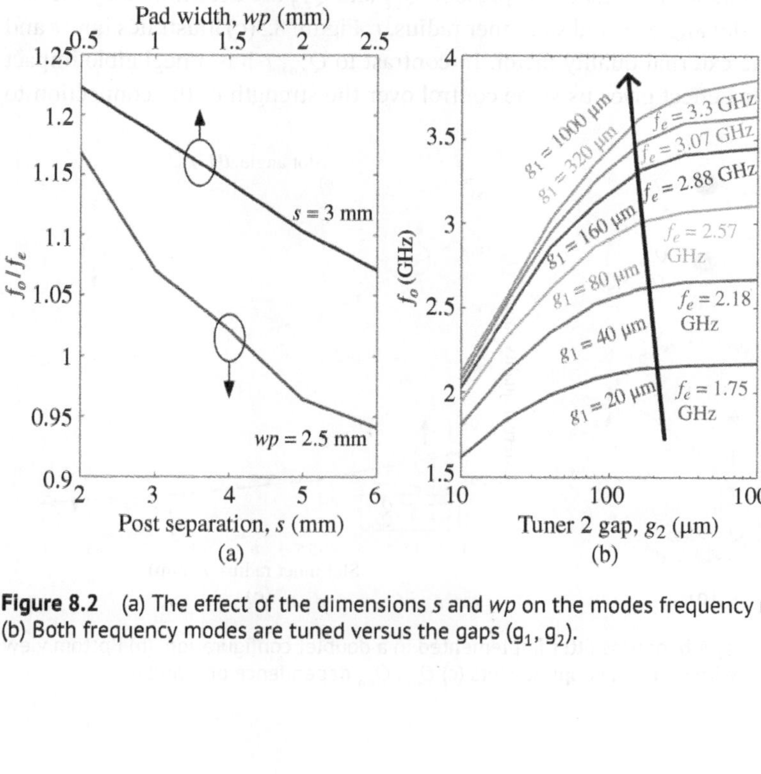


**Figure 8.2** (a) The effect of the dimensions $s$ and $wp$ on the modes frequency ratio. (b) Both frequency modes are tuned versus the gaps ($g_1$, $g_2$).

**Table 8.1** Summary of dimensions of the designed filters in millimeters.

| Dimension | $a$ | $b$ | $c$ | $s$ | $h$ | $w_p$ | $w_r$ | $\phi_v$ | $\phi_p$ |
|---|---|---|---|---|---|---|---|---|---|
| Value (mm) | 20 | 3.6 | 2.6 | 3 | 5.08 | 2.5 | 0.15 | 0.7 | 0.8 |

This dual-mode resonator is designed on a Rogers TMM3 substrate that is 5.08 mm thick and has the dimensions listed in Table 8.1. A 0.75 mm thick Rogers 4003 substrate with 17.5 μm thick copper is used to cut the tuners.

#### 8.1.1.2 Filter Design

A reconfigurable bandpass filter is implemented using the resonator previously described. This resonator is well suited for the doublet configuration depicted in Figure 8.3(a). A second-order bandpass filter with a finite transmission zero can be achieved with this topology. The following two requirements must be met in order to have a transmission zero [109]: First, only one of the topology's four connections needs to be polarity opposite. Second, there should be a discrepancy between the external quality factor values ($Q_{e,e}$, $Q_{e,o}$). In addition to the value of $f_o$, the difference between $Q_{e,e}$ and $Q_{e,o}$ determines the location of the transmission zero.

This topology is achieved by coupling both modes to the input/output ports via a grounded coplanar waveguide (GCPW) on the cavity's bottom, as illustrated in Figure 8.3(b). Figure 2.23 illustrates the coupling slots that are placed where the magnetic field for both modes is present. $Q_{e,e}$ and $Q_{e,o}$ are determined by the slot width, *ws*, slot angle, $\theta$, and slot inner radius, $r$. Figure 8.3(c) illustrates how $r$ and $\theta$ affect the external quality factor. In contrast to $Q_{e,o}$, $r$ has a negligible impact on $Q_{e,e}$. This effect gives us some control over the strength of the connection to

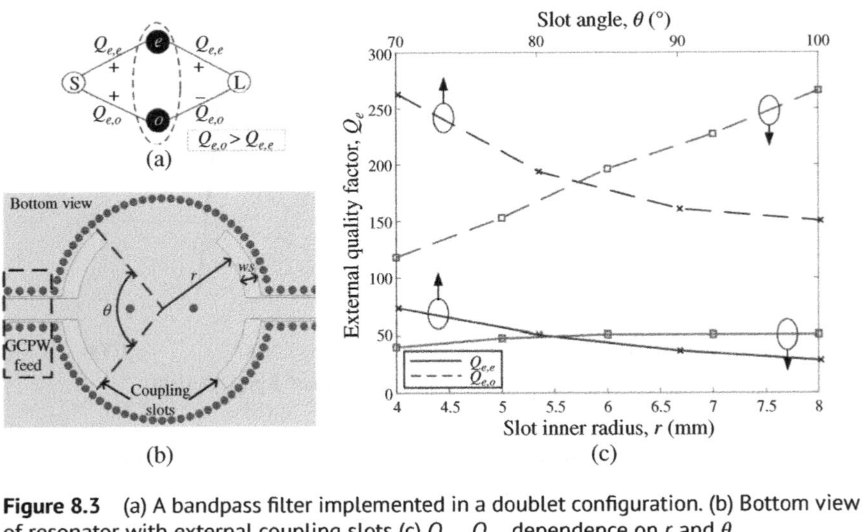


**Figure 8.3** (a) A bandpass filter implemented in a doublet configuration. (b) Bottom view of resonator with external coupling slots (c) $Q_{e,e}$, $Q_{e,o}$ dependence on $r$ and $\theta$.

each mode separately. The input/output GCPW here has a 1.94 mm-wide signal line and a 0.2 mm-wide gap. The dimensions chosen for the external coupling are: $r = 8$ mm, $\theta = 100°$ and $ws = 1.5$ mm.

#### 8.1.1.3 Experimental Validation

The tuning ranges and the modes quality factors ($Q_{u,e}$, $Q_{u,o}$) were assessed by fabricating and measuring a weakly coupled resonator design. To generate the vertical motion needed to tune both resonant modes, each tuner was connected to a linear actuator. Tuner 1 was used to keep $f_e$ at a specific value for each measurement, while Tuner 2 was used to adjust $f_o$ around it. Figure 8.4(a) displays a sample of

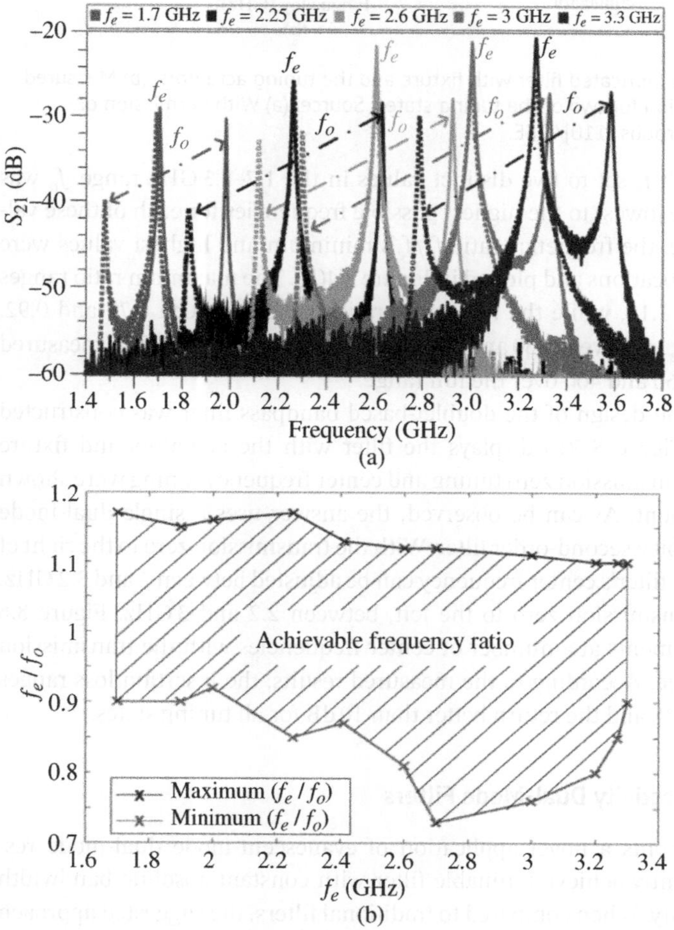


**Figure 8.4** A weakly coupled measurement of the dual-mode resonator (a) $S_{21}$ with $f_o$ tuned to minimum (dashed line) and maximum (solid line) frequencies for each value of $f_e$. (b) Mode frequency ratio versus $f_e$ over the tuning range.

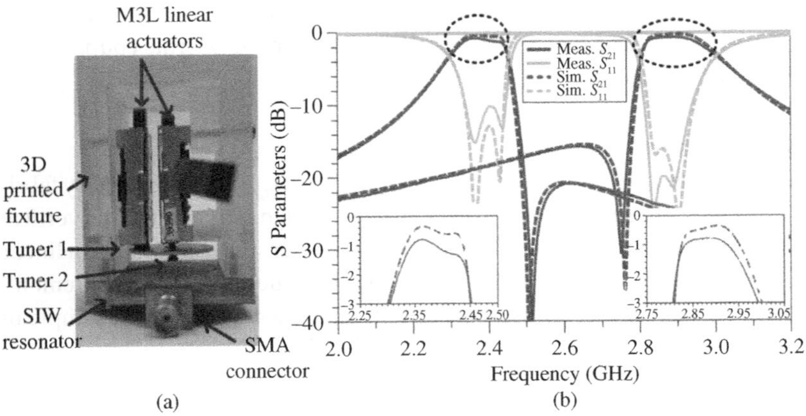


**Figure 8.5** (a) The fabricated filter with fixture and the tuning actuators. (b) Measured results and simulation for two of the tuning states. Source: (a) With permission of Abdelfattah and Peroulis [110]/IEEE.

these results, with $f_e$ set to five distinct values in the 1.7–3.3 GHz range. $f_o$ was adjusted from the lowest to the highest possible frequencies for each of these values. Furthermore, the frequency ratio $f_o/f_e$'s minimum and highest values were noted at various locations and plotted in Figure 8.4(b). The maximum ratio ranges between 1.1 and 1.18, while the minimum ratio ranges between 0.72 and 0.92. The extracted $Q_{u,e}$ is between 150 and 400 over the full tuning range. The measured $Q_{u,o}$ is between 260 and 400 over the full range.

Additionally, the design of the doublet-based bandpass filter was constructed and measured. Figure 8.5(a) displays the filter with the actuators and fixture assembly. Both transmission zero tuning and center frequency tuning were shown in the measurement. As can be observed, the answer uses a single dual-mode resonator to exhibit a second-order filter. With the transmission zero to the right of the passband, the filter's center frequency can be adjusted between 2 and 3.2 GHz, and with the transmission zero to the left, between 2.2 and 3 GHz. Figure 8.6 displays measurements at a number of center frequencies with the transmission zero on either side. According to the measured results, the insertion loss ranges from 0.65 to 1.1 dB, and the return better than 10 dB for all tuning states.

### 8.1.2 High Selectivity Dual-Mode Filters

This section presents a novel application of evanescent-mode dual-mode resonators to efficiently achieve a tunable filter with constant absolute bandwidth and high selectivity. When compared to traditional filters, the suggested approach accomplishes both objectives without requiring more tuning elements or larger size. The reflection zeros and transmission zeros are generated by the

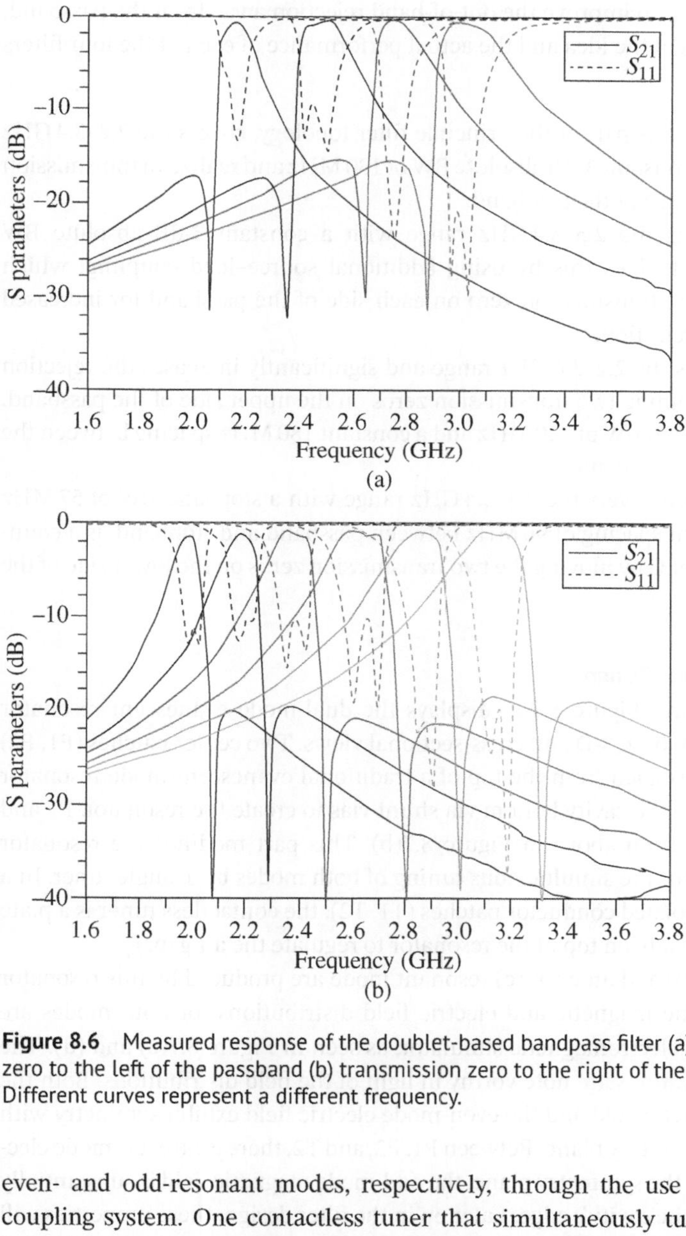


**Figure 8.6** Measured response of the doublet-based bandpass filter (a) transmission zero to the left of the passband (b) transmission zero to the right of the passband. Different curves represent a different frequency.

even- and odd-resonant modes, respectively, through the use of an L-shaped coupling system. One contactless tuner that simultaneously tunes both modes is built into each resonator. The dual-mode resonators and their tuners can generate a regulated separation between the transmission zeros and reflection zeros across the tuning range with the suggested structures. These filters use methods to either increase selectivity and widen the rejection bandwidth close to the passband or to improve the out-of-band rejection away from the passband. A quick synopsis of the idea and the actual performance of each of the four filters is given below:

1) Filter A, which is part of the principle filter topology, covers the 2.2–3.4 GHz range with a constant 3-dB absolute BW of 130 MHz and realizes a transmission zero on each side of the passband.
2) Filter B spans the 2.5–3.4 GHz range with a constant 3-dB absolute BW of 130 MHz. It does this by using additional source–load coupling, which adds one more transmission zero on each side of the passband for increased out-of-band rejection.
3) Filter C spans the 2.2–3.6 GHz range and significantly increases the rejection level by combining two transmission zeros on the upper side of the passband. It has a stopband BW of 120 MHz and a constant 180 MHz spacing between the passband and stopband.
4) Filter D, which covers the 2.3–3.4 GHz range with a stopband BW of 57 MHz and a constant spacing of 96 MHz between passband and stopband, is a counterpart of filter C by moving the two transmission zeros on the lower side of the passband.

#### 8.1.2.1 Resonator Design

***Resonator Structure*** Figure 8.7(a) displays the dual-mode evanescent resonator employed here in both 3-D and cross-sectional views. Two center patches (P1, P2) are electrically isolated from the top of a traditional evanescent-mode resonator and connected to the cavity bottom via shunt vias to create the resonator. P1 and P2 are depicted from above in Figure 8.7(b). This part modifies the resonator design to allow for the simultaneous tuning of both modes by a single tuner. In a pattern of two isolated conductor patches (T1, T2), the contactless tuner is a plate that moves vertically on top of the resonator to regulate the air gap, $g_t$.

Both an odd (o) and an even (e) resonant mode are produced by this resonator arrangement. The magnetic and electric field distributions for both modes are shown using an electromagnetic simulator, as seen in Figure 8.7(c) and (d). The following observations are noteworthy in light of the field distributions. Both the odd-mode magnetic field and the even-mode electric field exhibit symmetry with respect to the symmetry plane. Between P1, P2, and T2, there is no even-mode electric field. Across the symmetry plane, the odd-mode magnetic field is substantially weaker close to the cavity's edge. Later on in the filter design, these properties will be used.

As illustrated in Figure 8.8(a), the two modes can be represented as TEM-mode transmission-line (TL) resonators with distributed coupling. The total capacitance between P1 or P2 and ground is denoted as $C_1$. The total capacitance between P1

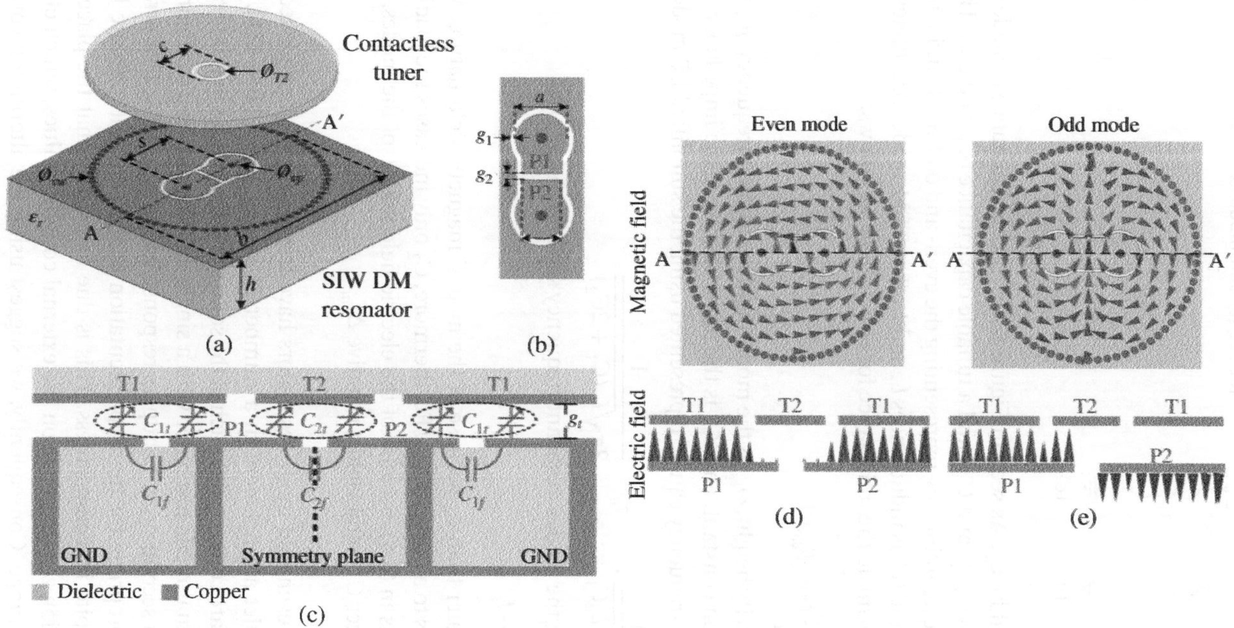


**Figure 8.7** Dual-mode resonator (a) 3D view with critical dimensions (b) top view (c) cross-section (A–A′) with highlighted capacitors. Electric and magnetic fields for the (d) even and (e) the odd modes. The symmetry plane is orthogonal to A–A′.

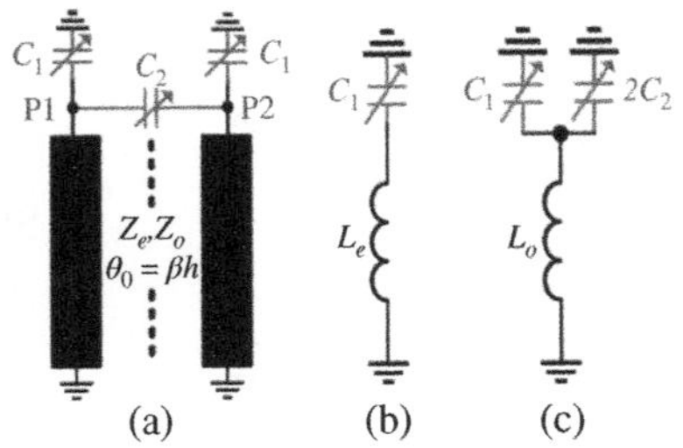


**Figure 8.8** (a) TEM-mode TL equivalent model for DM resonator, (b) even-mode LC model, (c) odd-mode LC model.

and P2 is denoted by $C_2$. As seen in Figure 8.7(c), each capacitance is made up of a fixed capacitance ($C_{1f}$ and $C_{2f}$) and a tunable capacitance ($C_{1t}$ and $C_{2t}$). The TL resonators can be simplified by representing the even- and odd-mode inductances in Figure 8.8(b) and (c) as inductances $L_e$ and $L_o$. The TEM-modes parameters can be used to approximate the inductance for either mode as follows:

$$L_{e,o} = \frac{Z_{e,o}}{\omega_{e,o}} \tan\left(\beta_{e,o} h\right) \tag{8.3}$$

where $h$ is the cavity height, $\omega_{e,o}$ is the mode angular resonant frequency, $\beta_{e,o}$ is the mode propagation constant, and $Z_{e,o}$ is the mode characteristic impedance. Each mode's resonant frequency can be represented using the simplified LC models as

$$f_e = \frac{1}{2\pi\sqrt{L_e C_1}}, \qquad f_o = \frac{1}{2\pi\sqrt{L_o(C_1 + 2C_2)}} \tag{8.4}$$

We also define the modes' resonant frequency separation ($\Delta f_m$) as

$$\Delta f_m = f_o - f_e \tag{8.5}$$

$L_e$ is always greater than $L_o$ because of the modes' magnetic field nature. When $C_2 = 0$, this leads to a positive $\Delta f_m$. Furthermore, $C_2$ only increases the loading of the odd-mode, as in (8.4), because of the electric field nature of the modes. Consequently, a bigger $C_2$ could yield a negative $\Delta f_m$.

***L-Coupling*** The even/odd mode resonators have frequently been employed in a traditional doublet arrangement in a number of earlier works [109–111]. Both resonant modes are connected to each of the source and load nodes in this arrangement. A single dual-mode resonator with a single doublet coupling configuration could generate a second-order bandpass response with a single TZ. This arrangement is not appropriate for the implementation of the constant absolute bandwidth filter, despite its compactness. This is due to the fact that the placement of the transmission zero depends on the external coupling values, which change over the tuning range. Consequently, we suggest using the alternate L-coupling depicted in Figure 8.9(a) in this study. In this arrangement, just the source is coupled to the odd mode, whereas the source and load are both linked to the even

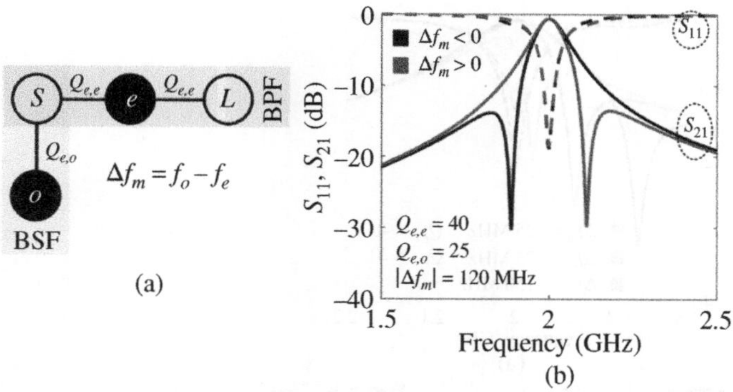


**Figure 8.9** The L-coupling delivering the dual-mode resonator (a) coupling diagram (b) frequency response.

mode. This corresponds to a BPF-BSF (first-order bandpass–bandstop) cascade. As can be seen from the S-parameter response in Figure 8.9(b), the odd mode results in a transmission zero, but the even mode produces a first-order bandpass response. Both the transmission zero locations and the center frequency in this setup depend on $f_e$ and $f_o$. When $\Delta f_m$ is positive or negative, the transmission zero might be found above or below the passband.

The impacts of $\Delta f_m$ and the external quality factor values ($Q_{e,e}$ and $Q_{e,o}$) are examined in Figure 8.10, with $f_e$ fixed at 2 GHz. It is demonstrated that the placement of the transmission zero is controlled by $\Delta f_m$. Additionally, as would be expected from a traditional BPF, $Q_{e,e}$ regulates the passband BW. $Q_{e,o}$ regulates the transmission zero BW, as would be expected from a BSF. It is important to note, nevertheless, that the in-band return loss level is impacted by both $\Delta f_m$ and $Q_{e,o}$. This adds more degrees of freedom.

As illustrated in Figure 8.11, the L-coupling can be physically achieved by means of coupling vias that link to the input/output microstrip feed lines after passing through the resonator. The via may couple to both modes or just the even mode, depending on where it is located. The distribution of magnetic fields in the resonator helps to explain this. The via's coupling to the odd mode undergoes self-cancellation when it is positioned on the symmetry plane, leading to coupling to only the even mode. The position of the via, which is determined by the distance, $d_c$, from the center, determines the strength of the external coupling. Depending on the cavity structure, the $Q_{e,e}$ and $Q_{e,o}$ dependency on $d_c$ may vary.

#### 8.1.2.2 Constant Absolute Bandwidth BPF Filter

We could use two resonators to generate a second-order response with two transmission zeros since a single resonator with the L-coupling can provide a

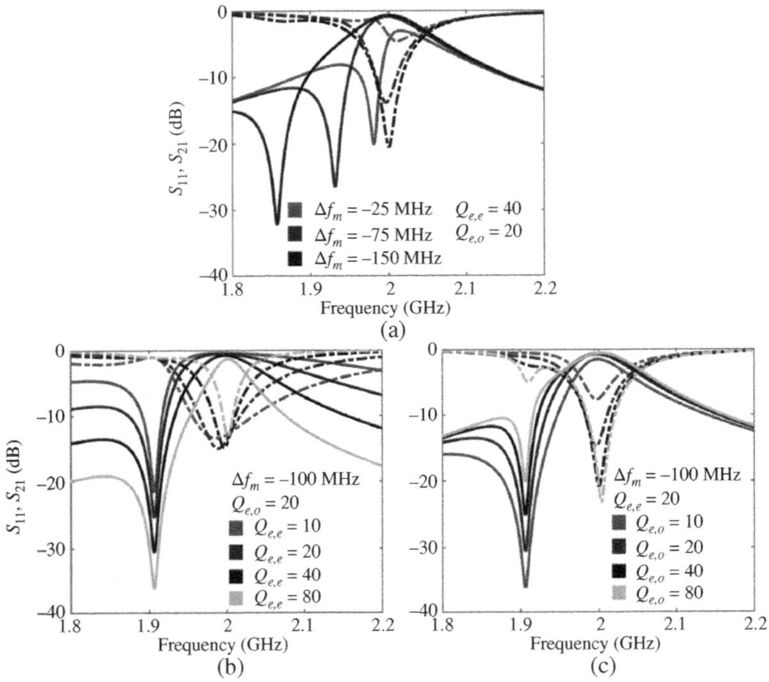


**Figure 8.10** The frequency response dependence on (a) $\Delta f_m$, (b) $Q_{e,e}$, and (c) $Q_{e,o}$.

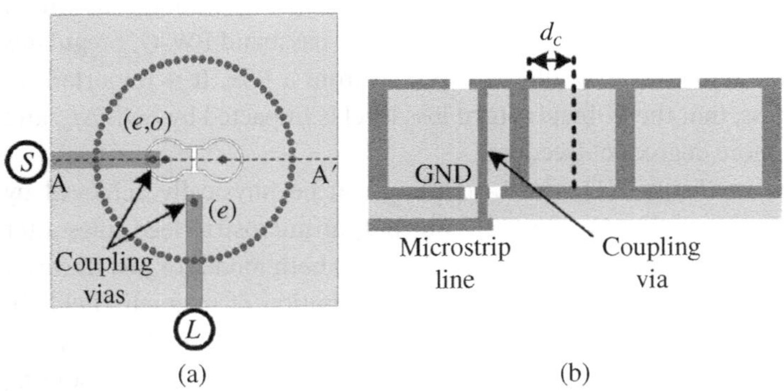


**Figure 8.11** The implemented L-coupling filter (a) top view (b) cross-section view (A–A′).

first-order response with a transmission zero. With the BPF architecture depicted in Figure 8.12(a), this may be easily achieved. In order to achieve a filter response with customizable notches, this topology was first presented in [112] and implemented using single-mode SIW resonators and standard microstrip lines. The coupling of both even-modes ($e_1, e_2$) via the inter-resonator coupling ($k_{12,e}$) creates

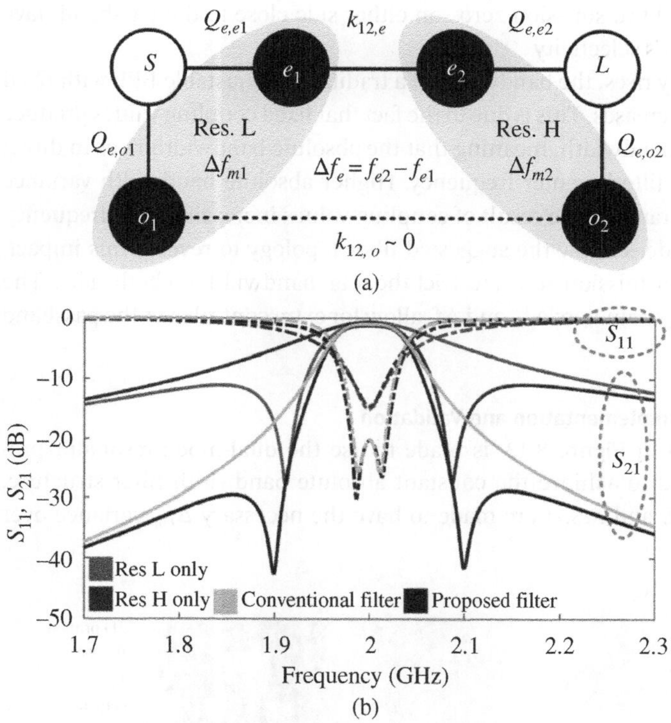


**Figure 8.12** The second-order bandpass filter topology using dual-mode resonators (a) coupling diagram (b) filter response with 80 MHz absolute bandwidth at 2 GHz, $Q_{e,e} = 29$, $Q_{e,o} = 20$, $k_{12,e} = 0.028$, $\Delta f_m = 100$ MHz $\Delta f_e = 25{:}6$ MHz.

the passband when the topology is implemented using dual-model resonators. Additionally, one resonator must have a negative value of $\Delta f_m$ and the other must have a positive value in order to generate a transmission zero on each side of the passband. (*Res.L*) is the resonator with negative $\Delta f_m$, and (*Res.H*) is the other one. It may be necessary for $e_1$ and $e_2$ to have an offset in their resonant frequencies in order to provide a certain filter response. Therefore, we use ($\Delta f_e = f_{e2} - f_{e1}$) to define the even-mode frequency separation. Figure 8.12(b) shows a BPF response with 3-dB ABW of 80 MHz and return loss better than 20 dB as an example of the topology. The response is overlaid with the response of *Res.L* and *Res.H* individually. This illustrates how the passband is formed within the two resonators' overlapping passband. Additionally, it demonstrates a correlation between the transmission zero locations and the individual resonators' transmission zeros. A traditional second-order Chebyshev filter response with the same bandwidth and return loss is also used to compare the response. This demonstrates unequivocally how the additional transmission zeros on either side close to the passband have increased the filter's selectivity.

As the frequency rises, the bandwidth of a traditional adjustable BPF with fixed coupling values increases. This is due to the fact that fixed coupling values produce a fixed fractional bandwidth, meaning that the absolute bandwidth rises in direct proportion to the filter's center frequency. Higher absolute bandwidth variance across the tuning range is the result of coupling values increasing with frequency in practice. Two ideas allow the suggested filter topology to reverse this impact. The first is the transmission zeros, restrict the total bandwidth on both sides. The second is that the parameters $\Delta f_m$ and $\Delta f_e$ allow for extra control over the passband shape.

#### 8.1.2.3 Filter A: Implementation and Validation

Filter A, depicted in Figure 8.13, is made to use the dual-mode resonator previously mentioned to achieve the constant absolute bandwidth filter structure. Individually, *Res.L* and *Res.H* are made to have the necessary $\Delta f_m$ variance over

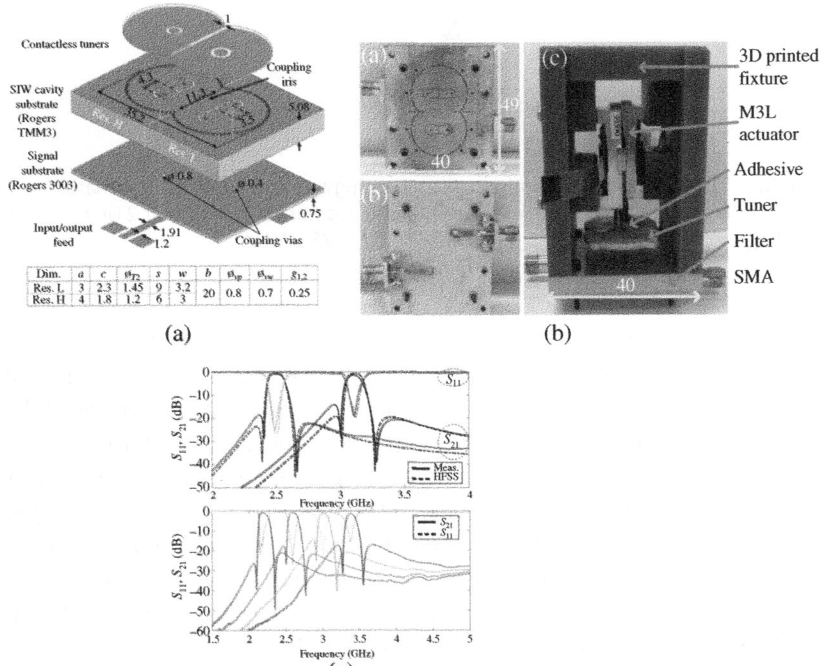


| Dim. | $a$ | $c$ | $\phi_{T2}$ | $s$ | $w$ | $b$ | $\phi_{vp}$ | $\phi_{vw}$ | $g_{1,2}$ |
|---|---|---|---|---|---|---|---|---|---|
| Res. L | 3 | 2.3 | 1.45 | 9 | 3.2 | 20 | 0.8 | 0.7 | 0.25 |
| Res. H | 4 | 1.8 | 1.2 | 6 | 3 | | | | |

**Figure 8.13** The structure of Filter A (a) implemented with linear actuators as tuners (b). Measured results in (c) show a practically constant absolute bandwidth across the tuning range. Source: (b) With permission of Abdelfattah and Peroulis [60]/IEEE.

the tuning range. The table in Figure 8.13 also displays the dimensions for each resonator. The coupling iris between the resonators creates the inter-resonator coupling. The symmetry line, where the odd-mode magnetic field is weakest, is where the coupling iris is situated. This guarantees that, as required by the topology [37], $k_{12,o}$ is near zero. The design of the external coupling vias is described in Figure 8.11. Each resonator has different coupling via spacing ($d_c$) and coupling via diameter in order to realize the required coupling values.

Printed Circuit Board (PCB) technology was used in the production of Filter A. Figure 8.13 displays a photo of the completed assembly and the created filter. The tuners' vertical movement is provided by two linear actuators. The filter center frequency can be adjusted between 2.2 and 3.4 GHz. Tuning $g_t$ between 80 μm and 2 mm covers the tuning range. The 3-dB *ABW* was kept at 130 ± 1 MHz in all tuning states. The insertion loss is 1.36 dB and the filter return loss is better than 15 dB. The unloaded quality factor of the resonators is predicted to be between 200 and 400. Both the upper and lower transmission zeros provide a maximum rejection level between 25 and 50 dB.

#### 8.1.2.4 Filter B: Implementation and Validation

Filter A measurements show that the out-of-band rejection is restricted to 30 dB at frequencies above the passband. By altering the topology to incorporate deliberate source-to-load coupling ($k_{SL}$), as illustrated in Figure 8.14(a), this rejection could be strengthened. One extra transmission zero is produced on each side of the passband by this additional coupling path. The filter selectivity won't be impacted by the new transmission zeros because they won't be close to the passband. They will, however, enhance the out-of-band rejection on both sides. By merely altering each microstrip feed line before its coupling via, this coupling might be realized. Filter B is developed with dimensions similar to Filter A and the updated microstrip

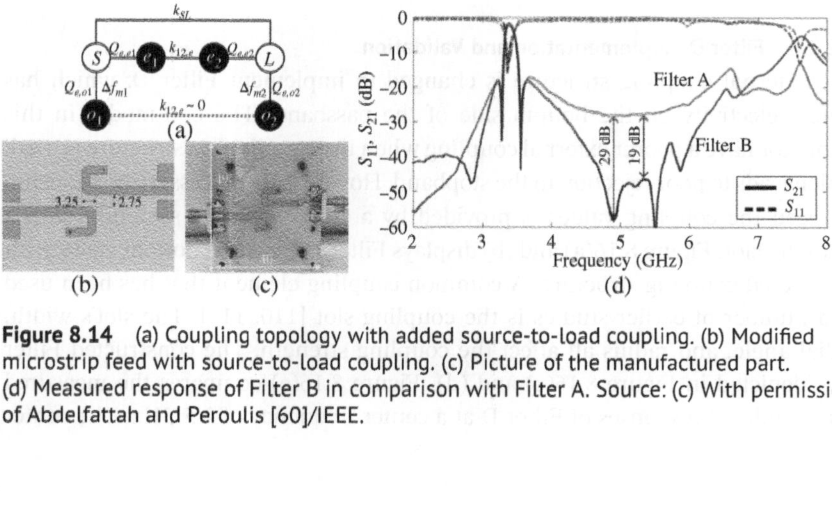


**Figure 8.14** (a) Coupling topology with added source-to-load coupling. (b) Modified microstrip feed with source-to-load coupling. (c) Picture of the manufactured part. (d) Measured response of Filter B in comparison with Filter A. Source: (c) With permission of Abdelfattah and Peroulis [60]/IEEE.

feed structure depicted in Figure 8.14(b) to demonstrate this design adjustments. The length of the overlap and the distance between the two feeding lines determines the strength of the source-to-load coupling. Additional information about source-to-load coupling extraction may be found in [113].

Filter B was fabricated and measured. Figure 8.14(c) depicts the modified feeding pattern. Filter B keeps the *ABW* at 130 ± 1 MHz across the 2.5–3.4 GHz range. The return loss is better than 10 dB, while the insertion loss ranges from 1.37 to 1.97 dB. As illustrated in Figure 8.14(d), the benefit of the source-to-load coupling is demonstrated by the 19–29 dB improvement in out-of-band rejection at higher frequencies (about 5 GHz). The spurious bands of Filters A and B are observed by measuring them up to 8 GHz.

#### 8.1.2.5 Filter C: Implementation and Validation

In Filter C, both transmission zeros are combined above the passband frequency. Consequently, $\Delta f_m$ is positive for both resonators. This means that both resonators can be identical in design.

Filter C's top and bottom views are displayed in Figure 8.15(a) and (b). Both resonators are the same, but the filter is developed using the same stack-up as Filter A. To maintain a high degree of rejection on both sides of the passband, the filter additionally uses a source-to-load coupling, as seen in Figure 8.15(b). The images of the constructed filter are displayed in Figure 8.15(c) and (d). The center frequency tuning of the manufactured filter was found to be between 2.2 and 3.6 GHz. Figure 8.15(e) presents a comparison of simulations and measurements for the state $f_c = 3.2$ GHz. Additionally displayed are the measurements under various tuning states. The filter's return loss is better than 16 dB, and its insertion loss ranges from 0.9 to 1.6 dB. The stopband rejection level is more than 28 dB. The 3-dB BW is between 88 and 139 MHz.

#### 8.1.2.6 Filter D: Implementation and Validation

The external coupling structure is changed to implement Filter D, which has better selectivity on the bottom side of the passband. The odd-modes in this resonator have a poorer external coupling when the coupling via structure is used, which leads to poor rejection in the stopband. However, the necessary relationship between the coupling values is provided by a GCPW feeding structure with a coupling slot. Figure 8.16(a) and (b) displays Filter D's top and bottom views with the altered coupling structure. A common coupling element that has been used in a number of earlier studies is the coupling slot [110, 114]. The slot's width, radial angle, and radius all affect the coupling strength. The constructed Filter D is depicted in Figure 8.16(c) and (d). Figure 8.16(e) compares the measured and simulated responses of Filter D at a center frequency of 2.5 GHz. The center

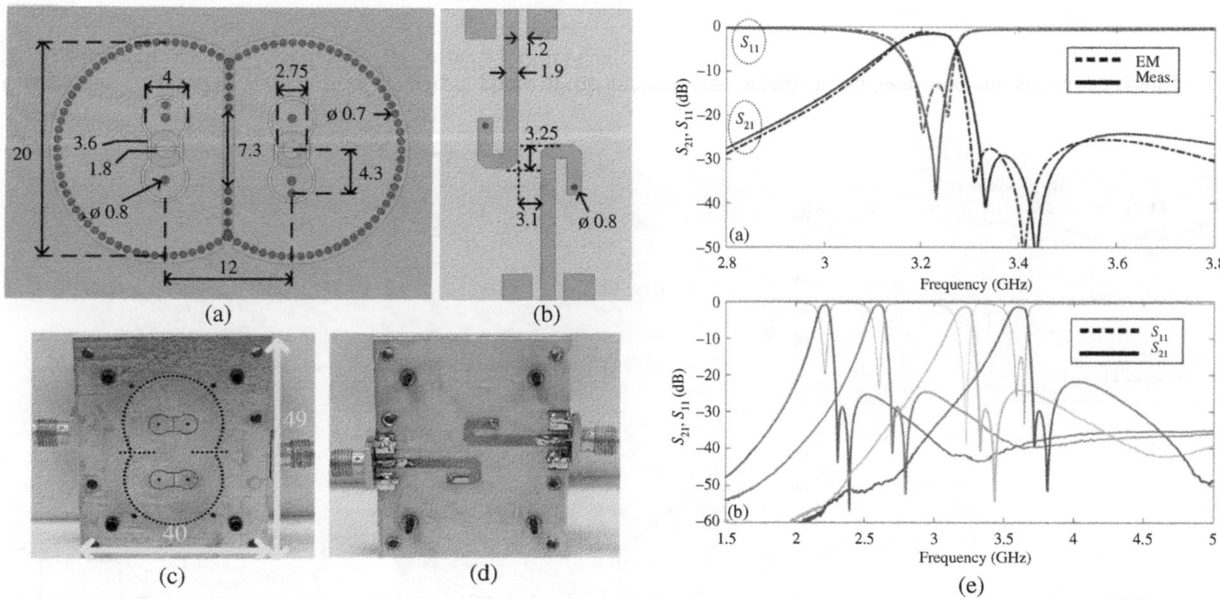


**Figure 8.15** Filter C (a) top view (b) bottom view (c, d) the implemented prototype. (e) Measured and simulated results for a single tuning state, and measured tunable states. Source: (c, d) With permission of Abdelfattah and Peroulis [60]/IEEE.

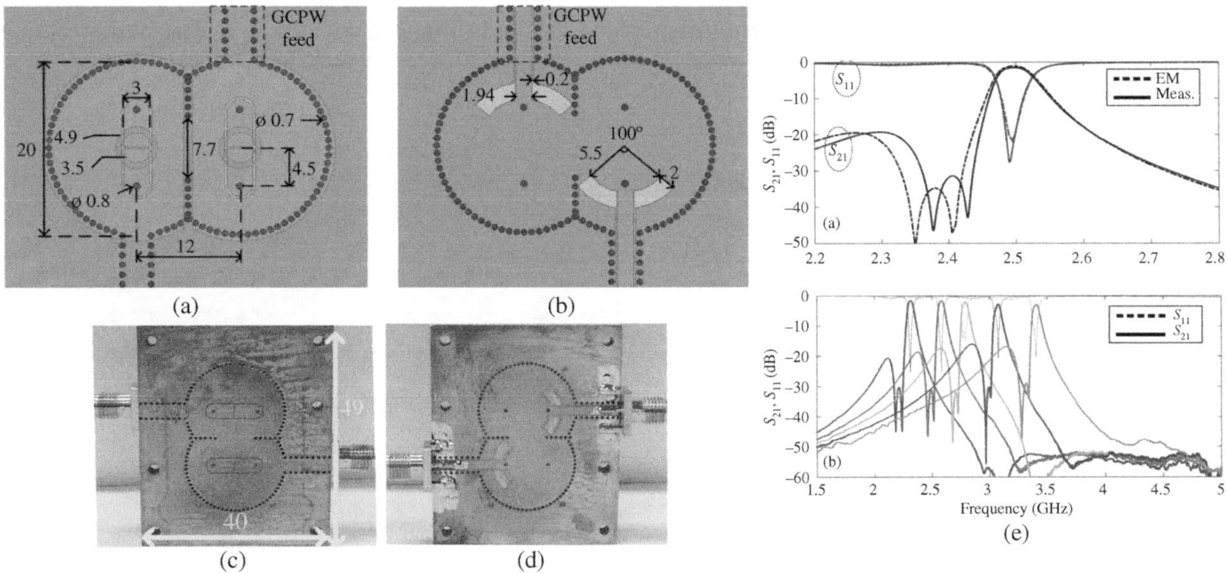


**Figure 8.16** Filter D (a) top view (b) bottom view (c, d) photo of the implemented prototype. (e) Measured and simulated results. Source: (c, d) With permission of Abdelfattah and Peroulis [60]/IEEE.

frequency of the filter is between 2.3 and 3.4 GHz. While the return loss is better than 12 dB, the insertion loss ranges from 1.5 to 2.8 dB. With a stopband bandwidth of 57 MHz and $f_c - f_S = 96$ MHz, the stopband rejection level is greater than 28 dB. The 3-dB BW passband is between 49 and 59 MHz.

### 8.1.3 Four-Wedge Evanescent-Mode Resonator

A novel dual-mode, dual-band resonator and filter structure that offers several poles and passbands per resonator with adjustable center frequencies are introduced in this section. The new structure comes from a single physical resonator that is comparable in size to the resonators illustrated in earlier chapters. It has a single fundamental resonance and two higher-order resonances. Many beneficial filtering capabilities are possible, and the coupling to and between these three resonances can be regulated within certain bounds. With just the physical volume of two resonators, such a device may, for instance, be used to switch between a relatively wide band second-order filter and a relatively narrow band fourth-order filter. By adjusting the higher-order modes to the band of interest and the basic modes away from it, this conversion could be achieved. This idea is illustrated in Figure 8.17 below. The left side of Figure 8.18 shows a model of a strongly loaded coaxial cavity with the loading post cut into four wedges.

#### 8.1.3.1 Bandpass Filter Design

In order to design a filter with a specific type of resonator, external coupling and inter-resonator coupling values available from that resonator with particular coupling structures must be known and understood. Below we discuss the necessary coupling structures.

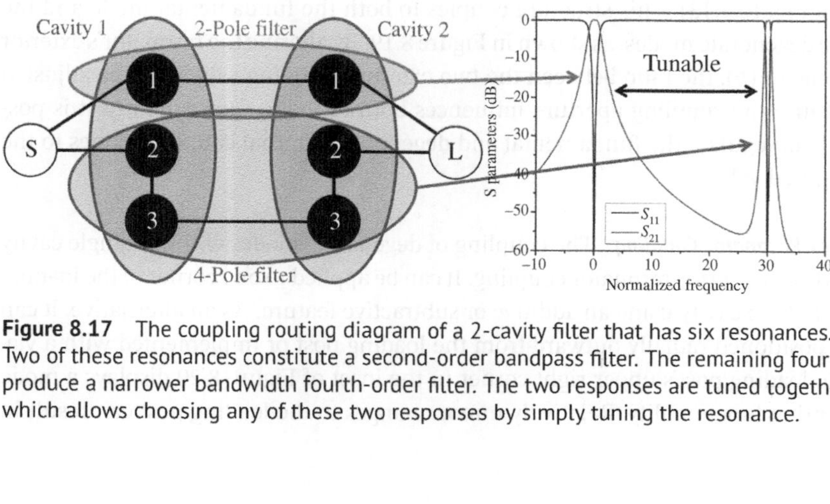


**Figure 8.17** The coupling routing diagram of a 2-cavity filter that has six resonances. Two of these resonances constitute a second-order bandpass filter. The remaining four produce a narrower bandwidth fourth-order filter. The two responses are tuned together which allows choosing any of these two responses by simply tuning the resonance.

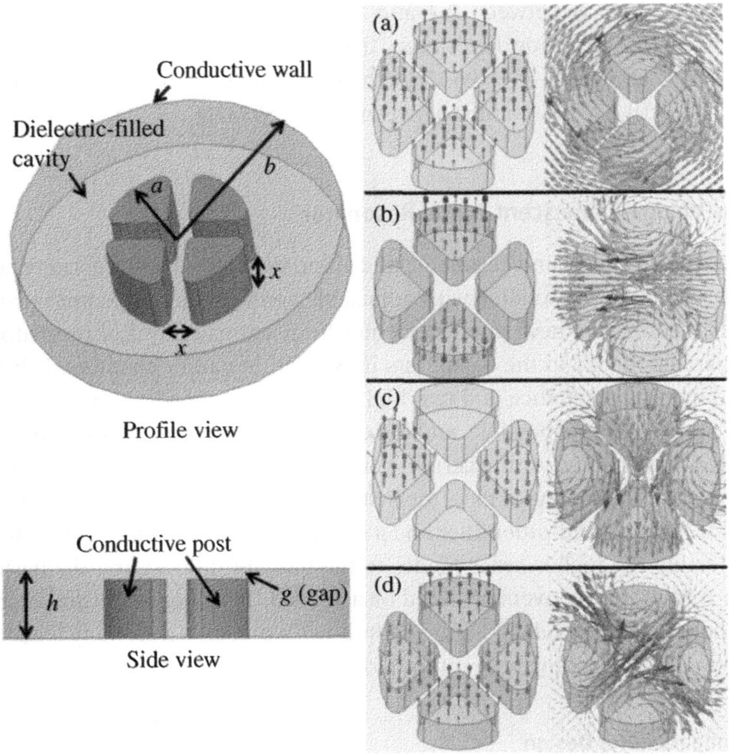


**Figure 8.18** Trimetric and cross-section views of the four-wedge evanescent-mode resonator. Electric and magnetic fields (left and right, respectively) (a) the fundamental mode, (b) and (c) degenerate modes, and (d) first spurious mode.

***External Coupling*** A CPW structure serves as the basis for this external coupling (see Figure 2.13). This structure couples to both the fundamental mode and the two degenerate modes, as shown in Figure 8.19. By altering the resonator's exterior diameter ($b$), the ratio between the two external coupling values can be adjusted because the coupling aperture influences both of them. Consequently, it is possible to regulate the fundamental and degenerate external coupling values to the required value.

***Intra-Resonator Coupling*** The coupling of degenerate modes within a single cavity is known as intra-resonator coupling. It can be applied to one corner of the loading post of the cavity using an additive or subtractive feature. As an alternative, it can be positioned radially outward from the loading post or implemented with a via. The loading post's upper right corner in the inset of Figure 8.20 displays a mode splitting functionality. Relatively minor changes to the loading post can result in

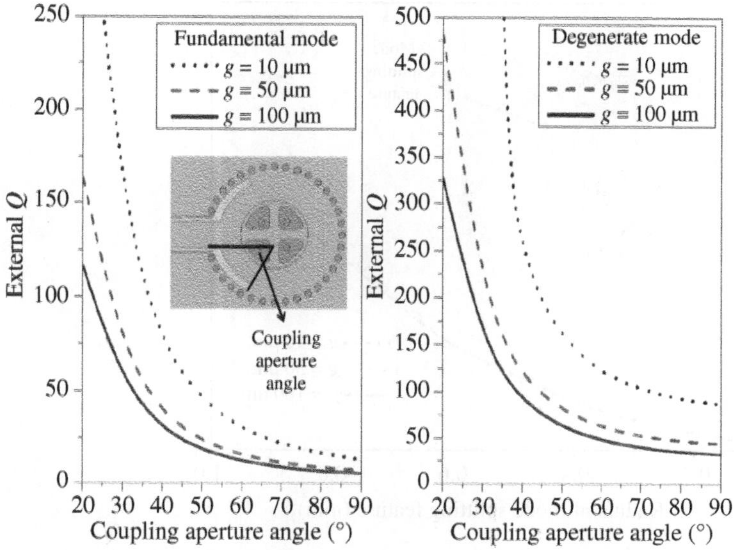


**Figure 8.19** External quality factor as a function of coupling size for the different modes of the four-wedge evanescent-mode resonator.

a broad range of intra-resonator coupling values without significantly altering the fundamental mode frequency, as seen in Figure 8.20. It is worth noting here that the coupling is a weak function of the gap (g) between the loading post and top wall of the cavity, allowing g to be used as a design parameter for external coupling.

***Inter-Resonator Coupling*** Waveguide irises can be used to achieve the inter-resonator coupling $k_{inter}$ between cavity resonators. An illustration of a waveguide iris that couples two resonators is shown in Figure 8.21 (inset).

Keep in mind that $k_{inter}$ also depends on any iris perturbations and the distance between the coupled cavities. Furthermore, since the fundamental and degenerate modes of a pair of resonators occur at different frequencies, the frequency dependence of the coupling iris can be altered by adding low-pass or high-pass structures in parallel with the iris [114]. This effectively alters the ratio of inter-resonator coupling for the fundamental and degenerate modes. When n is an even number, waveguide irises can be used to create n-pole filters, which employ two degenerate modes per resonator. The following part provides an illustrative design example.

#### 8.1.3.2 Design Example

Figure 8.22 illustrates the implementation of the resonator principle in a two-resonator filter. An intriguing feature of the design is that, as was first

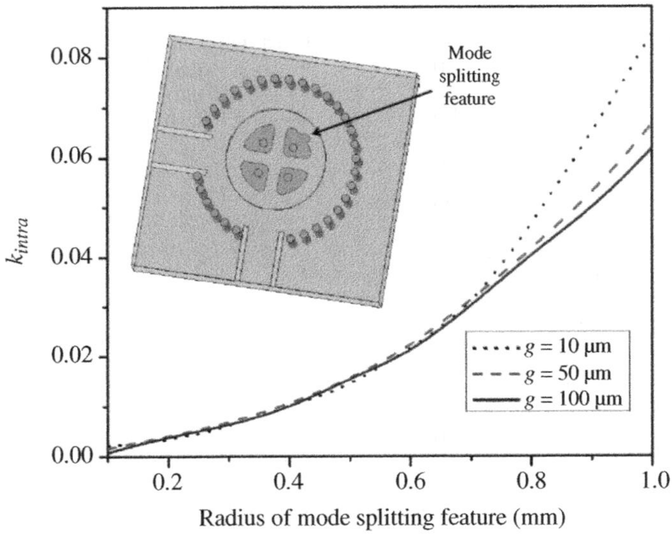


**Figure 8.20** Intra-resonator coupling, $k_{intra}$, as a function of the radius of mode-splitting post feature.

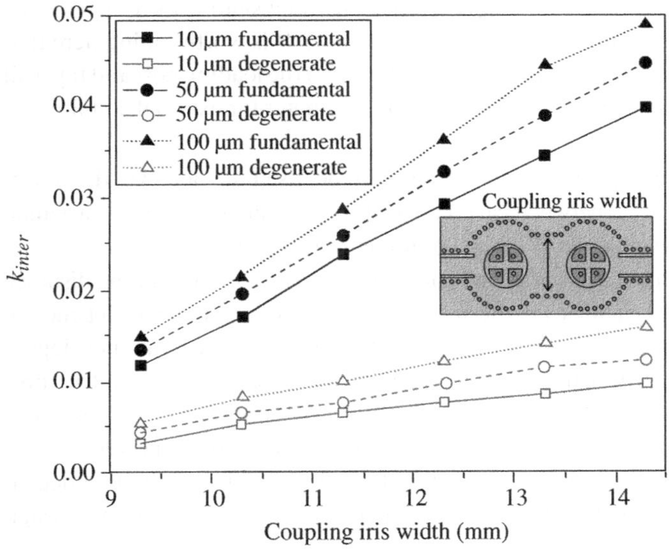


**Figure 8.21** The effect of the width of the coupling iris on the inter-resonator coupling, $k_{inter}$.

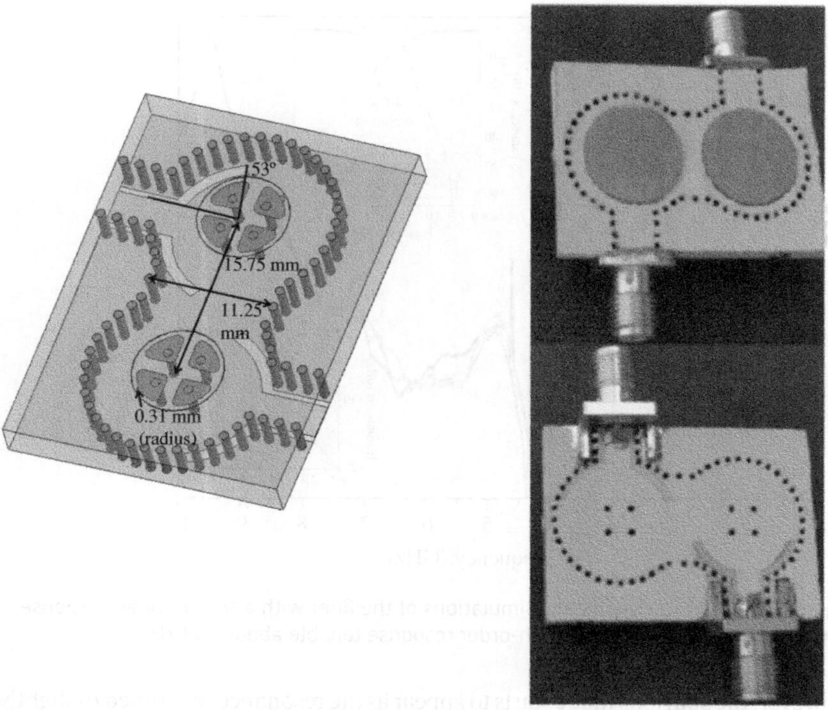


**Figure 8.22** The 3D model of the double-resonator, dual-band filter (left), and a picture showing both sides of the implemented double-resonator, dual-band bandpass filter. Source: With permission of Naglich et al. [106]/IEEE.

demonstrated in [115], the resonators are physically rotated by 180° with respect to one another. Creating the filter in this manner has two positive outcomes. First, the apertures are farther apart. As a result, there is less source-to-load coupling and greater attenuation further from the filter's passbands. The second benefit of rotating the resonators 180° relative to one another is that the degenerate and fundamental modes in both cavities constructively couple, while the subsequent spurious mode cancels because the magnetic fields of the spurious mode in the two resonators are pointing in the opposite directions.

The measured and simulated results for the double-resonator, dual-band filter with a 2-pole response and 4-pole filter displayed in Figure 8.23. The 4-pole response has a 3 dB fractional bandwidth of 2.1% and an insertion loss of 3.34 dB (extracted Q = 400), while the 2-pole response has a 3.26 dB insertion loss and a 5.9% 3 dB fractional bandwidth. As previously mentioned, when the resonators are positioned as they are in Figure 8.22, the first spurious mode cancels, and in two of the three states displayed in Figure 8.23, the first spurious mode is not visible.

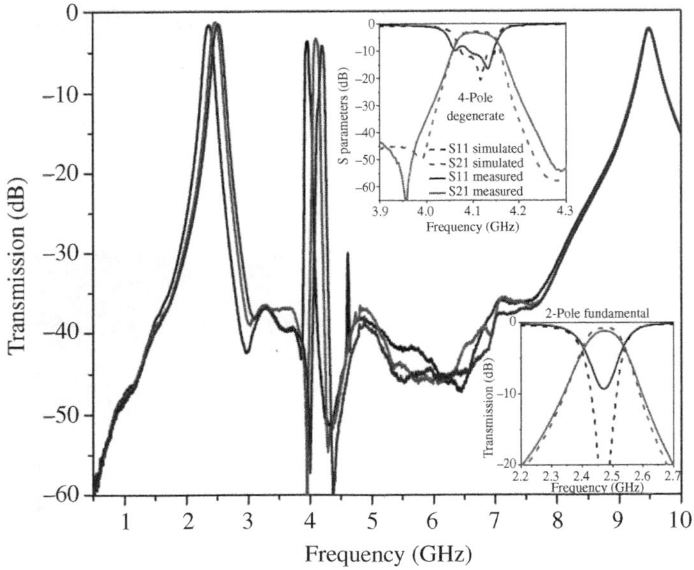


**Figure 8.23** Measurements and simulations of the filter with a second-order response tunable about 2.5 GHz and a fourth-order response tunable about 4.1 GHz.

However, the spurious mode starts to appear as the resonances are tuned to slightly offset frequencies as can be seen in the lowest-frequency response in Figure 8.23.

Measurement data up to 10 GHz are also displayed in Figure 8.23 to illustrate the structure's second spurious mode. The geometry between the loading post and the cavity's exterior wall corresponds to this mode, which is caused by the cavity's static properties. Although this design did not focus on it, the removal or modification of the dielectric material between the loading post and the cavity's outer wall can further push the spurious mode higher in frequency. Even though the mode appears to be on the verge of being critically coupled, which seems to allow for a triple-band response, this is a coincidence in the design. Nevertheless, future research may focus on ways to control the coupling for this mode independently of the coupling for other filtering modes. Without the addition of an additional tuning element, the mode would not be tunable in this setup.

## 8.2 Tunable Half-Mode SIW Filter

This section discusses a novel frequency-tunable half-mode evanescent-mode filter with a constant bandwidth that does not require a lossy tuning component on the coupling structures. A recently created static coupling structure that can

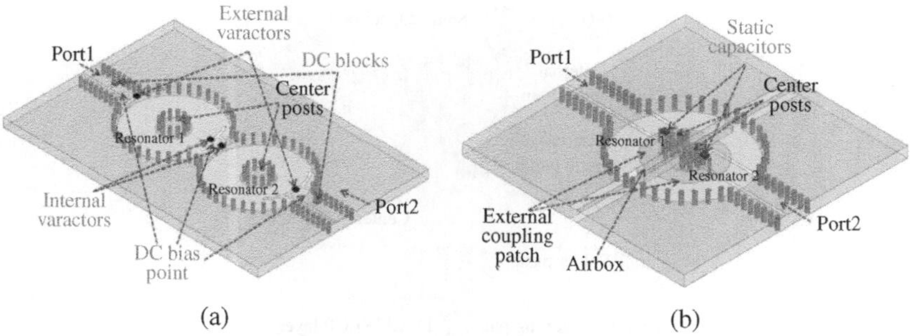


**Figure 8.24** (a) Conventional frequency-tunable evanescent-mode filter with a varactor to maintain a constant bandwidth, and (b) the novel second-order tunable evanescent-mode filter.

have specific coupling values across the frequency tuning range was created in order to avoid the need for tuning components on the coupling structures.

A traditional second-order frequency-tunable evanescent-mode filter and the one suggested by this work are both depicted in Figure 8.24. Several tuning devices are used in the conventional one to control coupling values. Therefore, when the center frequency changes, the coupling parameters can be changed to achieve a constant bandwidth. Two half-mode frequency-tunable evanescent-mode resonators comprise the Cbehyshev filter suggested by this work. The filter's bandwidth remains constant across its frequency tuning range thanks to the addition of new static external and inter-resonator coupling mechanisms.

The new external coupling structure utilized in this filter design is depicted in Figure 8.25(a). The post patch on the resonator's bottom side is surrounded by a coupling patch. This is an important modification compared to the conventional design. Since the capacitance between the post patch and the flexible layer with copper cladding varies, the resonant frequency can be changed by adjusting the space between them. In the meanwhile, there are variations in the capacitance between the Liquid Crystal Polymer (LCP) layer and the coupling patch. We can get the required external coupling values for a constant absolute bandwidth by controlling this capacitance. Using a piezoelectric actuator, which causes the LCP layer to move vertically, these capacitance values can be adjusted. The external coupling value, $k_{ex}$, is displayed for different coupling patch sizes and capacitors in Figure 8.25(b). The capacitor at the top of the filter can be controlled to compensate how quickly $k_{ex}$ decreases with frequency. By adjusting the coupling patch's diameter, we may control the magnitude of $k_{ex}$, which is inversely proportional to frequency when the static capacitor is 2 pF. The external coupling patch's diameter of 10.4 mm and the static capacitance of 2 pF have been selected. These values provide the appropriate coupling across the frequency tuning range.

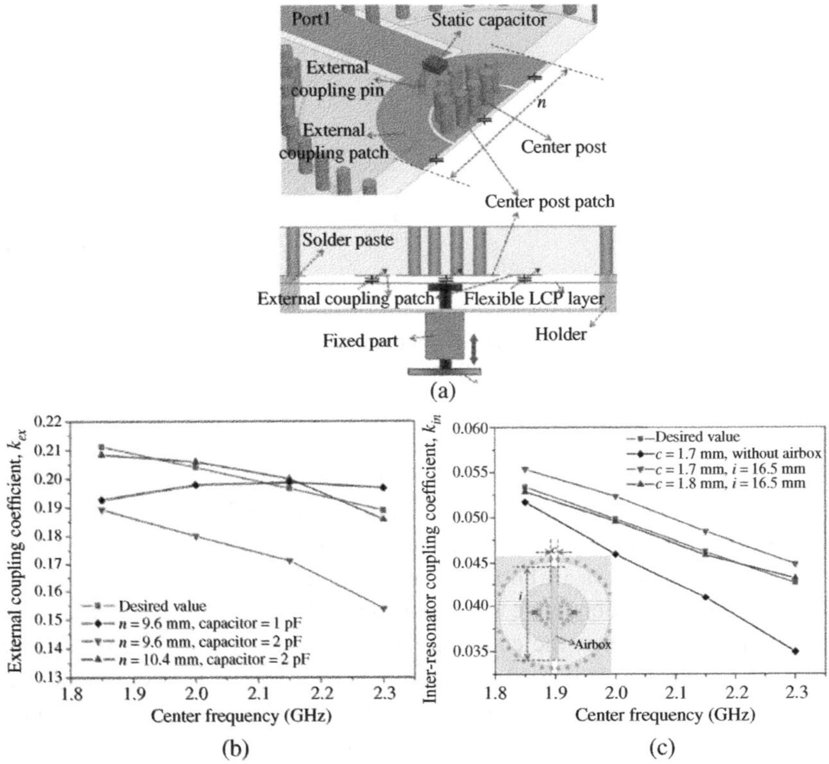


**Figure 8.25** (a) An isometric view and the cross-sectional view of the external coupling structure and the resonator discussed in this section. (b) The external coupling and (c) the inter-resonator coupling coefficients.

For the inter-resonator connection, a similar strategy is applied. The new inter-resonator coupling structure is depicted in Figure 8.25(c). $c$ is the distance between the two resonators. The line with the diamond symbols represents $k_{in}$, when the two resonators were 1.7 mm apart. It is demonstrated that the rate at which $k_{in}$ decreases with frequency is significantly higher than the desired rate, which is represented by the red line (noted as 'desired value' and visible in color in the eBook). By eliminating the dielectric between the two resonators, this rate can be regulated. This is the same as putting an air box between them. The green line (top line in the graph and visible in color in the eBook) illustrates the scenario in which an air box measuring 16.5 mm in length is placed and the distance between the two resonators is 1.7 mm. The rate at which $k_{in}$ decreases with frequency is about the same as the required rate. It is still necessary to modify the magnitude of $k_{in}$, which can be done by varying the distance between the two resonators.

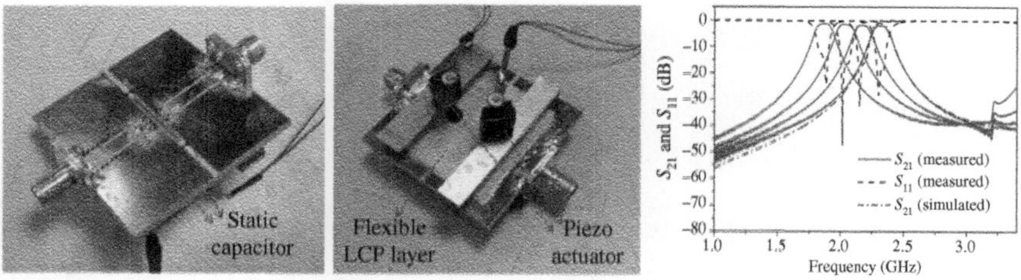


**Figure 8.26** A picture of the fabricated filter, in addition to simulated and measured frequency responses. Source: With permission of Nam et al. [112]/IEEE.

The required separation and $k_{in}$ correspond well when the separation is 1.8 mm (Figure 8.25(c)).

The design method followed here can be applied to filters of higher order. For instance, the structures shown here can be cascaded for achieving higher-order constant-bandwidth responses.

The structure of the manufactured filter and the measured frequency response along with the simulated one at the center frequency of 2.3 GHz are displayed in Figure 8.26. With a bandwidth of 136–142 MHz, the center frequency can be tuned from 1.85 to 2.3 GHz. There is a 1.3 dB minimum insertion loss at 1.85 GHz and a 2.0 dB maximum insertion loss at 2.3 GHz. The filter's absolute bandwidth is almost constant throughout the frequency range.

# 9

# Advanced Evanescent-Mode Filter Structures

## 9.1 Bandpass-to-bandstop Reconfigurable Filter

Filters occupy a relatively large amount of physical space in modern wireless systems compared with other typical radio components, most of which are now easily integrated on chip. In many cases, it would be impractical to implement both tunable bandpass filter (BPF) banks and tunable bandstop filter (BSF) banks in a system due to size constraints. A possible solution to this problem is to implement sets of coupled resonators whose response can be dynamically switched between a bandpass shape and a bandstop shape, saving space while providing added flexibility to a system. This response reconfiguration concept is shown in Figure 9.1(a). Figure 9.1(b) shows that Chebyshev bandpass, elliptic bandpass, and bandstop responses can be obtained from the same set of resonators if their coupling values are tunable. Note that Figure 9.1(b) is theoretical, and the ability to achieve all three responses in Figure 9.1(b) would require components that have wider tuning ranges than what is commercially available today. However, with the wide array of switches and varactors that are available today, the theory presented below can be used to implement a filter that can transform between any single BPF shape and any single BSF shape. Figure 9.2 shows the coupling diagram of the theoretical structure needed to implement the responses shown in Figure 9.1(b). The dashed lines represent tunable or switchable coupling values, and the solid line represents static coupling.

This section shows the theory of narrow-band, prescribed-response, bandpass-to-bandstop electronically reconfigurable filters and demonstrates a novel, high-$Q$, tunable, evanescent-mode cavity design with new capabilities not found in previous designs.

*Tunable Evanescent-Mode Filters: Principles, Implementation, and Applications*,
First Edition. Dimitrios Peroulis and Mohammad Abu Khater.

Companion website: www.wiley.com/go/Peroulis/TunableEvanescent-ModeFilters

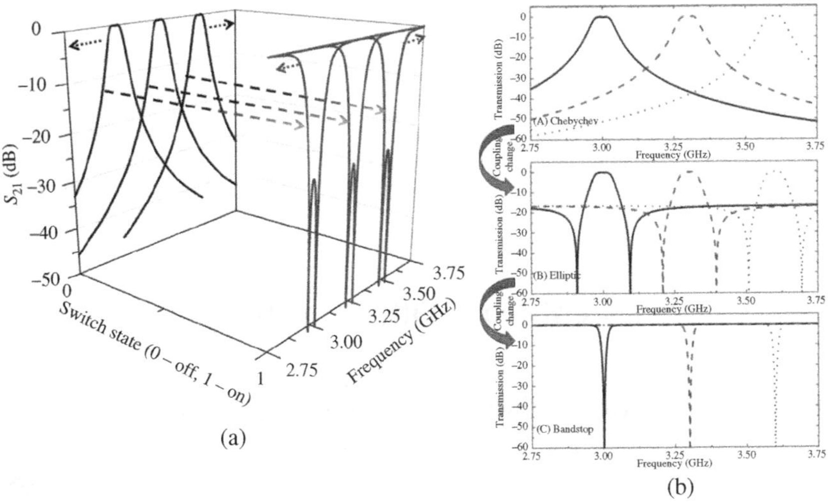


**Figure 9.1** (a) A concept of a bandpass filter that can switch its response to that of a bandstop filter by toggling a switch. (b) Simulated results showing possible $S_{21}$ responses of a set of two tunable resonators with tunable external and/or source-to-load coupling values.

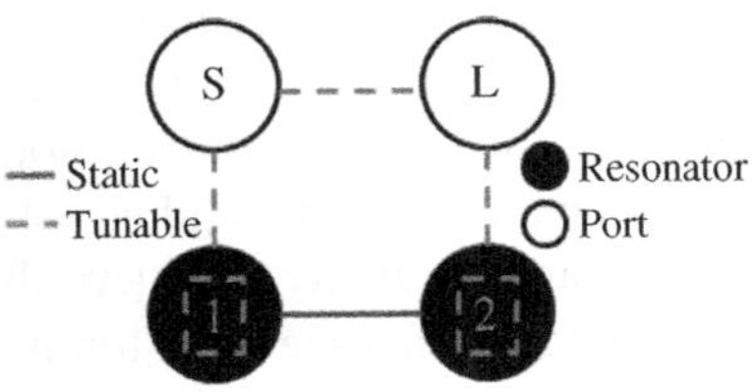


**Figure 9.2** Coupling diagram needed for implementation of a filter with responses shown in Figure 9.1(b). Note that the resonant frequencies of the two resonators are independently tunable.

### 9.1.1 Bandpass-to-bandstop Filter Theory

The same geometry resonators can be used for a BPF or a BSF if the coupling relationships between the source, load, and resonators are changed appropriately. For the second-order filter analysis in this chapter, input and output external coupling values are equal to each other, the resonator self-coupling values have the same magnitudes but opposite signs. Moreover, we can exploit certain reciprocities such as the source-to-load coupling being the same as the load-to-source coupling to simplify the generalized second-order filter $(N+2) \times (N+2)$ coupling matrix $M$ to

$$\mathbf{M} = \begin{bmatrix} 0 & M_{01} & 0 & M_{03} \\ M_{01} & M_{11} & M_{12} & 0 \\ 0 & M_{12} & -M_{11} & M_{01} \\ M_{03} & 0 & M_{01} & 0 \end{bmatrix} \tag{9.1}$$

where the subscripts 0 and 3 correspond to the source and load, respectively, and subscripts 1 and 2 correspond to the first and second resonators, respectively. It can be shown with coupling matrix methods that the coupling matrix for a typical second-order Butterworth BPF is

$$\mathbf{M}_{\mathbf{BPF}} = \begin{bmatrix} 0 & 0.8409 & 0 & 0 \\ 0.8409 & 0 & 0.7071 & 0 \\ 0 & 0.7071 & 0 & 0.8409 \\ 0 & 0 & 0.8409 & 0 \end{bmatrix} \tag{9.2}$$

and the coupling matrix for a second-order Butterworth BSF is

$$\mathbf{M}_{\mathbf{BSF}} = \begin{bmatrix} 0 & 1.189 & 0 & 1 \\ 1.189 & 0 & 0 & 0 \\ 0 & 0 & 0 & 1.189 \\ 1 & 0 & 1.189 & 0 \end{bmatrix} \tag{9.3}$$

Compared to the BSF in (9.3), the coupling matrix in (9.2) has two main differences. Typical BSFs have a source-to-load coupling ($M_{03}$) value equal to unity, which is commonly implemented as a 90° transmission line. Also, common BSFs have an inter-resonator coupling ($M_{12}$) value of zero, indicating that BSFs place the resonators in a shunt configuration, while BPFs, with a nonzero $M_{12}$ and zero-valued $M_{03}$, place the resonators in a series configuration. In addition to these two differences, the $M_{01}$ value changes slightly between the bandpass and bandstop cases. $M_{01}$ becomes 1.189 in the normalized frequency Butterworth bandstop case, compared with 0.8409 for the Butterworth bandpass case.

It is straightforward to change the zero-valued $M_{03}$ of a BPF to a value of 1 through the use of switches with high isolation, which switch in a 90° transmission line between the resonators. In addition, a uniquely designed coupling structure will be shown below that allows for the switching of $M_{01}$ between the bandpass and bandstop modes of the filter. However, changing from a nonzero $M_{12}$ value to an $M_{12}$ value of zero is difficult in practice, since it is difficult to fully block the coupling mechanism between resonators. Fortunately, it is possible to synthesize a bandpass-to-bandstop reconfigurable filter with a nonzero $M_{12}$ value if the resonators of the filter can be tuned asynchronously [116].

#### 9.1.1.1 Coupling Structure to Switch $M_{01}$ and $M_{03}$ Simultaneously

The coupling structure designed for bandpass-to-bandstop reconfiguration is a transmission line of electrical length $L$ degrees over a coupling aperture in the wall of the first resonator, followed by a switch, followed by a transmission line of length $270 - 2L$ degrees, followed by a switch, followed by a transmission line of length $L$ degrees over a coupling aperture in the wall of the second resonator. A model of such a structure can be seen in Figure 9.3(a), where the resonators have been omitted for clarity.

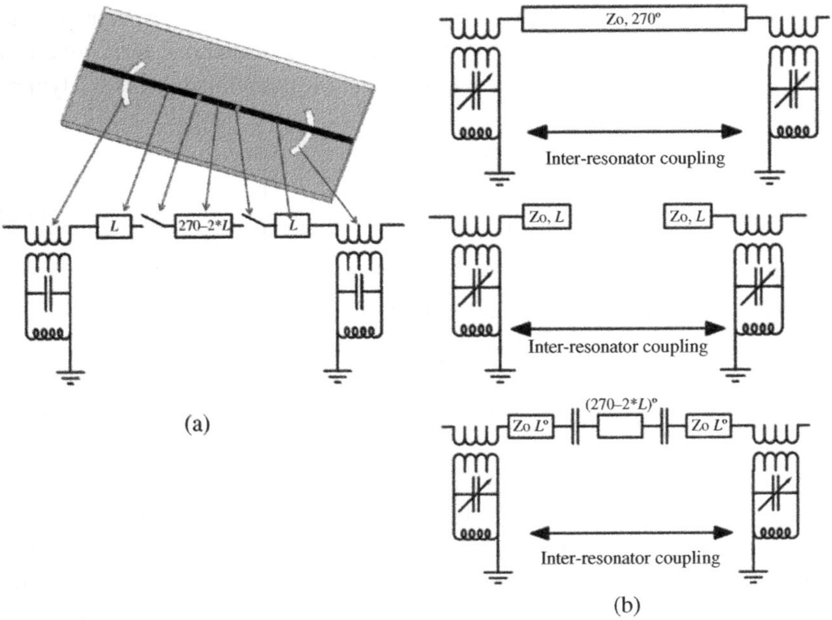


**Figure 9.3** (a) The coupling structure designed for bandpass-to-bandstop reconfiguration. (b) Top: coupling structure in the bandstop mode of the filter. Middle: coupling structure in the bandpass mode with very high isolation switches. Bottom: coupling structure in the bandpass mode with medium isolation switches, enabling an elliptic response.

This coupling structure has three modes of operation. First, in the bandstop mode of the filter, the switches are closed, and the coupling structure becomes a 270° transmission line over coupling apertures in the resonators. A transmission line has a positive coupling value in this analysis. This positive insertion phase allows for a bandstop response with a nonzero $M_{12}$ value. The coupling apertures in the walls of the resonators are sized to provide the appropriate $M_{01}$ value for the bandstop mode of the filter. An equivalent circuit model of this mode of operation can be seen at the top of Figure 9.3(b).

The coupling structure's second mode of operation requires very high isolation switches. In this mode, the switches are open, and the coupling structure becomes a transmission line of length $L$ degrees over a coupling aperture in the wall of the first resonator. The transmission line between the two switches is not coupled to; therefore, the source-to-load coupling is removed. Symmetry allows similar coupling into resonator 2 as resonator 1. The transmission lines of length $L$ are approximately open at their interior ends if high isolation switches are utilized. The length of the transmission line from the coupling aperture to the open end transforms the impedance seen at the coupling aperture. Therefore, $M_{01}$ also changes accordingly, and different values are possible between the bandpass and bandstop modes of the filter. An equivalent circuit for this mode of operation can be seen in the middle of Figure 9.3(b).

The coupling structure's third mode of operation requires a switch with a medium level of isolation, similar to the level found in common solid-state radio frequency (RF) switches. Variable isolation switches or varactors could also be used for even more flexibility. In this mode, the switches are in their off-state but have significant capacitance. An equivalent circuit for this mode of operation can be seen in the bottom of Figure 9.3(b). For a certain range of small switch capacitance values, the capacitance shown in the figure adds approximately two more 90° phase shifts into the circuit. Considering that 270 electrical degrees are already present from the transmission lines, the total insertion phase in this mode of the coupling structure becomes 450°. This is a significant result because the 450° of insertion phase is the sign opposite of the 270° of insertion phase obtained in the bandstop mode of operation, and this allows for negative source-to-load coupling. Negative source-to-load coupling results in an elliptic response [117, 118], providing a third useful mode of operation for the proposed filter.

### 9.1.2 Bandpass-to-bandstop Reconfigurable Filter Design

An evanescent-mode resonator-based bandpass-to-bandstop reconfigurable filter is fabricated with a static inter-resonator coupling structure for measurement. A model of the fabricated filter can be seen in Figure 9.4. The evanescent-mode cavity resonators are designed in a 3.175-mm-thick Rogers TMM3 printed circuit board material and defined by plated vias. The cavities have semi-circular apertures cut into their surface to enable magnetic field coupling. Single, weakly coupled resonators are fabricated out of the same TMM3 board as the filter cavities presented in this chapter, and the unloaded quality factor is measured to vary from 608 to 650 across the frequency range from 2.75 to 3.75 GHz.

A copper membrane is laminated onto the substrate to close the cavity and form the bottom plate of the capacitor inside of the resonator. Piezoelectric actuators are attached to the copper membrane to allow for electronically controllable deformation of the membrane. The cavity feeds are fabricated on a 1.27-mm-thick Rogers TMM10i material and laminated to the cavities. The feeds are fabricated on this high-dielectric material to reduce the size of the 270° transmission line for the source-to-load coupling described above.

The external coupling structure in the previous section can provide two independent values for the external couplings, one for the BSF and the other for the BPF. The denormalized inter-resonator coupling, $k_{12}$ value, however, is constant.

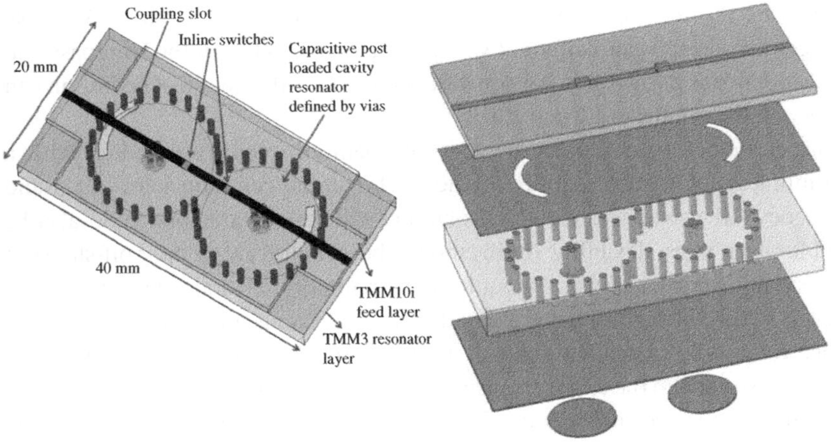


**Figure 9.4** Filter model showing geometry and materials used in the fabricated filter, in addition to a layer-by-layer expanded view of filter. From top to bottom, the layers are TMM10i, thin copper, TMM3, and thin copper. The disks represent the actuators. Note that there are switches in line with the transmission line shown above on the top layer.

The external couplings are thus chosen such that they deliver the sought BSF and BPF responses.

### 9.1.3 Measured Results

A labeled photograph of the device can be seen in Figure 9.5. The ideal synthesized response in Figure 9.6(a) shows a theoretical second-order Butterworth bandpass filter with no stray coupling between source and load ports. It is valuable to do a measurement of the fabricated filter without switches in the

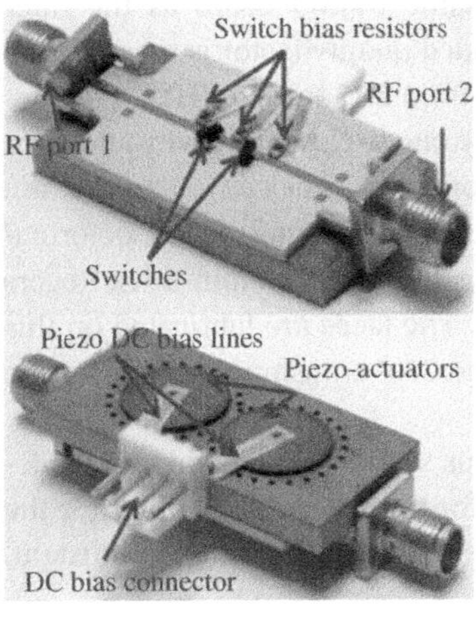


**Figure 9.5** Labeled photograph of both sides of the fabricated filter. For measurements, a network analyzer was connected to the RF ports, and DC power supplies were connected to the DC bias connector. The DC traces were fabricated on a 25.4-μm liquid crystal polymer (LCP) substrate.

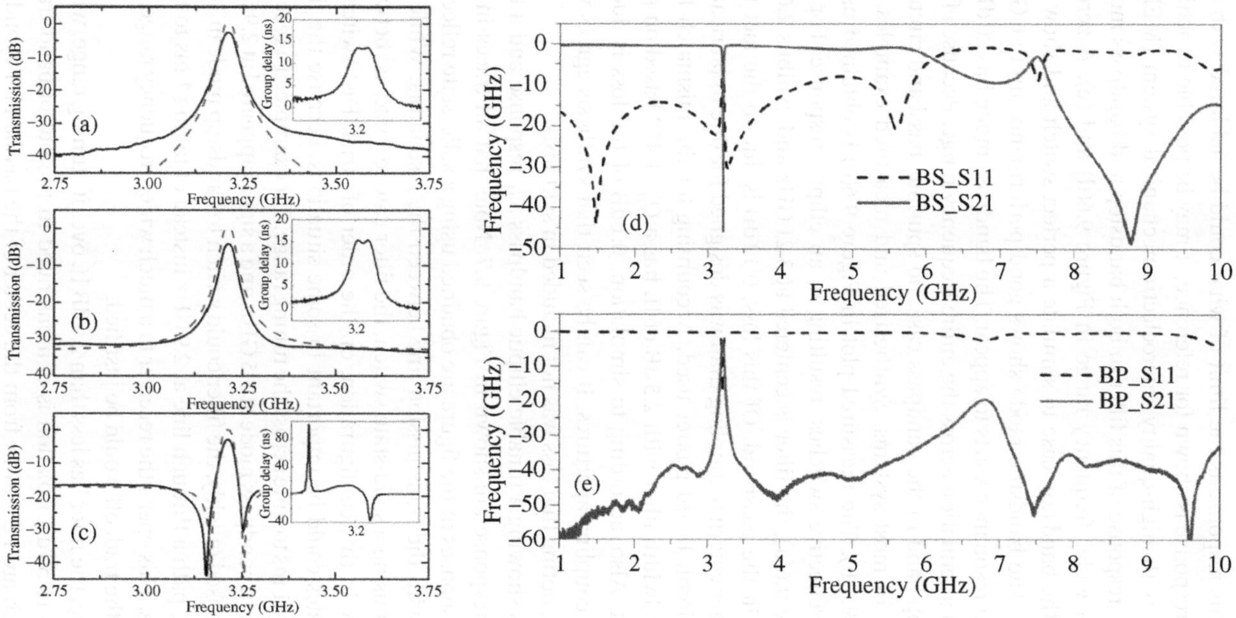


**Figure 9.6** Measured and synthesized $S_{21}$ responses of the filter in the bandpass mode. (a) Response with no switches (open) compared with the ideal Butterworth filter. (b) Response with absorptive switches compared with the ideal self-equalized filter. (c) Response with reflective switches compared to ideal elliptic filter. Note that loss was not included in synthesis. Measured $S_{11}$ and $S_{21}$ responses of the filter in the bandpass and bandstop modes from 1 to 10 GHz. (d) Bandstop mode. (e) Bandpass mode.

source-to-load coupling structure to observe the amount of coupling due to radiation and substrate paths for comparison. This measurement can also be seen in Figure 9.6(a). This response is the limit of what could be achieved with this design. While this response is shown for reference, it may be possible to achieve a similar response with high-quality microelectromechanical system (MEMS) switches [119]. The response of this filter in both bandstop and bandpass modes can be seen over a wider frequency range in Figure 9.6(d) and (e). A zero Ω resistor is used in the bandstop case to simulate a perfect switch and show the limit of the design. The bandstop mode shows good performance up to 6 GHz, where an unwanted resonance starts to appear. The bandpass mode has 20 dB or greater out-of-band attenuation across the entire frequency range. Because of the extended upper stop band in the bandpass case, no spurious passband circuitry would be required in most systems. Synthesized and measured examples can be seen in Figure 9.6(b). The measured plot in Figure 9.6(c) is obtained using medium isolation reflective switches, resulting in an elliptic response. For the ideal case in Figure 9.6(a), the filter is centered at 3.21 GHz and exhibits 2.6 dB insertion loss (IL) in the passband. Of this loss, 0.12 dB is due to the fact that the filter has stronger external coupling than was designed for due to fabrication tolerances of the circuit board plotter used. Accounting for the mismatch loss, a 1% fractional bandwidth filter with 2.5 dB of IL has a $Q_u$ of 457 based on gain bandwidth product. Also, according to simulation, 0.1 dB of the loss is due to radiation from the coupling apertures. It can be seen that synthesis agrees well with measurement, noting that loss was not included in synthesis.

Measured results showing a tunable elliptic bandpass $S_{21}$ response and a tunable bandstop $S_{21}$ response are shown in Figure 9.7. Note that all traces in the plot are $S_{21}$. The responses in the figure are obtained using a solid-state reflective RF switch in line with the transmission line between the resonators. While the loss is higher when using a solid-state switch, the filter can be switched between modes very quickly in this configuration, on the order of 10 ns. Fast switching speed between modes could be important in some situations, because the filter takes approximately 1 ms to tune across the tuning range. For example, if a receiver were using the filter in bandpass mode at 3.5 GHz, and a signal appeared at 2.9 GHz that needed to be sensed quickly, the filter could switch to bandstop mode in 10 ns to make the receive path a through line at 2.9 GHz instead of taking 1 ms to tune that BPF to 2.9 GHz. This opens the receiver to a much wider frequency range, but in some situations the tradeoff would be justified.

In bandpass mode, the filter has less than 2.9 dB IL over its tuning range, which is 2.9–3.6 GHz. Up to 50 dB isolation is demonstrated in the out-of-band transmission nulls, which are 1.5% away from the center of the passband. Beyond the transmission zeros (TZs), the out-of-band attenuation is 17 dB. In bandstop mode, the filter produces up to 45 dB isolation at its center frequency and has the same

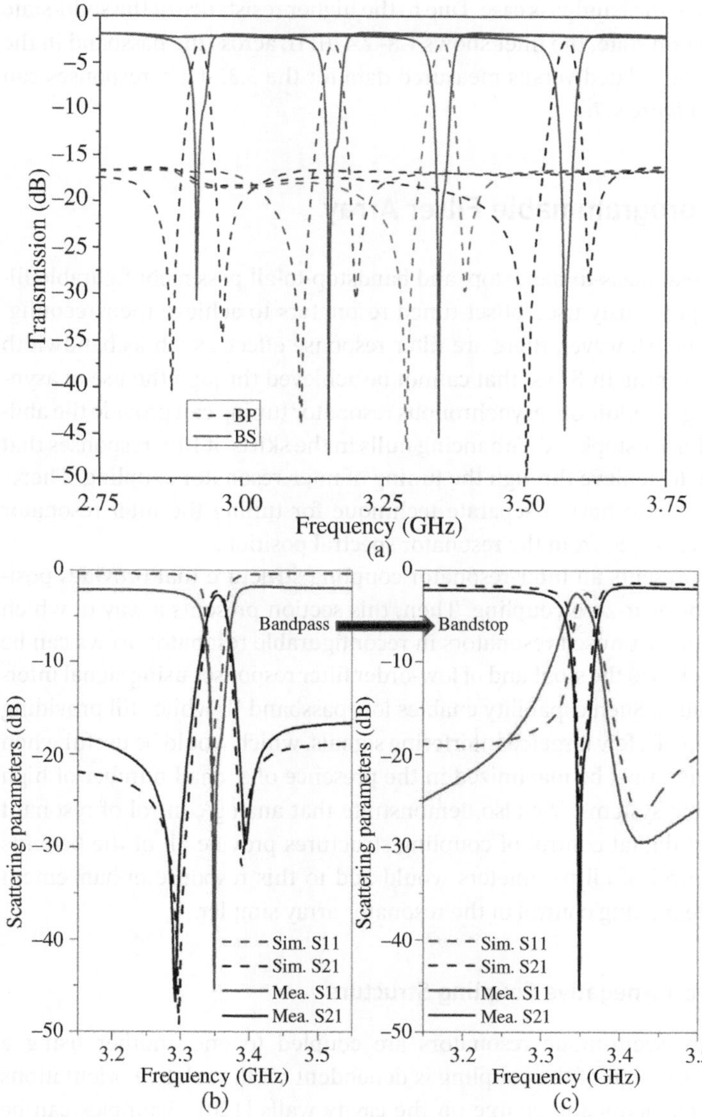


**Figure 9.7** (a) Measured $S_{21}$ response of the filter in bandpass and bandstop modes, tuned across the available tuning range. Simulated versus measured data for the 3.35-GHz responses in (a). (b) Bandpass data. (c) Bandstop data.

tuning range as in the bandpass case. Due to the higher resistance of the solid-state switches in their on state, the filter shows 1.8–2.4 dB IL across the passband in the bandstop state. Simulated versus measured data for the 3.35-GHz responses can also be seen in Figure 9.7.

## 9.2 Field-programmable Filter Array

The bandstop, bandpass-to-bandstop, and bandstop-to-all pass reconfigurable filters presented previously used offset-tuned resonators to achieve their reconfiguration capability. However, there are filter response effects, such as bandwidth and filter order tuning in BPFs, that cannot be achieved through the use of asynchronous tuning. In addition, asynchronous resonator tuning can provide the ability to add multiorder stopband-enhancing nulls in the skirts of BPF responses that are not possible to achieve through the tuning of inter-resonator coupling. Therefore, it is desirable to have a separate technique for tuning the inter-resonator coupling that is isolated from the resonator spectral position.

This section presents an inter-resonator coupling structure that provides positive, negative, or near-zero coupling. Then, this section presents a way in which adjacent, previously unused resonators in reconfigurable resonator arrays can be used to add notches to the stopband of low-order filter responses using signal interference techniques. Such capability enables low passband IL while still providing high attenuation of a few targeted interfering signals, which would be useful when system sensitivity must be maximized in the presence of a small number of high power interfering systems. We also demonstrate that analog control of resonant frequencies and digital control of coupling structures provide all of the benefits that analog control of all parameters would add to this response enhancement technique while making control of the resonator array simpler.

### 9.2.1 Positive-to-negative Coupling Structure

When two evanescent-mode resonators are coupled to one another using a waveguide iris, the sign of the coupling is dependent on the relative orientations of the loading elements and charge on the cavity walls [120]. Examples can be seen in Figure 9.8(a) and (b). In [121], it was shown that a probe connecting the top wall of the iris to the bottom wall of the iris over a distance can provide negative inter-resonator coupling relative to the coupling of the waveguide iris through effectively reversing the direction of the coupled current. In the proposed concept, several of these probes are used inside each waveguide iris. A switch is placed between one end of each probe and the iris wall, expanding the concept to provide coupling value switching capability. Such a coupling structure can be seen in Figure 9.8(c). With the switches off, the coupling structure is approximately

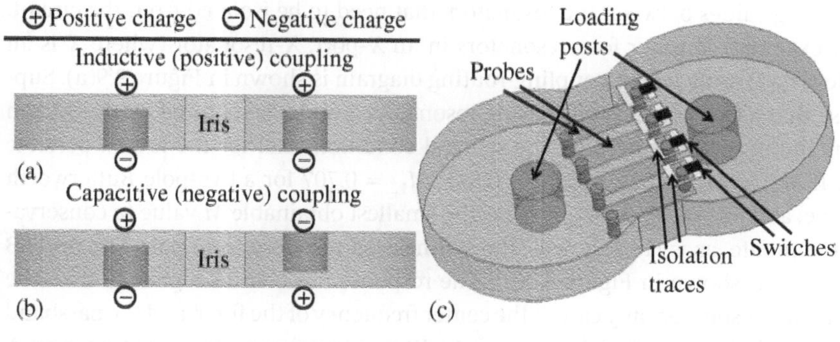


**Figure 9.8** (a) Side view, positive coupling (like charge on each cavity side). (b) Side view, negative coupling (opposing charge on each cavity side). (c) Switchable coupling structure for positive-to-negative coupling.

equivalent to a waveguide iris with positive coupling as in Figure 9.8(a). With the switches on, the negative coupling provided by the probes overtakes the positive coupling of the waveguide iris as long as the probes are sufficiently large. With some switches on and some off, other positive, negative, and near-zero coupling values result. The structure effectively twists the configuration in Figure 9.8(a) into the configuration in Figure 9.8(b) as more switches are turned on. In addition, the current distribution across the width of the iris varies, so altering switch states in different probe locations results in different coupling values. Therefore, $2^N$ coupling states exist, where $N$ is the number of probes. This allows each coupling structure to be addressed as a digital $N$-bit word, with each bit representing the state of a switch, on or off. Although discrete, the coupling values available can be varied enough to facilitate the design of highly reconfigurable filters.

### 9.2.2 Response Enhancements Enabled By Positive-to-negative Inter-Resonator Coupling

When used in a multi-resonator array, positive-to-negative inter-resonator coupling enables reconfiguration between several filter response types that are not possible otherwise. In addition, it can enhance more standard responses and increase isolation between resonators that are desired to be isolated. Such capability is shown in the following subsections.

#### 9.2.2.1 Zero Net Coupling State Enhancement Using Destructive Interference

Due to manufacturing tolerances and environmental effects, the zero net coupling state will often not provide exactly zero coupling. However, the structure is resilient against these nonidealities in rectangular arrays of resonators due to destructive interference paths within the circuit. The destructive interference provided by these additional paths can be enhanced through intelligent selection of coupling values between the resonators that need to be isolated from the circuit. For example, consider four resonators in an $X$-port, $X$-resonator (where $X$ is an integer $\geq 4$) array whose coupling-routing diagram is shown in Figure 9.9(a). Suppose that a two-pole response using resonators 1 and 2 is required by the system containing the filter because the spectral conditions call for the lowest possible IL. If $M_{A1} = M_{B2} = M_{C3} = M_{D4} = 0.841$, $M_{12} = 0.707$ for a two-pole Butterworth shape, and $M_{13} = M_{34} = M_{24} = 0.2$ is the smallest obtainable $M$ value (a conservative estimate, as is later shown), the synthesized response from port A to ports B and C is as shown in Figure 9.9(b). The response has 0.66 dB higher IL than the ideal two-resonator-only case at the center frequency of the filter, and the passband shape is slightly changed due to the −14 dB transmission coefficients from port A to port C and port B to port D. While this may be acceptable in many applications, optimizing the destructive interference between the paths shown in Figure 9.9(c) can further isolate resonators 3 and 4. If $M_{34}$ is set to

$$M_{34} = \frac{M_{A1}^4 + M_{13}^2}{M_{12}} \tag{9.4}$$

destructive interference is maximized at the center frequency of the filter, and the result is the response in Figure 9.9(d) with 0.34 dB IL. In the example above, $M_{34} = 0.764$ is the ideal value. While the proposed coupling structure cannot dial in an exact $M_{34}$ value, setting $M_{34}$ to the closest available value still significantly

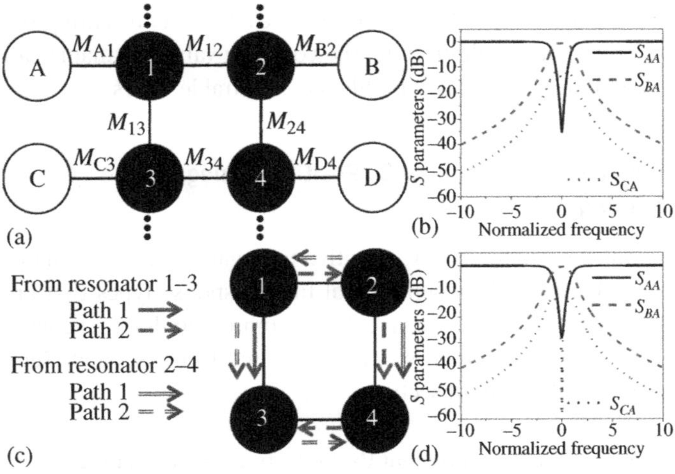


**Figure 9.9** (a) Coupling-routing diagram of four resonators in an array. Resonators are black, and ports are white. (b) Response when $M_{13} = M_{34} = M_{24} = 0.2$. (c) Paths for destructive interference between resonators 1 and 3 and 2 and 4. (d) Improved response when $M_{34} = 0.764$.

aids in isolating adjacent resonators. If the resonators are tunable, tuning the to-be-isolated resonators far from the passband further increases isolation.

#### 9.2.2.2 Local Stopband Attenuation Enhancement Technique

Adjacent resonators that are not in use to create the current passband response can be exploited to enhance the filter response by adding TZs to the filter stopband. TZs locally enhance the stopband isolation in a filter response. This is especially valuable in low-order filters due to their more gradual stopband slopes but is also applicable to all filter orders and shapes. For example, consider the coupling-routing diagram shown in Figure 9.10(a) with six resonators and two ports. Such a coupling-routing diagram could represent a reconfigurable resonator array with six resonators or a subset of resonators in a larger array.

If the coupling to and between resonators 1, 2, 5, and 6 is instead set to nonzero values, resonant destructive interference paths can be created at targeted frequencies. This concept is shown in Figure 9.10(b). In Figure 9.10(b), resonators 3 and 4 produce a two-pole response at $f_0$. Energy at frequencies $f_1$ and $f_2$, stopband frequencies where increased attenuation is desired, leaks through the path that contains resonators 3 and 4 due to the finite stopband attenuation of the two-pole response. An equal magnitude and opposite phase signal at $f_1$ can be leaked through the adjacent path that contains resonators 1 and 2 if the coupling values and/or resonant frequencies associated with resonators 1 and 2 are set properly, creating a resonant TZ path. A similar effect can be achieved at $f_2$ using the adjacent path through resonators 5 and 6. A particularly attractive trait of the

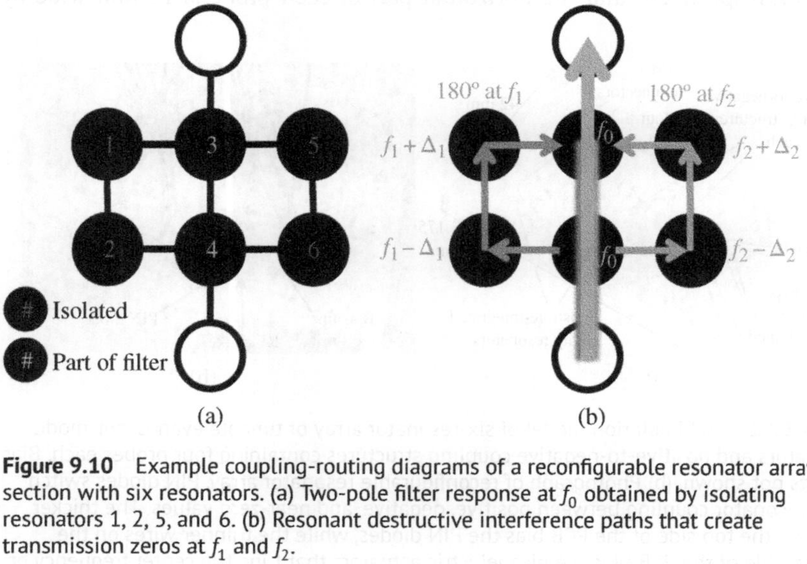


**Figure 9.10** Example coupling-routing diagrams of a reconfigurable resonator array section with six resonators. (a) Two-pole filter response at $f_0$ obtained by isolating resonators 1, 2, 5, and 6. (b) Resonant destructive interference paths that create transmission zeros at $f_1$ and $f_2$.

proposed method to create TZs is that the zeros are created on demand using paths composed of multiple resonators as opposed to the conventional method of using reactive cross-coupling [122–124]. Using paths composed of multiple resonators enables the design of the shape, tuning range, and bandwidth of TZs, similar to how these parameters can be designed in conventional multi-resonator BSFs.

### 9.2.3 Resonator Array Design and Fabrication

Four devices are fabricated to demonstrate the positive-to-negative inter-resonator coupling structure and its response enhancement capabilities. The devices are composed of piezoelectrically tuned evanescent-mode resonators. The first two devices are six-resonator arrays with ports at each resonator and positive-to-negative coupling structures used to couple all resonators to all adjacent neighbors. A simulation model of these structures can be seen in Figure 9.11(a), and a photograph can be seen in Figure 9.11(b). Each positive-to-negative coupling structure contains four probes. Each probe uses a positive intrinsic negative diode (PIN) diode as a switch in the first design. The second design uses zero Ω resistors and open circuits for comparison. The PIN diode version would be useful to dynamically adapt to signals in the field, while the zero Ω resistor version would be useful as a post-mass production shape-customizable filter since the surface-mount zero Ω resistors can be easily added or removed with solder. The structure volume is 63 mm × 42 mm × 3.175 mm. The resonators and loading posts in all designs have radii of 7.5 and 2.1 mm, respectively, and the horizontal part of each probe is 1.2 mm wide by

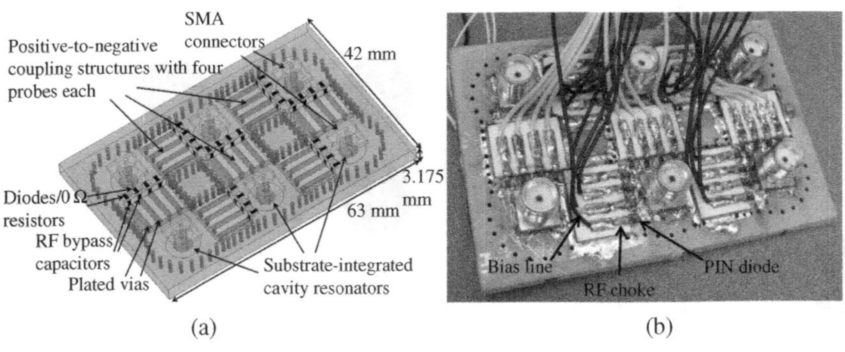


**Figure 9.11** (a) Simulation model of six-resonator array of tunable evanescent-mode resonators and positive-to-negative coupling structures containing four probes each. Bias circuits not shown. (b) Photograph of reconfigurable resonator array. PIN diodes switch inter-resonator coupling between positive, negative, and near-zero values. The thicker wires on the top side of the PCB bias the PIN diodes, while the thinner wires on the bottom side of the PCB bias the piezoelectric actuators that tune the center frequency of the resonators.

7.8 mm long. The vertical parts of the probes are 0.8-m-diameter plated vias on one side and vertically mounted 68 pF RF bypass capacitors on the other side. Figure 9.12(a) shows closer views of the positive-to-negative inter-resonator coupling structures and supplementary direct current (DC) bias circuit board. The DC bias board uses a 120 nH inductor in each bias circuit to block RF signal from propagating into the bias generator. Since the four PIN diodes in each inter-resonator coupling structure are connected in parallel, bias resistors on a separate breadboard are also used to equate the current in each branch of the parallel circuit. Another view of the positive-to-negative inter-resonator coupling structure and supplementary DC bias board is shown in Figure 9.12(b).

A closer view of the loading post, which also forms the external coupling structure, is shown in Figure 9.12(c). The top row of images in Figure 9.12(c) show the dimensions of the loading post and its positional relationship with the Subminiature Type A (SMA) connector port above each resonator. The bottom row of images shows the equivalent circuit of the external coupling circuit/loading post. The central plated via in the loading post connects to an isolated pad on the copper surface of the cavity substrate that is soldered to the center conductor of the SMA connector above each resonator. The ground housing of the SMA connectors is soldered to the rest of the copper surface of the cavity substrate. The three outer viae in the loading post connect it to the copper surface of the cavity substrate and provide a DC path to ground. At RF, these three viae provide significant inductance to ground. Together, the vias in the post forms an inductive impedance transformer. The radius and number of these vias can be adjusted to obtain various external coupling values.

The third and fourth fabricated structures are sets of two resonators with single positive-to-negative coupling structures between them. One of these structures has an inter-resonator coupling structure of the same dimensions as those of the six-resonator arrays discussed above and weak external coupling. This structure is used to extract measured inter-resonator coupling values for use when switching coupling values in the larger array. The other structure with two resonators is designed with an inter-resonator coupling structure of the same concept as the others, but slightly different dimensions and near-critical external coupling. This structure is designed to work with different types of switches. The purpose of the design with those switches is to show that the positive-to-negative inter-resonator coupling structures described in this chapter can be used with minimal increase in filter IL when modern, low-loss switches are integrated into the design.

### 9.2.4 Measured Results

During device operation, the SMA connectors that are not used as input and output ports are left open (unterminated). This is possible because the highly capacitively

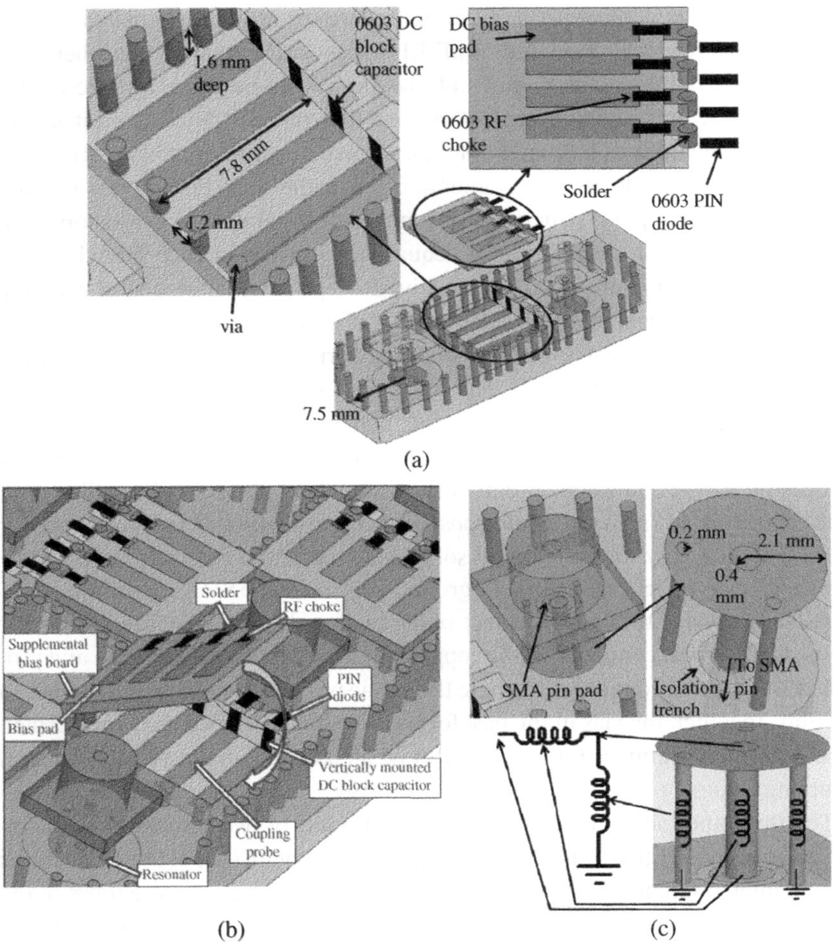


**Figure 9.12** (a) Model of inter-resonator coupling structure and supplementary DC bias board. The DC bias board is soldered to the iris in the fabricated structure but raised in this figure for clarity. (b) Model of an inter-resonator coupling structure with the supplementary bias board rotated off of the cavity substrate to show internal structure. As PIN diodes are switched on, waveguide iris current is redirected, changing coupling values. (c) External coupling circuit/loading post and equivalent circuit. Together, the vias in the loading post acts as an inductive impedance transformer.

loaded resonators used in this design have a very low characteristic impedance. Therefore, termination with an open-circuit approximates termination with a 50 Ω load. Leaving the port open can save volume and cost if such a filter were used in a production application.

Measurements of the two weakly coupled resonators can be seen in Figure 9.13(a) along with switch states and extracted $M_{12}$ values. We observe that coupling values that would be useful in critically coupled responses, isolation paths between desired and undesired resonators, and cross-coupling paths with negative coupling can be obtained from the same electronically reconfigurable inter-resonator coupling structure. The measured coupling values are plotted graphically in Figure 9.13(b). Figure 9.13(c) shows measured 1.25% fractional bandwidth two-pole responses from resonators 3 and 4 of the six-resonator array that uses PIN diodes to electronically switch coupling. Measured IL ranges from 3.11 to 3.96 dB over the tuning range. A similar two-pole response from the device that uses zero Ω resistors to mechanically switch coupling is also shown at 1.8 GHz to show the tradeoff between electronic reconfiguration capability and IL. The diodes add 1.06 dB IL. The silicon construction of these diodes is part of the reason for this additional loss. Lower-loss diodes in the same package size are also available, but they have larger off-state capacitance that would limit the spurious-free range of the field programmable filter array (FPFA). Another possible reason is that the complete recommended bias circuit in the diode's data sheet could not be implemented in the FPFA prototype shown here due to space constraints. Lower-loss switches such as RF MEMS devices could possibly reduce switching element IL .

Figure 9.14(a) shows measured 1.02% fractional bandwidth four-pole responses from resonators 1–4 of the six-resonator array that uses zero Ω resistors. TZs are added and removed by toggling $M_{34}$ between near-zero and negative values. IL ranges from 4.07 to 5.31 dB with TZs and 4.81 to 6.38 dB without TZs. The measured IL is within 1.3 dB of fullwave simulations. The first spurious response is at 3.86 GHz and is at −20.2 and −48.2 dB levels in the two- and four-pole cases, respectively. The spurious responses for both a two-pole response and a four-pole response can be seen in Figure 9.14(b). The spurious modes are a result of the electrical length of the probes in the inter-resonator coupling structure, and their frequency does not change when the resonators are tuned across their tuning range.

Measured superposed results of the fabricated filter that show the destructive interference stopband enhancement technique can be seen in Figure 9.15(a). One adjacent path through two resonators is activated in order to obtain the plots shown. A tunable two-pole notch response can be seen at various locations in the stopband of the two-pole passband response. The final two resonators of the device are isolated. When the notch is significantly far away from the passband, the filter exhibits 2.32 dB IL and a 1.2% fractional bandwidth. Passband insertion loss increases as the notch is brought close to the passband due to the skirt reactance of the notch. Figure 9.15(a) shows that when the notch is placed 20 MHz away from the passband at 1.8646 GHz, the minimum IL of the filter degrades by 1 dB and the passband shape becomes distorted. The passband response is designed expecting 50 Ω terminations, and the input impedance of the notch at the passband frequency is not 50 Ω when it is tuned close to the passband. In order to mitigate this distortion, the passband response can be

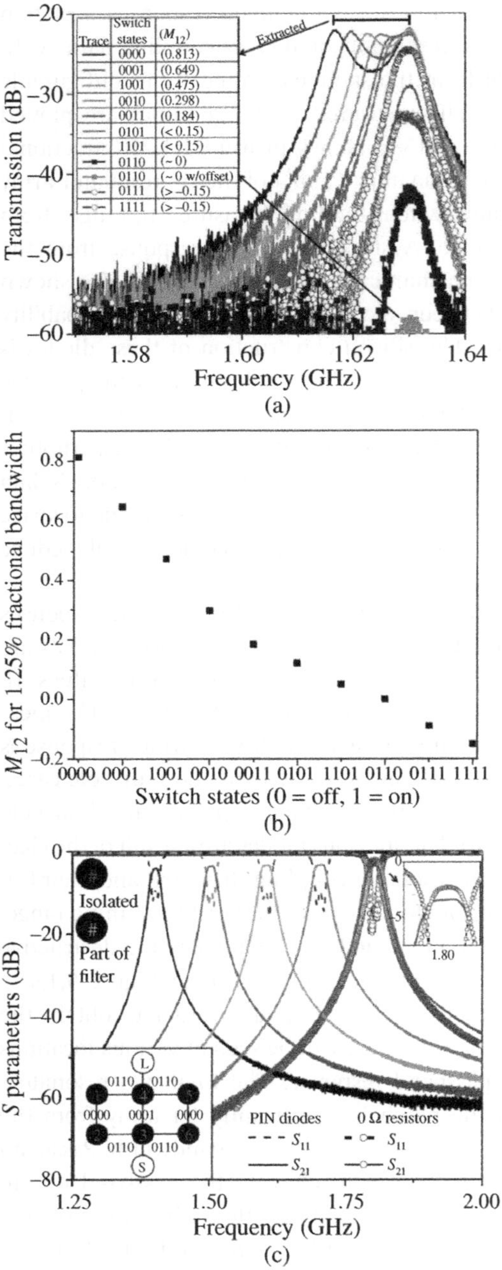


**Figure 9.13** (a) Measured weakly coupled two-pole responses used for $M_{12}$ coupling value extraction. The table shows PIN diode states, where 1 is on and 0 is off, and extracted $M_{12}$ values for a 1.25% fractional bandwidth. Offset tuning the resonators results in minimal coupling. (b) Measured extracted inter-resonator coupling values (a) for different switch/PIN diode states. (c) Measured two-pole responses obtained by isolating four resonators with the shown switch states. Solid lines show responses of the PIN diode version. Lines marked with white circles show a response of the 0 Ω resistor version for comparison.

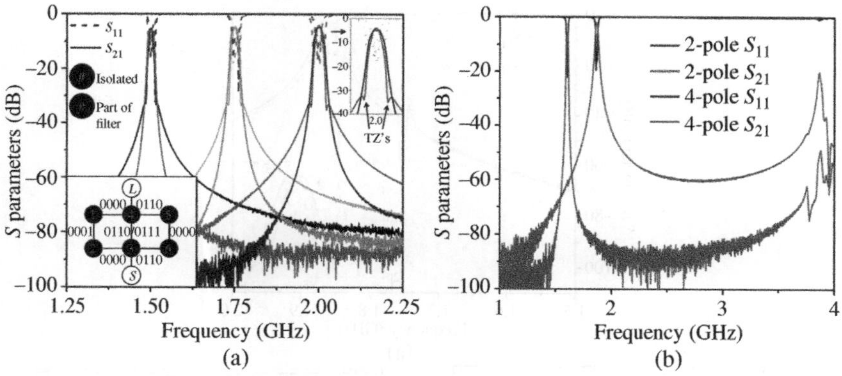


**Figure 9.14** (a) Measured four-pole responses with and without TZs obtained by isolating two resonators and toggling M34 between near-zero and negative values. All responses are from the zero Ω resistor version. (b) Measured two-pole and four-pole responses showing the first spurious response in both states.

designed to expect a complex load [125]. However, the response would become distorted when the notch is far from the passband in this case unless the external coupling could also be tuned, which is not a capability of the fabricated structure.

Figure 9.15(b) shows measured superposed responses when two adjacent paths are activated to create two-pole notches in the stopband of the two-pole passband response. The filter passband exhibits 2.93–2.30 dB IL and a 1.2% fractional bandwidth over the tuning range shown. The two notches can be placed on opposite sides or the same side of the passband. In addition, Butterworth or equi-ripple notches are possible. Figure 9.15(c) shows the response from Figure 9.15(b) with a passband at 1.7 GHz along with a synthesized response.

The plots shown in this chapter are representative of numerous filter states that can be generated with the six-resonator array, but they are not all-encompassing. The combination of many digitally addressable coupling states and tunable resonators provides the means to create a wide array of filter responses. However, there are still limitations to the responses of the FPFA prototype due to discretization of the inter-resonator coupling states, lack of source-to-load coupling, and static external coupling. While the discrete inter-resonator coupling structure used in the FPFA prototype did not limit its capability to add notches to the stop band at arbitrary frequencies, it does limit the number of possible passband shapes. Each unique shape and each order of each shape require different inter-resonator coupling values. Therefore, as FPFAs grow in number of resonators, more inter-resonator coupling values will be needed in order to fully utilize the capability of a large FPFA. The design of inter-resonator coupling structures that can provide many specific discrete values will be challenging because the position of each current probe affects multiple coupling states. In these cases,

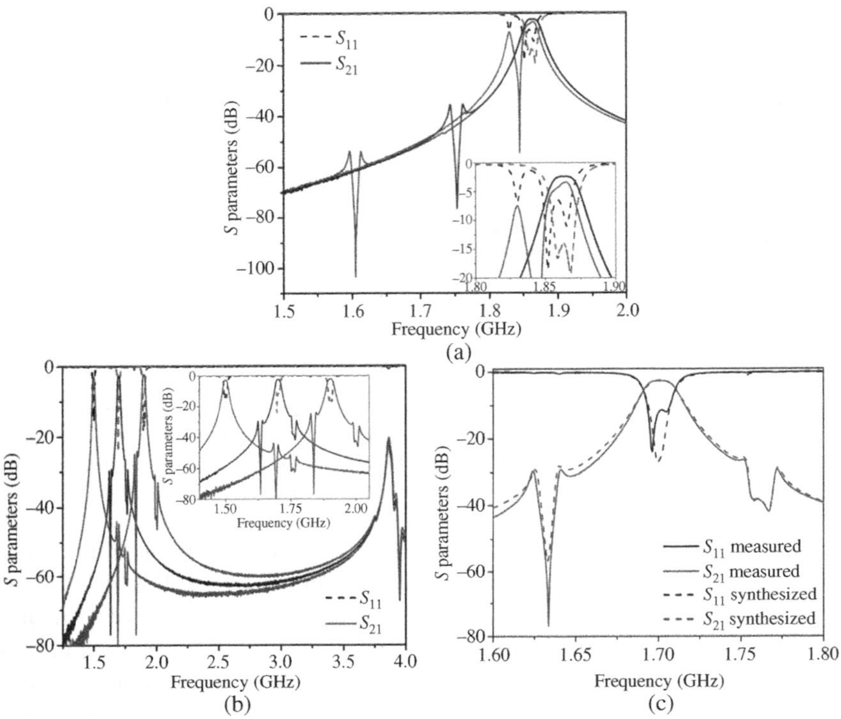


**Figure 9.15** (a) Three superposed measured results showing a two-pole passband response with one adjacent resonant path creating a tunable notch in the filter stopband. (b) Wideband measurement showing first spurious resonances and three superposed two-pole passband responses with two adjacent resonant paths creating tunable notches in the filter stopbands. Inset shows closer view of the frequency range near the passband frequencies. Therefore, the addition of notches in the skirt of the passband does not change the first spurious resonance frequency. (c) Measured versus synthesized response.

designing coupling structures with quasi-continuous tuning through the use of many discrete closely spaced coupling values may be the best design option. In contrast to these cases, the design of structures that switch between a few specific inter-resonator coupling values is useful for future systems that need a more limited amount of reconfiguration capability. For example, the inter-resonator coupling structure shown in this section has 10 unique coupling values. One specific value was designed for a two-pole response, two for a four-pole response, one for zero coupling, and two for negative coupling. However, the remaining four unique values were left to be what naturally occurred after the first six values were obtained due to the difficulty in optimizing for 10 specific values. Still, the FPFA was able to produce a multitude of responses.

# 10

# Passive Applications

## 10.1 Impedance Tuner

Impedance tuners are key devices in adaptive radio frequency (RF) systems. They are needed in many applications, such as source-pull, load-pull, and noise parameter characterization, antenna impedance matching, and reconfigurable power amplifiers (PAs). Impedance tuners are typically designed to cover as much of the Smith chart as possible and have low loss as well as high linearity and power-handling levels.

Conventional impedance tuners are often realized as motorized mechanical devices integrated in either coaxial or waveguide structures. These are usually low-loss tuners with good range of impedance coverage. However, they are also heavy, bulky, and slow. Modern tuners employ electronically variable elements such as semiconductor varactors, microelectromechanical systems (MEMS), or ferroelectric capacitors in different topologies. Although each of these tuning technologies has its own advantages and drawbacks, a comprehensive solution for having electronically controlled tuning while maintaining low loss, high linearity, and high power handling is challenging.

In this section, an impedance tuner, based on evanescent-mode (EVA) technology, is introduced. Due to its high-$Q$ resonators, the measured power loss of the proposed tuner is very low (approximately 0.8 dB in S-band). Also, the measured third-order input intermodulation intercept point (IIP3) is +64.3 dBm, while the tuner handles up to 90-W input power. Both of these parameters are superior when compared with what semiconductor-based tuners provide.

### 10.1.1 Design and Fabrication

Design of this impedance tuner is based on a two-pole coupled-resonator structure in which each resonator is independently controlled. The considered proof-of-concept demonstration frequency is 3.3 GHz, while the frequency

*Tunable Evanescent-Mode Filters: Principles, Implementation, and Applications*, First Edition. Dimitrios Peroulis and Mohammad Abu Khater.

Companion website: www.wiley.com/go/Peroulis/TunableEvanescent-ModeFilters

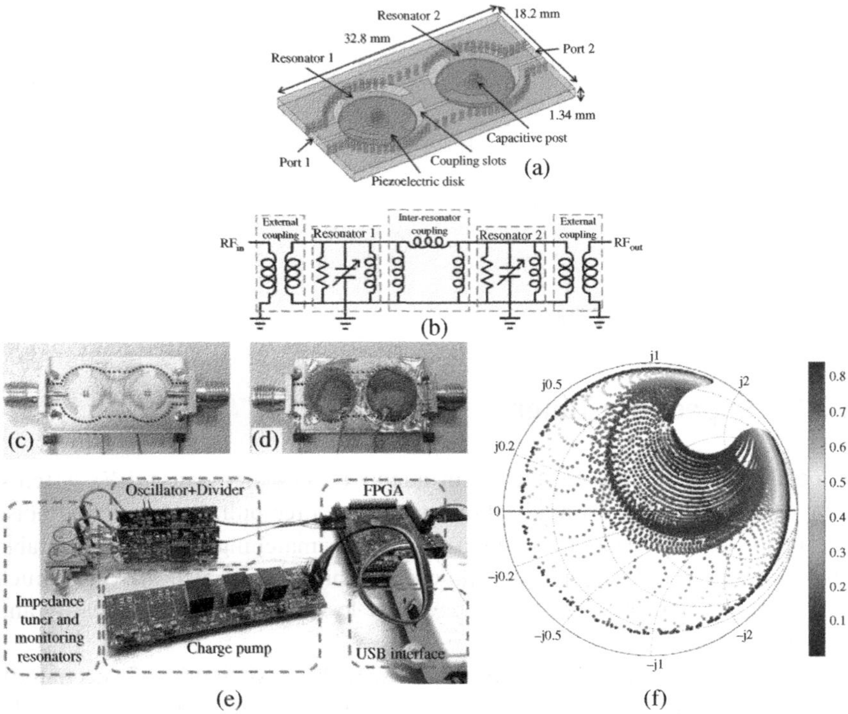


**Figure 10.1** (a) Layout of the designed 3.3-GHz tuner using evanescent-mode resonators. (b) Equivalent circuit. (c) Top view of the fabricated 3.3-GHz impedance tuner showing the resonators and the coupling slots. (d) Bottom view of the tuner with the piezoelectric disks. (e) Entire tuner system including the bias board and the closed-loop control unit. (f) Measured input impedance coverage and power gain (in linear scale) of the fabricated tuner at 3.3 GHz.

bandwidth directly affects the tuner coverage range. Figure 10.1(a) shows the structure of the designed tuner, while Figure 10.1(b) represents the equivalent circuit of the tuner. On the top side, external and inter-resonator coupling slots have been implemented, while the bottom side includes two piezoelectric disks that change the gap size over the resonators' posts.

If the frequency of each resonator changes individually, a wide range of impedances is observed from the tuner input. The operating principle of this tuner can be explained as follows: When the second resonator gap is constant, varying the gap size of the first resonator changes $S_{11}$ on a curve in the Smith chart. Hence, by also changing frequency of the second resonator, wide coverage of the Smith chart can be obtained.

The designed tuner has been fabricated as shown in Figure 10.1(c) and (d). Since piezoelectric disks have displacement hysteresis and may also drift with time/temperature, a closed-loop monitoring and control system, explained in detail in Chapter 12, is also employed to stabilize the tuning. Figure 10.1(e) shows the entire tuner system including the tuner itself, the monitoring resonators, the feedback loop, and the power board.

### 10.1.2 Measured Results

The Smith chart coverage of the fabricated tuner has been experimentally evaluated for 9,840 states (different sets of values of the gaps in the resonators). As Figure 10.1(f) shows, the input impedance covers most of the Smith chart. The smaller measured coverage close to the Smith chart border is due to the minimum gap size achieved in practice, which seems to be larger than the targeted 10-μm gap due to the manual fabrication process. Also, the depicted density of measured points in different regions of the Smith chart is only related to how applied voltages and therefore the gap sizes have been distributed in the matrix laboratory (MATLAB) code. Measurements also show very good stability and reproducibility of impedance points thanks to the presence of the closed-loop control system.

The power gain (ratio of the power dissipated in the load to the power delivered to the input of the tuner) has been calculated using the equation below, assuming that the tuner is terminated to a matched load,

$$G = \frac{|S_{21}|^2}{1 - |S_{11}|^2} \tag{10.1}$$

Figure 10.1(f) also represents (in linear-scale colors) the measured power gain for each covered point on the Smith chart. It is seen that the maximum power gain is 0.84 (−0.77 dB) at 3.3 GHz. Also, the impedance points that achieve a loss of lower than 1 and 1.5 dB are, respectively, 20.2% and 43.5% of the covered Smith chart shown.

Linearity of the tuner is also examined through measurement of the IIP3 using the two-tone test. Linearity depends on the gap size and, consequently, frequency of the resonators. Smaller gap size results in worse linearity due to stronger nonlinear electromechanical effects. Hence, IIP3 measurement is performed at 3 GHz using two tones with 1-MHz separation. With the measured main tone and third-order intermodulation terms of +25.3 and −52.75 dBm, respectively, the IIP3 is calculated to be +64.3 dBm.

Finally, power handling of the tuner is investigated through measuring the tuner's forward transmission scattering parameter ($S_{21}$) when the input power ($P_{in}$) is swept. Due to the presence of a PA in the measurement setup, a response calibration in the presence of the PA and a 30-dB high-power load (before the vector network analyzer [VNA] receiver) is conducted. This measurement is performed under two-port 50-Ω matched condition at 3 GHz. $S_{21}$ remains nearly constant. Another important point is that the input power to the tuner is increased up to about 90 W (maximum available power from the PA) with no failure sign observed, including gas micro-breakdown or damage of piezoelectric disks. The power handling test was repeated ten times, and the results are consistent.

Higher speed (nanosecond time) and higher frequency (mm wave) impedance tuners have been also presented using EVA resonators [126].

## 10.2 Single-ended (SE) and Balanced (BAL) Diplexers

Diplexers play a significant role in frequency-division duplexing (FDD) systems. There are many measures associated with the performance of diplexers, such as size, isolation, loss, integration, and tunability. As a result, designing a diplexer that meets all such measures is often challenging. Conventional architectures consist of two bandpass filters (BPFs) centered at two different frequencies, combined using a distribution network [127–129]. The size and performance overhead caused by the distribution network can be eliminated by using dual-mode diplexing [130–134]. On the other hand, EVA resonators often present a good compromise between volume and performance due to their compact size and high-quality factors.

Tunable diplexers, with various SE and BAL capabilities, with state-of-the-art performance are introduced. Furthermore, a unique dual-mode diplexing concept is discussed that reduces the required volume by half, compared to previous implementations. The presented implementations also allow for wideband tuning (around 50%) while maintaining low insertion loss and high isolation. These concepts are experimentally demonstrated by designing SE–SE, SE–BAL, and BAL–BAL diplexers. The main difference between the three diplexers is in the external coupling structure, which provides the SE and BAL ports. The dual-mode operation results in a dependent tuning between the two ports. The average measured insertion loss is 1.32 dB for the SE–SE, 1.95 dB for the SE–BAL, and 2.15 dB for the BAL–BAL. The diplexing structures for the three diplexers are approximately the same size. For the proposed SE–BAL and BAL–BAL diplexers, the in-band common-mode rejection (CMR) is better than 40 dB throughout the tuning range.

### 10.2.1 Dual-Mode Diplexing Concept

#### 10.2.1.1 Diplexing Architecture

The dual-mode diplexing concept is shown in Figure 10.2(a). Depending on the frequency of the incident wave at port 1, two modes can be excited in the common dual-mode resonator (resonator A). Mode II is coupled to resonator B (creating

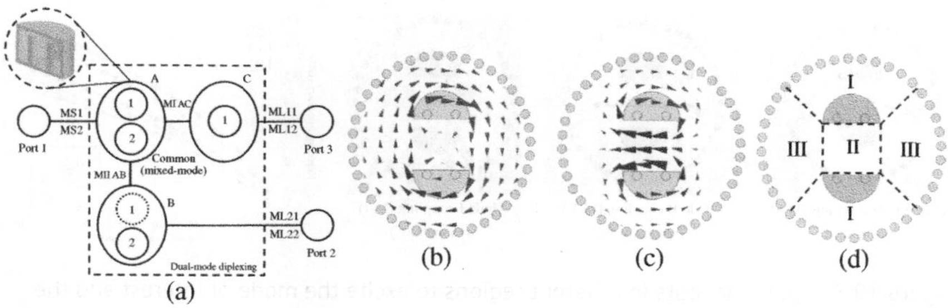


**Figure 10.2** (a) The architecture of dual-mode diplexer. The two channels are created by exciting two modes in a common dual-mode resonator. A cross section of a dual-mode split-post evanescent-mode resonator is shown in the inset figure. (b) Magnetic field distribution of the resonators in Mode I and (c) Mode II. (d) Resonator region classification based on magnetic field distribution of the modes.

high channel on port 2), while Mode I is coupled to resonator C (creating low channel on port 3). While resonator B can be designed to support only Mode II, it is designed as a dual-mode resonator to match the $Q$ in all resonators of the mode of interest. On the other hand, resonator C is not necessarily dual-mode because it has to support only Mode I. The resonators used here are split-post loaded substrate integrated waveguide (SIW) evanescent-mode cavities, shown in Figure 10.2(a) and discussed in Section 2.5.1. The magnetic field distributions of both modes are shown in Figure 10.2(b) and (c). Mode I, shown in Figure 10.2(b), can be excited from anywhere in the cavity except the area between the posts (region II in Figure 10.2(d)). On the other hand, Mode II, shown in Figure 10.2(c), can be excited only in between the posts or behind them (regions I and II in Figure 10.2(d)). The magnetic field distribution and the resonant frequency of each mode are affected differently by introducing coupling slots in the corresponding regions shown in Figure 10.2(d).

#### 10.2.1.2 Resonant Frequency Misalignment

Introducing a slot in different regions in Figure 10.2(d), shown in Figure 10.3, results in different resonant frequencies and different mode frequency ratio (MFR): $f_{modeI}/f_{modeII}$. Having different MFRs for each resonator gives the advantage of detuning the undesired mode in the high or low channels, which provides the required diplexing.

Figure 10.3 also shows the resonant frequency dependence of the two modes in each resonator on the capacitance gap (g). As illustrated in the figure, when the frequencies of Mode II (the desired mode) in resonators A and B are aligned, their undesired Mode I frequencies are misaligned. Likewise, the Mode I frequencies of resonators A and C are aligned, and the Mode II frequencies are misaligned. The frequency misalignment of the undesired modes provides some degree of isolation

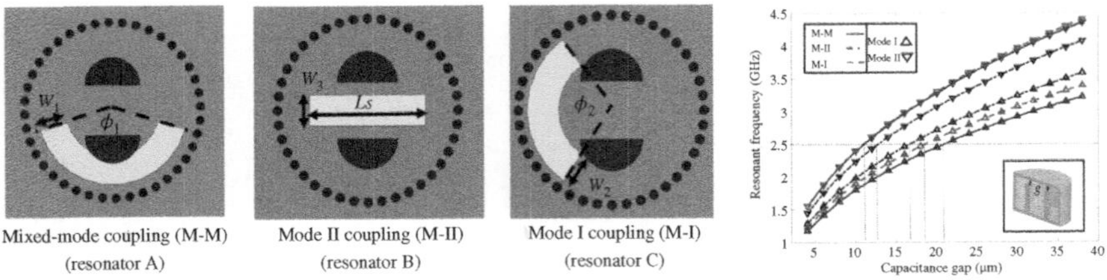


**Figure 10.3** Coupling slots in different regions to excite the mode of interest and the resonant frequency of the modes versus capacitance gap ($g$).

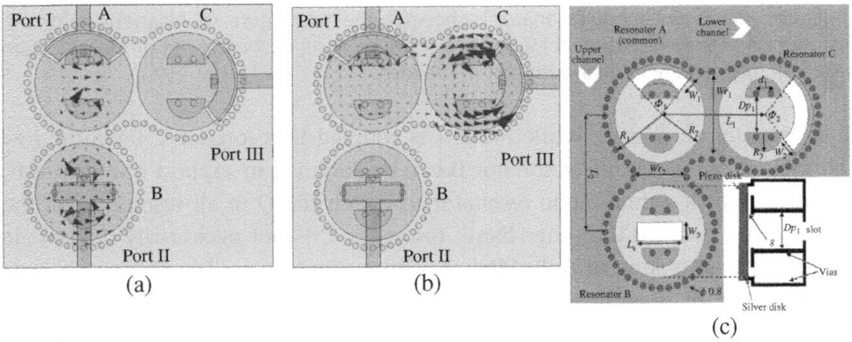


**Figure 10.4** Magnetic field distribution in the diplexer when (a) Mode II resonant frequency of resonator A is aligned with Mode II of resonator B, and (b) Mode I of resonator A is aligned with of Mode I of resonator C. (c) Structure of the proposed diplexing schematic, along with a cross section of resonator B.

between the output ports, as does the appropriate choice of external coupling, which will be discussed in the next section. This mechanism is illustrated further in Figure 10.4 that shows the magnetic field distribution in the diplexer when resonator A is aligned to Mode II resonant frequency of resonator B and to Mode I resonant frequency of resonator C.

The schematic of the proposed diplexer is shown in Figure 10.4(c). In order to design a dual-mode resonator (resonant frequency of Mode I and the MFR), we start by considering its single mode operation first. Then, adjusting $Dp_1$ primarily influences the resonant frequency of Mode II, while having a minor impact on the resonant frequency of Mode I. The external coupling is subsequently designed as shown in the following sections. Since the external coupling slots affect the resonant frequency, the dimensions $d_1$ and $Dp_1$ are adjusted to achieve the desired MFR and resonant frequency. Increasing $Dp_1$ decreases the MFR with a slight change in the resonant frequency. However, the MFR starts to increase as the post-vias becomes close to the cavity wall vias.

#### 10.2.1.3 Inter-Resonator Coupling

The inter-resonator coupling is calculated when each mode is aligned in frequency for the different resonators of the coupler. The inter-resonator coupling also plays a role in suppressing Mode II in the lower channel. This is due to the fact that Mode II is nearly zero in the direction of this coupling (Figure 10.2). Consequently, the coupling of Mode II in the lower channel is practically zero. As a result, making resonator C a single mode or dual mode does not further improve the isolation as shown in the following sections.

### 10.2.2 SE–SE Diplexer Implementation and Measurements

In order to achieve a SE or BAL port, the external coupling structure of a diplexer has to be designed accordingly. As a result, the external coupling structure for each one of the proposed diplexers is discussed first, followed by implementation details and measurements. Each channel of the diplexers is synthesized as a second-order BPF using the coupling matrix method. The employed coupling coefficients are based on a Chebyshev filter response (for 0.2-dB passband ripple). Then, the diplexers are simulated and optimized using simulations.

#### 10.2.2.1 External Coupling

The external coupling of resonator A couples into both modes, while resonators B and C primarily couple into Mode II and Mode I, respectively. This further improves isolation between the channels. The leakage in resonator B from Mode I results in a finite amount of coupling. However, Mode I in resonator A and Mode I in resonator B occur at different frequencies. As a result, the leaked coupling shows no significant effect on isolation. At each port, the external coupling aperture excites a single shorted transmission line (to maximize the magnetic field coupling), providing an SE operation. This structure is shown in Figure 10.5(a).

#### 10.2.2.2 Measured Results

The SE–SE diplexer is fabricated using multilayer printed circuit board (PCB) process, and the final product is shown in Figure 10.5(b) and (c). The total size of the diplexer is $0.45\lambda_g \times 0.48\lambda_g$ ($40 \times 43\ \mathrm{mm}^2$) ($\lambda_g$ at 2.1 GHz). The bottom of each resonator is covered with a flexible 1-mil-thick silver disk attached to a piezoelectric actuator using silver epoxy.

Figure 10.6 presents the measured and simulated S-parameters of the proposed SE–SE diplexer at the beginning, middle, and end of the tuning range. The lower channel can be tuned from 2.07 to 3 GHz, and the 3-dB fractional bandwidth changes from 4.35% to 5.9% within the tuning range. The upper channel is tuned (with the lower channel) from 2.71 to 3.9 GHz with a 3-dB fractional bandwidth changing from 2.5% to 3.6%.

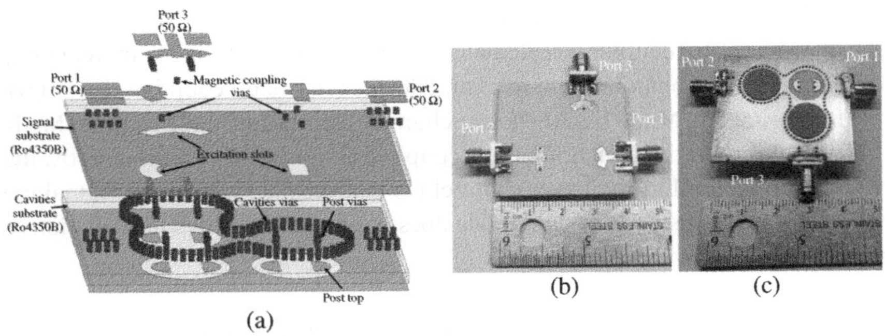


**Figure 10.5** (a) An exploded view of the model of the SE–SE diplexer. (b) Top and (c) bottom photograph of the SE–SE diplexer. (c) The piezoelectric tuner is removed from the common resonator to show the split-post structure.

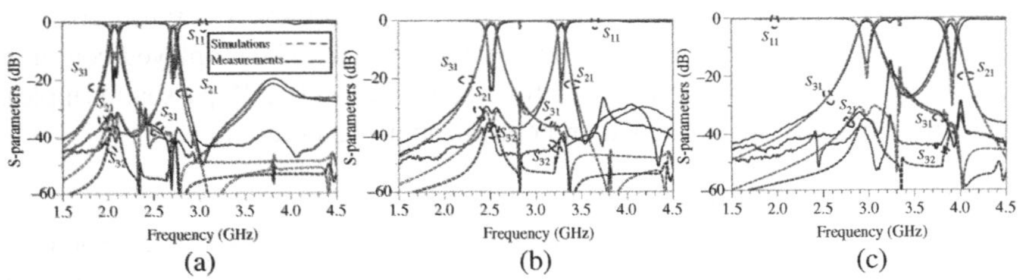


**Figure 10.6** Simulated and measured S-parameters of SE–SE diplexer.

### 10.2.3 SE–BAL Diplexer Implementation and Measurements

#### 10.2.3.1 External Coupling

The external coupling of the common resonator A is similar to that of the SE–SE ones.

For resonators B (M-II) and C (M-I), the external coupling is a transmission line, with the coupling aperture below its middle point (symmetry plane) as shown in Figure 10.7 [135]. An external excitation to the BAL port can be analyzed as a combination of differential- and common-mode excitations. A differential-mode excitation creates maximum current at the coupling point due to the electric wall at the symmetry plane. This maximizes the magnetic field coupling to the cavity. A common-mode excitation results in zero current due to the magnetic wall at the symmetry plane. This suppresses the coupling of the common mode into the resonator. These concepts are demonstrated in Figure 10.7.

Resonator C is designed with a single post (single mode) to increase the magnetic field at the coupling location, providing sufficient BAL coupling. This also eliminates the second mode in the resonator. Using dual-mode cavity does not provide sufficient BAL coupling because region III (in Figure 10.2) would have

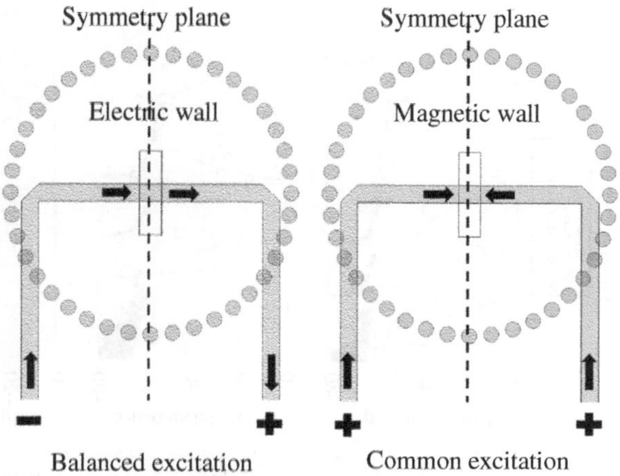


**Figure 10.7** Schematic of the balanced external coupling structure showing that balanced excitation creates an electric wall in the coupling slot, which maximizes the magnetic field coupling, while common excitation creates magnetic wall, which suppresses the coupling of the magnetic field.

weaker magnetic field compared to region I, and small slots are used for balanced excitation to decrease CMR.

#### 10.2.3.2 Measured Results

The SE–BAL diplexer is fabricated using a similar process as described in the SE–SE example. The total size of the diplexer is $0.7\lambda_g \times 0.6\lambda_g$ ($60 \times 50\,\text{mm}^2$) ($\lambda_g$ at 2.2 GHz). While the diplexing structure is almost the same size compared to the SE–SE diplexer, the external coupling is adding the extra size. A photograph of the implemented diplexer is shown in Figure 10.8. The piezoelectric tuners are identical to the ones used in the SE–SE diplexer. Each channel is measured separately by terminating the other channel in a matched load. First, with matching the upper channel (port 2), Figure 10.9(a) presents the measured port 3 (lower channel)

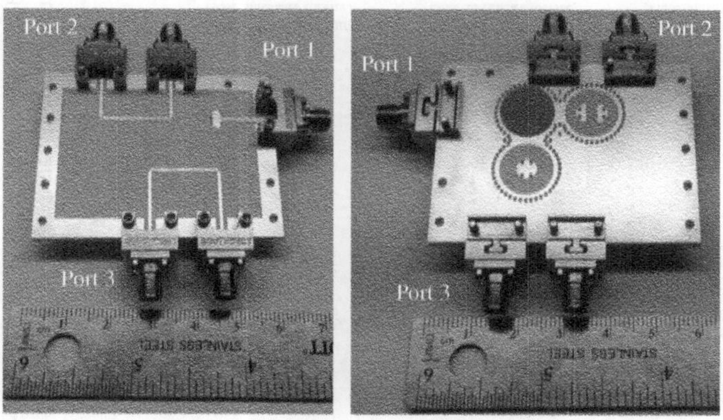


(a) (b)

**Figure 10.8** Photograph of the SE–BAL diplexer: (a) top view and (b) bottom view.

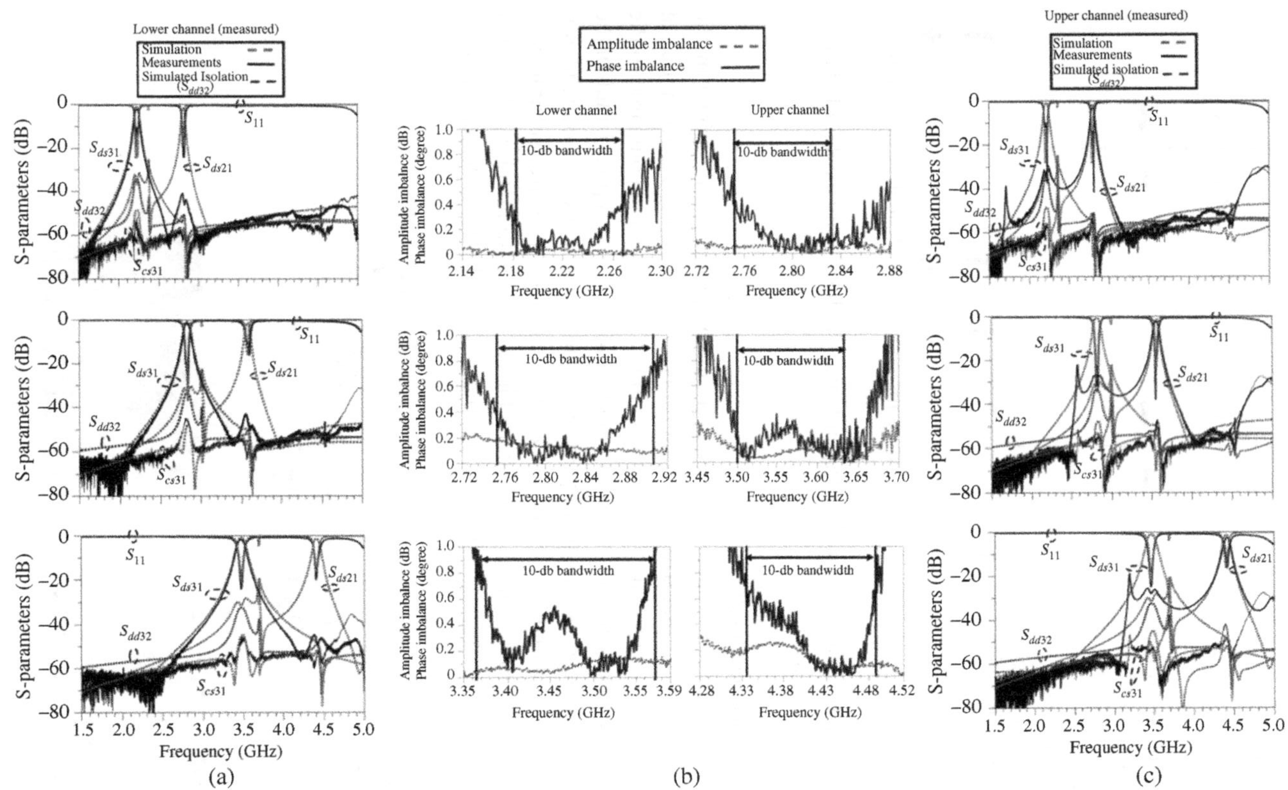


**Figure 10.9** Simulated and measured S-parameters of SE–BAL diplexer channels: (a) Lower channel is measured, and upper channel is terminated with 100 Ω differential load. (c) Upper channel is measured and lower channel is terminated with 100 Ω differential load. (b) The amplitude and phase imbalance within the 10-dB bandwidth associated with the tuning states shown in (a)(c).

S-parameters and the simulated S-parameters of the proposed SE–BAL diplexer at the beginning, middle, and end of the tuning range. The lower channel can be tuned from 2.2 to 3.46 GHz, and the 3-dB fractional bandwidth changes from 2.4% to 3% across the tuning range. Second, with matching the lower channel (port 3), Figure 10.9(c) presents the measured port 2 (upper channel) S-parameters and the simulated S-parameters of the diplexer at the same tuning states. The upper channel is tuned (with the lower channel) from 2.8 to 4.4 GHz with its 3-dB fractional bandwidth changing from 1.6% to 1.8%. Figure 10.9(b) shows the amplitude and phase imbalance within the 10-dB bandwidth associated with the tuning states shown in Figure 10.9(a) and (c). In all cases, the amplitude imbalance is less than 0.5 dB, and the phase imbalance is less than 0.7° for both channels. In all measured states, the maximum observed CMR is better than 44 dB for both channels.

Throughout the tuning range, the insertion loss varies between 2.2 dB at 2.2 GHz to 1.7 dB at 3.4 GHz for the lower channel and from 2.1 dB at 2.8 GHz to 2.7 dB at 4.4 GHz for the upper channel. The maximum leakage measured from port 1 to the upper channel ($S_{21}$) within the passband of the lower channel is 28 dB, while the maximum observed leakage to the lower channel ($S_{31}$) within the passband of the upper channel is 42 dB. To measure the isolation, four ports are needed for characterizing the two BAL ports, and a fifth port is needed to align the common resonator (port 1). Due to the four-port limitation in the measurement setup, only the simulated isolation is shown in Figure 10.9. The isolation is expected to be lower than the leakage because the signal path for isolation ($S_{32}$ or $S_{23}$) passes through three resonators, whereas the leakage path involves only two resonators. This is supported in the simulation results in Figure 10.9. As a result, the leakage values are considered as an upper bound for the isolation.

### 10.2.4 BAL–BAL Diplexer Implementation and Measurements

#### 10.2.4.1 External Coupling

The external couplings in the BAL–BAL diplexer for resonators B and C are identical to the external couplings in the SE–BAL case. The common resonator (resonator A) is differentially excited at region I (from Figure 10.2) such that it excites both modes.

#### 10.2.4.2 Measured Results

The BAL–BAL diplexer is fabricated using a multilayer PCB process. The total size of the diplexer is $0.5\lambda_g \times 0.44\lambda_g$ ($60 \times 50\,\text{mm}^2$) ($\lambda_g$ at 2 GHz). Also, the diplexing structure does not exceed in size the one in SE–BAL example. A photograph of the implemented diplexer is shown in Figure 10.10. A 60-mil-thick RO4350B substrate is employed as a cavity substrate, and a 10-mil-thick RO5880 substrate is employed as a signal substrate. The tuning is done using the same aforementioned manner.

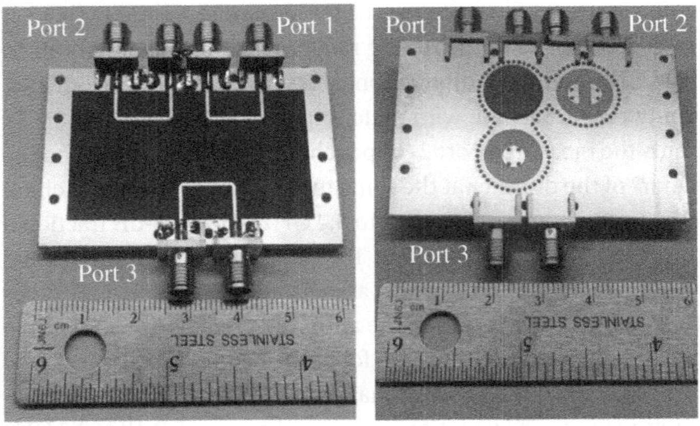


(a) (b)

**Figure 10.10** Photograph of the proposed BAL–BAL diplexer: (a) top view and (b) bottom view.

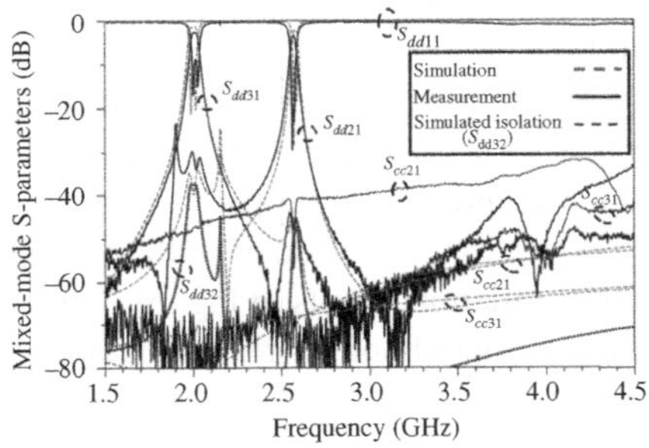


(a)

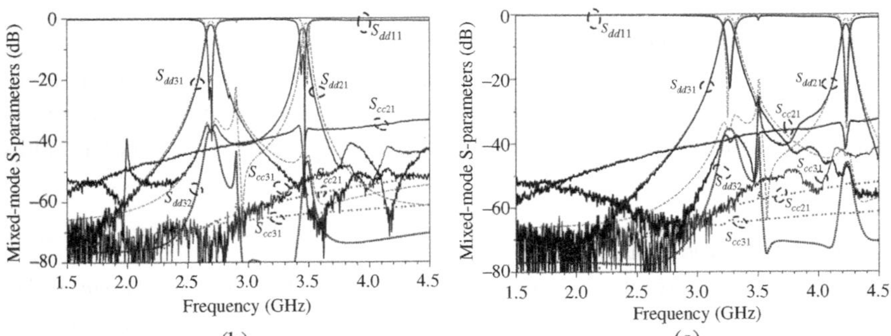


(b) (c)

**Figure 10.11** Simulated and measured S-parameters of BAL–BAL diplexer.

As in the SE–BAL example, each channel is measured separately by matching the other channel. However, here, for the sake of compactness, upper channel and lower channel results are plotted in one figure. Figure 10.11 presents the measured and simulated S-parameters of the BAL–BAL diplexer at the beginning, middle, and end of the tuning range. The lower channel can be tuned from 2 to 3.27 GHz, and the 3-dB fractional bandwidth changes from 2.3% to 2.8% within the tuning range. The upper channel is tuned (with the lower channel) from 2.56 to 4.23 GHz with a 3-dB fractional bandwidth changing from 1.2% to 1.4%. In all measured states, the maximum observed CMR is less than 35 dB for both channels. Throughout the tuning range, the insertion loss varies between 2.5 to 1.8 dB for the lower channel and from 3 to 2.4 dB for the upper channel. The same difficulty in measuring the isolation exists as mentioned above. However, based on the simulated isolation and the leakage between channels, the measured channel-to-channel isolation should be better than 35 dB, throughout the tuning range.

## 10.3 Tunable Filtering Rat-race Couplers Based on Half- and Full-mode Evanescent-mode Resonators

The rat-race coupler is a four-port antisymmetric RF network that can operate as an in-phase or out-of-phase power divider and power combiner by properly choosing the excitation ports [13]. As such, rat-race couplers are used in numerous RF front-end components such as PAs, BAL mixers, and antenna array feeding networks [13, 39, 136, 137].

In this section, tunable, filtering rat-race couplers based on full-mode and half-mode EVA are discussed. Half-mode resonators are discussed in Section 2.5.2. The design methodology relies on mapping the design of a standard BPF into the design of the coupler, achieving the desired filtering response. The EVA choice results in a tunable filtering coupler, while maintaining a wide spurious-free range. Compared to conventional planar rat-race couplers, the proposed full-mode and half-mode rat-race couplers achieve a 42% and 92% reduction in structure footprint, respectively. Throughout the tuning range, the couplers show an equal power splitting ratio and filtering profile with a measured insertion loss between 4.2–3.7 dB and 5–4.2 dB, respectively. Both couplers have measured isolation better than 30 dB. The half-mode rat-race coupler measurements show a very wide spurious-free range, which is more than $5.45 \times f_0$.

### 10.3.1 Design

For a desired filtering response, the rat-race coupler is synthesized as a second-order BPF, shown in Figure 10.12(b). Then, the coupling coefficients are mapped to the rat-race topology shown in Figure 10.12(c).

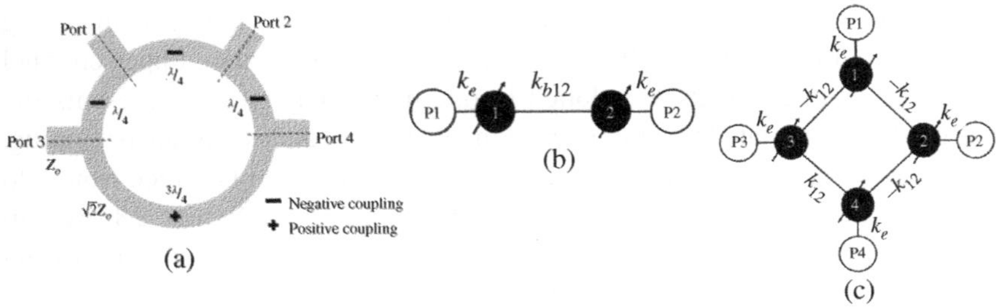


**Figure 10.12** (a) Conventional planar rat-race coupler. (b) Topology of tunable second-order bandpass filter. (c) Proposed tunable filtering rat-race coupler topology.

A conventional planar rat-race coupler is shown in Figure 10.12(a). The characteristic impedances of the transmission lines are obtained using the even–odd mode analysis [13]. For a rat-race coupler with $Z_0$ terminal impedance, all connecting transmission lines have $\sqrt{2}Z_0$ characteristic impedance. Any two adjacent ports are connected with a 90° transmission line, except ports 3 and 4. These ports are connected with a 270° transmission line.

Figure 10.13 shows a lumped-element equivalent of a coupled-resonator rat-race coupler. The resonators are coupled magnetically. The inter-resonator coupling value $k_{inter}$ can be obtained by modeling the magnetic coupling as admittance inverters and following a procedure similar to the one in [138].

A 90° transmission line with $Z_0$ characteristic impedance can be replaced by

$$k_{12} = \frac{\omega L}{Z_0} \tag{10.2}$$

As a result, the inter-resonator couplings are mapped accordingly.

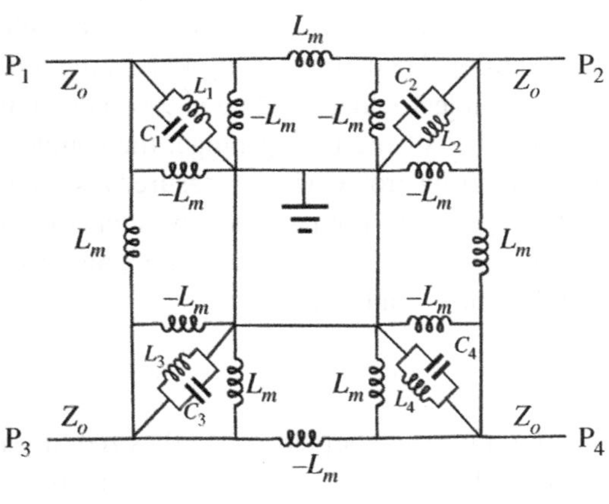


**Figure 10.13** Lumped-element equivalent of a coupled-resonator rat-race coupler.

### 10.3.2 Full-mode Rat-race Coupler

The 3D structure of the tunable rat-race coupler is shown in Figure 10.14. Negative inter-resonator coupling is realized using a conventional iris structure between the resonators. Positive coupling is realized using a line of vias connecting the top and bottom at the coupling region accompanied with coupling slots. This allows the current to transfer from the top ceiling of one of the cavities to the bottom of the other cavity, thus flipping the sign of the magnetic coupling [107]. The physical dimensions of the proposed coupler are shown in Figure 10.14(b). Figure 10.14(c) shows the piezoelectric actuator assembly of the cavity. Shown in Figure 10.14(d) is the magnetic field distribution in the proposed rat-race coupler when excited from port 4. As shown, the magnetic fields of the resonators attached to ports 2 and 3 are 180° out of phase. This eliminates coupling to port 1 (isolated port).

The proposed rat-race coupler is synthesized with a second-order BPF response. At the low end of the tuning range, the employed coupling coefficients are based on a Chebyshev filter response (with a 0.2-dB passband ripple).

#### 10.3.2.1 Experimental Validation

The rat-race coupler is fabricated using a commercial milling and plating system. A 60-mil-thick RO4003C substrate is employed. The total size of the coupler is $0.38\lambda_g \times 0.38\lambda_g$ ($54 \times 54\ \text{mm}^2$) ($\lambda_g$ is calculated at 1.3 GHz). A photograph of the implemented coupler is shown in Figure 10.15(a).

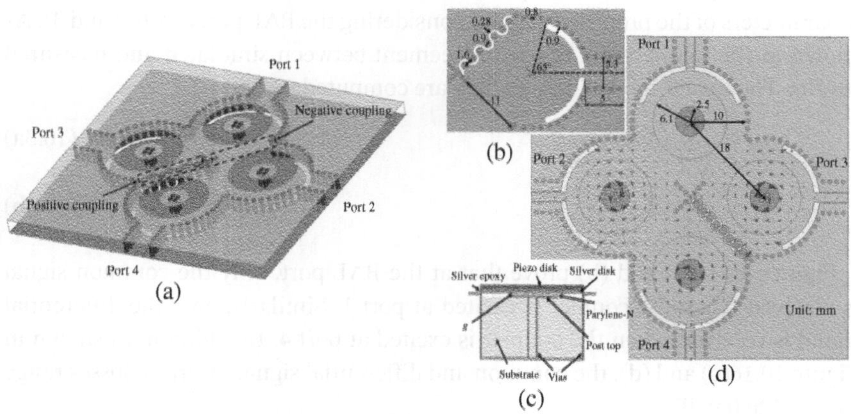


**Figure 10.14** (a) 3D structure of tunable filtering evanescent-mode rat-race coupler (tuning elements are not shown). (b) Bottom view of one of the resonators showing the coupler's physical dimensions. (c) Side view of one of the cavities showing the tuning element (piezoelectric disk) assembly on top of the cavity. (d) Bottom view of the rat-race coupler showing the magnetic field distribution in the four resonators when port 4 is excited.

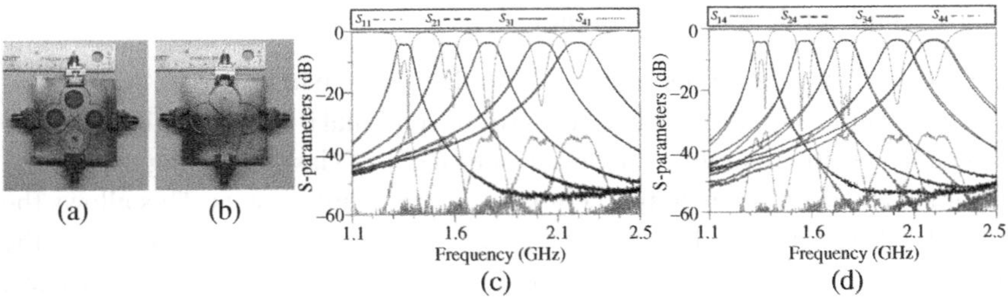


**Figure 10.15** Photograph of the rat-race coupler: (a) top view and (b) bottom view. Measured S-parameters of the fabricated rat-race coupler when excited from (c) port 1 and (d) port 4.

Figures 10.15(c) and (d) present the measured S-parameters of the proposed rat-race coupler at five different states within the designed tuning range. For clarity, only measured results are shown. Figure 10.15(c) shows the measured S-parameters when the coupler is excited at port 1, while Figure 10.15(d) shows the measured S-parameters when the coupler is excited at port 4. As shown, the power is split equally between ports 2 and 3. The coupler is tunable from 1.3 to 2.21 GHz (70% tuning range). Throughout the tuning range, the measured insertion loss changes from 4.2 to 3.7 dB, the 3-dB fractional bandwidth changes from 5.5% to 6.46%, and the isolation is better than 30 dB.

In order to show the amplitude and phase balance at ports 2 and 3 when the coupler is excited at ports 1 and 4, both ports 2 and 3 are considered as BAL (differential) port terminals. Figure 10.16 presents the measured and simulated S-parameters of the proposed coupler considering the BAL port (ports 2 and 3). As shown in the figure, there is good agreement between simulated and measured results. The mixed-mode S-parameters are computed using

$$S_{ds1} = \frac{1}{\sqrt{2}}\left(S_{21} - S_{31}\right) \tag{10.3a}$$

$$S_{cs1} = \frac{1}{\sqrt{2}}\left(S_{21} + S_{31}\right) \tag{10.3b}$$

Figures 10.16(a) and (b) prove that at the BAL port, only the common signal is received when the coupler is excited at port 1. Similarly, only the differential signal is received when the coupler is excited at port 4. In addition, as shown in Figure 10.16(c) and (d), the common and differential signal insertion losses range from 1.4 to 0.9 dB.

### 10.3.3 Half-mode Rat-race Coupler

#### 10.3.3.1 Half-mode Structure

Half-mode EVA resonators are briefly discussed and shown in Figure 2.24.

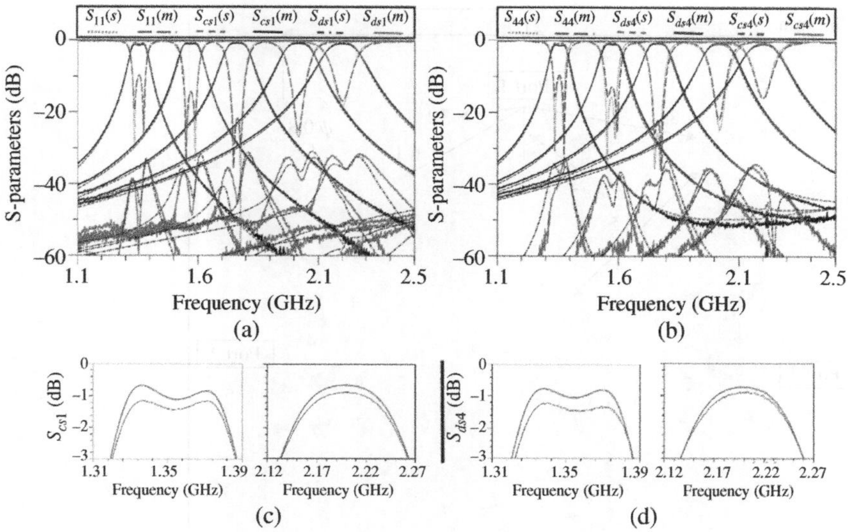


**Figure 10.16** (a) Measured (*m*) and simulated (*s*) S-parameters of the evanescent-mode rat-race coupler. Port 1 is excited, and both ports 2 and 3 are considered as common port terminals. (b) Measured (*m*) and simulated (*s*) S-parameters of the proposed rat-race coupler. Port 4 is excited, and both ports 2 and port 3 are considered as differential port terminals. (c) Detail of the passband insertion loss (from (a)) at the low and high ends of the tuning range. (d) Detail of the passband insertion loss (from (b)) at the low and high ends of the tuning range.

Aside from the use of half-mode resonators, the overall structure of the rat-race coupler is similar to that of the full mode. For example, through exciting the rat-race coupler from port 4, the magnetic field distribution of the resonators is shown in Figure 10.17. As shown, the magnetic fields in the port 2 and 3 resonators are out of phase, which eliminates coupling to port 1 (isolated port). The physical dimensions of the proposed coupler are shown in the figure.

#### 10.3.3.2 Experimental Validation

An exploded view of the rat-race coupler is shown in Figure 10.18(a). The final prototype is fabricated through four main steps. First, a PCB is fabricated using a commercial milling and plating system. A 100-mil-thick Roger TMM3 substrate is employed. The top of each post is coated with a 4-μm-thick parylene-N layer to prevent shorting with the copper ceiling. Second, the top of the cavity is covered by a 1-mil-thick Roger ULTRALAM 3850HT, which is a liquid crystal polymer (LCP) coated with 18-μm-thick copper layers on both sides. The top copper is etched away to cover the cavity ceiling from the other side without shorting the positive inter-resonator coupling structure. It is attached to the top of the PCB using silver epoxy. Third, a piezoelectric actuator is attached on the LCP layer above

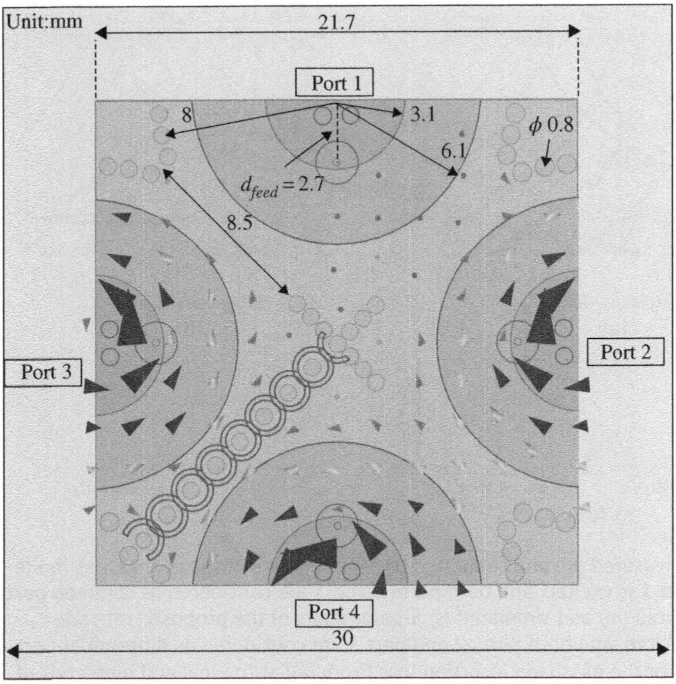


**Figure 10.17** Bottom view of the half-mode evanescent-mode rat-race coupler showing the magnetic field distribution in the four resonators when port 4 is excited. The physical dimensions of the coupler are shown in the figure.

each half-mode resonator using epoxy. Finally, the ports are attached to the bottom using four connectors. A photograph of the implemented coupler is shown in Figure 10.18(b).

The measured four-port S-parameters of the proposed half-mode rat-race coupler are shown in Figure 10.19 at five successive states. Figure 10.19(a) shows the measured S-parameters when the coupler is excited at port 4, while Figure 10.19(b) shows the measured S-parameters when the coupler is excited at port 1. As shown, the coupler is designed to have an equal power dividing ratio between ports 2 and 3, and it is tunable from 0.885 to 1.65 GHz (86% tuning range). Throughout the tuning range, the measured insertion loss and the 3-dB fractional bandwidth range from 5–4.2 dB and 6.8–8%, respectively. The isolation is better than 30 dB. The absolute phase error remains within 6°.

In order to clearly compare the measured and simulated S-parameters, both ports 2 and 3 are considered as terminals of a BAL port. Figure 10.20(a) shows the measured and simulated S-parameters at the BAL port when the coupler is excited from port 4. As expected, the common signal is rejected and the differential signal

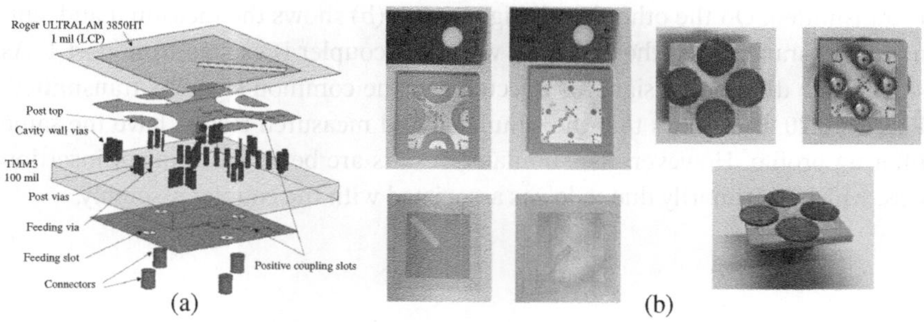


**Figure 10.18** (a) An exploded view of the model of the rat-race coupler. (b) Photograph at different stages of the assembly processes of the rat-race coupler.

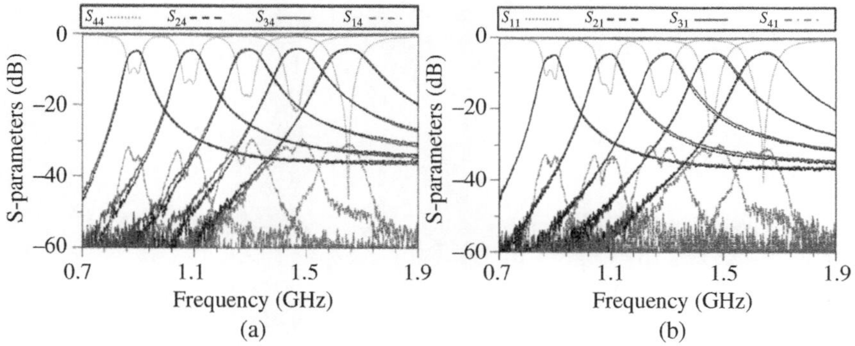


**Figure 10.19** Measured S-parameters of the fabricated HM-EVA-B rat-race coupler when excited at (a) port 4 and (b) port 1.

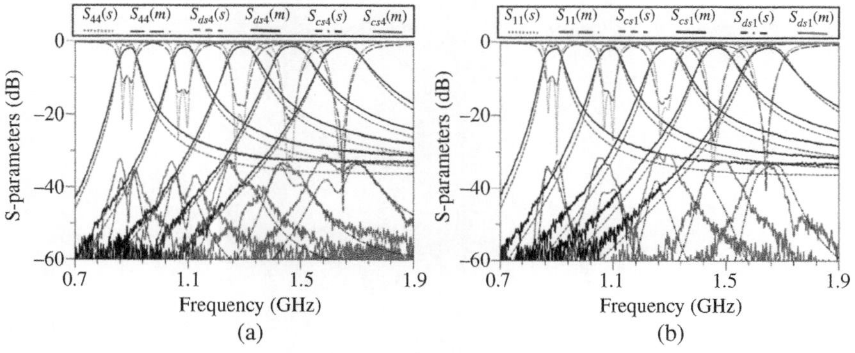


**Figure 10.20** (a) Measured (*m*) and simulated (*s*) S-parameters of the proposed HM-EVA-B rat-race coupler. Port 4 is excited, and both ports 2 and port 3 are considered as differential port terminals. (b) Measured (*m*) and simulated (*s*) S-parameters of the proposed rat-race coupler. Port 1 is excited, and both ports 2 and 3 are considered as common port terminals.

is transmitted. On the other hand, Figure 10.20(b) shows the measured and simulated S-parameters at the BAL port when the coupler is excited from port 1. As shown, the differential signal is rejected and the common signal is transmitted. Figure 10.20 also shows that the simulated and measured results have the same filtering profile. However, the simulated results are better in terms of insertion loss, which is primarily due to losses associated with the coupler assembly.

# 11

# Active Applications

## 11.1 Co-Design of Power Amplifiers and High-$Q$ Filters

High power amplifiers (PAs) are typically followed by post-selection bandpass filters in modern wireless transmitter front ends. Moreover, the recently developed digital (burst-mode) transmitters require a low-loss bandpass filter at the PA output to reconstruct the amplitude information from pulse-width or delta-sigma modulation while maintaining high efficiency and linearity [139–141].

Conventionally, the PA and filter are designed independently based on 50-Ω system impedance, as illustrated in Figure 11.1(a). Nevertheless, given the non-negligible insertion loss of a narrowband filter due to the limited quality factor, the overall performance of a PA-filter module may be degraded significantly. For example, a 1-dB insertion loss results in a drop of PA efficiency from 80% to approximately 60%. Furthermore, additional loss may be introduced from the interconnecting line and mismatch between the PA and filter. To minimize the loss term, the PA output matching network can be entirely eliminated by designing the filter input impedance to directly match the transistor output, which usually needs a non-50-Ω complex impedance, as shown in Figure 11.1(b). This co-design approach leads to a reduced circuit complexity, smaller size/volume, minimized loss, and enhanced overall efficiency. Specifically, it offers an effective solution for applications where small size is a top priority, such as pico and femto base stations [142–144].

It is also important to note that filter bandwidths need to fit those of the signals, which can be below 1%. Therefore, resonators with high quality factors are of great importance for such output filters to ensure low insertion loss and outstanding overall system performance.

In this chapter, a co-design approach is expanded to multi-pole filters. The synthesis of a second-order Chebyshev bandpass filter with non-50-Ω input impedance is utilized, as discussed in Section 1.2.3.5. We also present the detailed

*Tunable Evanescent-Mode Filters: Principles, Implementation, and Applications*,
First Edition. Dimitrios Peroulis and Mohammad Abu Khater.

Companion website: www.wiley.com/go/Peroulis/TunableEvanescent-ModeFilters

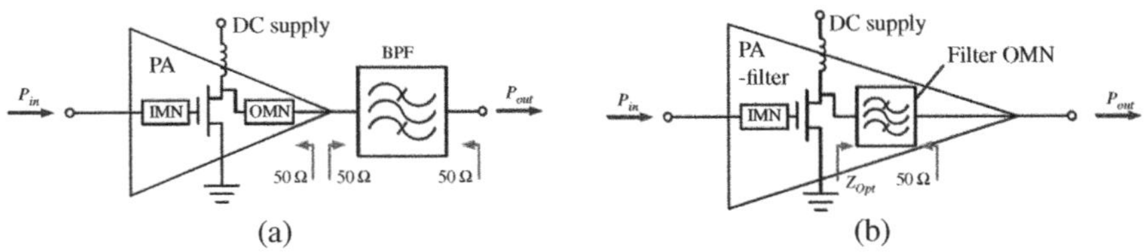


**Figure 11.1** Power amplifier and bandpass filter in cascade: (a) conventional topology and (b) co-designed module. IMN: Input Matching Network, OMN: Output Matching Network.

procedure for realizing the PA-filter module, including filter-response shaping and harmonic matching to enhance efficiency. The implemented circuit exhibits a Chybeshev filter response with efficient measured performance throughout the passband. Moreover, a stand-alone PA and filter are developed and cascaded as a control structure for performance comparison. It is seen that the overall efficiency of the conventional PA-filter cascade is highly sensitive to the interconnection line length. This practical issue can be completely avoided by using the co-design approach.

### 11.1.1 Filter Design

From the coupling matrix theory, the normalized input impedance of a second-order filter, regardless of its shape, is given by

$$Z_{in} = \frac{M_{12}^2 + jM_{11}M_{2L}^2}{M_{S1}^2 M_{2L}^2} \tag{11.1}$$

While $M_{11}$ typically equals zero, complex sources (PA transistor) require detuning the first resonator, which in turn requires changing the input external coupling. As a result,

$$Z_{in\ Complex} = R + jX = \frac{M_{12}^2 + jM'_{11}M_{2L}^2}{M_{S1}'^2 M_{2L}^2} \tag{11.2}$$

To find the new coupling values, the following relationships apply

$$M'_{S1} = \frac{M_{S1}}{\sqrt{R}}, \quad \text{and} \quad M'_{11} = \frac{X}{R}M_{S1}^2 \tag{11.3}$$

For example, the filter can be designed to have the second-order 15-dB-equiripple Chebyshev response with the normalized input impedance of 0.5 + j0.5 within the passband by setting $M'_{S1}$ = 1.467 (instead of 1.037) and $M'_{11}$ = 1.075 (instead of 0).

### 11.1.2 Transistor Characterization

To experimentally demonstrate the co-design concept of a PA and filter, a GaN transistor (Cree CGH40010F) is selected as the power device of this circuit module. The transistor is characterized using load-pull simulations. The multi-harmonic load-pull simulation results are presented in Figure 11.2. Figure 11.2(a) shows the optimal fundamental impedances at 3 GHz for achieving maximum power added efficiency (PAE) and maximum output power, respectively, while the contours of PAE and $P_{out}$ are also plotted. For PA design at such high frequencies, optimal impedance matching of higher-order harmonics is very important to ensure high PA efficiency. Figure 11.2(b) and (c) show the PAE contours of the second and third harmonics, indicating the regions in which the filter's harmonic impedances need to be located to ensure a high PAE. The load-pull simulation provides the reference impedance for designing the filter as the output matching network of the PA.

### 11.1.3 Matching Filter Design

Based on the filter synthesis theory presented earlier, a 3%-bandwidth, 15-dB equi-ripple, second-order Chebyshev bandpass filter is designed and realized with evanescent-mode resonators, as shown in Figure 11.3(a). Figure 11.3(b) illustrates the cross-section view of a single EVA resonator and the internal electromagnetic field distribution. The final geometrical dimensions of the resonator are also listed in Figure 11.3.

In conventional PA design, the impedance of the output matching network is normally set to the point that leads to either maximum PAE ($Z_{opt,PAE}$) or maximum power ($Z_{opt,Power}$). It can be seen from Figure 11.2 that ($Z_{opt,PAE}$) and ($Z_{opt,Power}$) share the same real part ($\approx$11 Ω in this design), while their imaginary parts are both very small (Im($Z_{opt,PAE} \approx 6$ Ω), Im($Z_{opt,Power} \approx 0$ Ω)). Therefore, the filter's

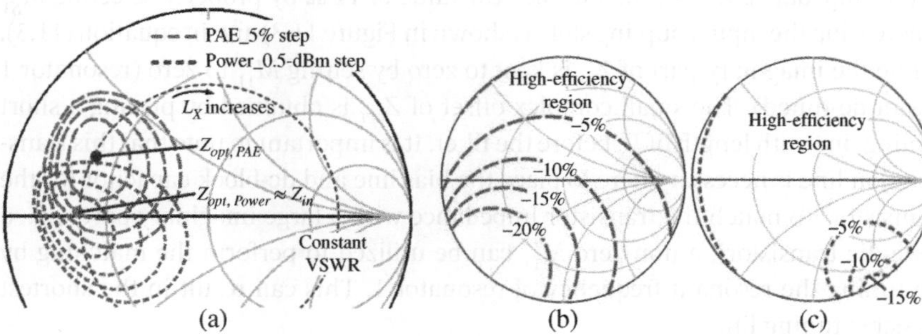


**Figure 11.2** Load-pull simulation results of the CGH40010 GaN transistor at 3.0 GHz: (a) fundamental impedance, (b) second-harmonic impedance, and (c) third-harmonic impedance.

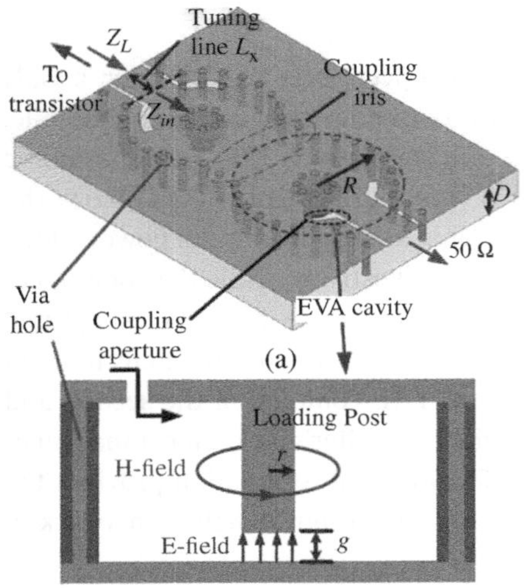


Evanescent-mode cavity parameters

| Parameter | Value |
|---|---|
| Cavity radius $R$ | 7.2 mm |
| Cavity depth $D$ | 3.15 mm |
| Post radius $r$ | 1.2 mm |
| RF gap $g$ | 20 μm |
| Simulated $Q_u$ | 680 |
| Equivalent inductance | 1.4 nH |
| Equivalent capacitance | 2.0 pF |

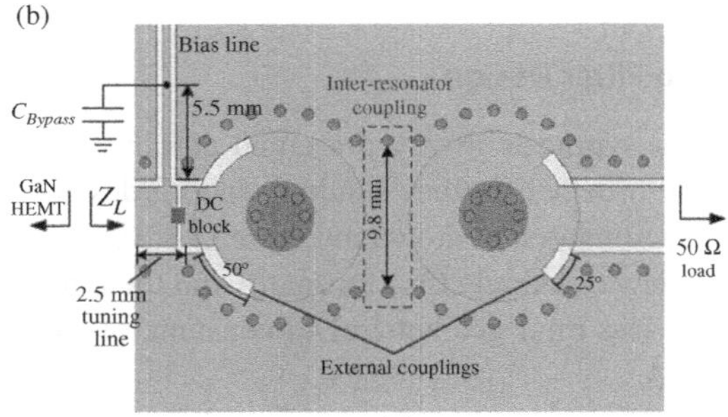


**Figure 11.3** Evanescent-mode cavity filter: (a) 3-D illustration, and (b) description of the EVA resonator. (c) Finalized output matching network with the 2-pole evanescent-mode cavity filter.

input impedance $Z_{in}$ is tuned to the real value of 11 Ω by properly selecting $M'_{S1}$ (increasing the input coupling slot, as shown in Figure 11.3) and in equation (11.3), while the imaginary part of $Z_{in}$ is kept to zero by setting $M'_{11}$ to zero (resonator 1 is not de-tuned). The small complex offset of $Z_{opt}$ is obtained by placing a short tuning line with length of $L_x$ before the filter. It is important to note that this transmission line is necessary here to place the bias line and dc-block capacitor for the transistor. To match the transistor impedance with a large imaginary part (e.g., a bare-die transistor), a non-zero $M'_{11}$ can be utilized to perform the matching by de-tuning the resonant frequency of resonator 1. This can result in the shortest possible tuning line.

#### 11.1.3.1 Fundamental and Harmonic Matchings

The matching filter theory states that a given filter response can be achieved for any given impedance of the source. In this co-design case, however, the filter's input port is not matched to a static source impedance but the transistor's output that exhibits a frequency-dependent impedance behavior. Thus, the filter's input impedance needs to be properly selected to result in a desired frequency response of the entire PA-filer module. However, it is found in this design that neither of the conventional optimal points ($Z_{opt,PAE}$ and $Z_{opt,Power}$) leads to an expected Chebyshev filter response of the entire module. An optimized filter response can be obtained by optimizing the tuning line length ($L_x$). The load impedance $Z_L$ is eventually set to a point in between $Z_{opt,PAE}$ and $Z_{opt,Power}$ ($L_x = 2.5$ mm), leading to a flat pass-band and symmetrical skirt. Figure 11.4 shows the input impedance of the designed filter, extracted from full-wave simulation. It can be seen that optimized impedance matching is achieved with this filter.

In addition to fundamental impedance matching, harmonic impedance matching is also important for achieving high efficiency. As shown in Figure 11.4, the second harmonic impedance ($Z_{2f_0}$) of the initial filter does not fall in the high efficiency region. Therefore, the bias line is utilized to further tune the harmonic impedance. This bias line is realized using a high-impedance (75 Ω) short-ended transmission line appended to the transmission line between transistor and filter. By properly selecting the bias-line length, the second harmonic impedance is tuned to the high-efficiency region while the fundamental impedance almost remains the same, as shown in Figure 11.4. The third harmonic impedance is

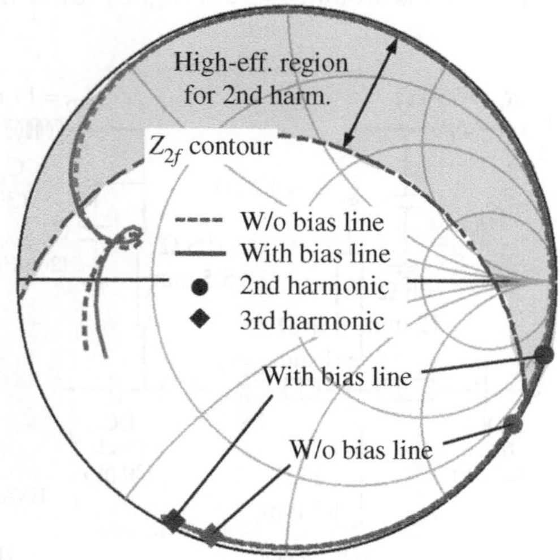


**Figure 11.4** Simulated $Z_L$ versus frequency provided by the filter.

located in the optimal region for both the initial and bias-line-tuned filters, compared to the load-pull result shown in Figure 11.2.

### 11.1.4 PA Design

To complete the design of the entire circuit, the input matching network is realized using a single open-ended stub, which provides an input impedance of $3 - 5j\ \Omega$ at 3.0 GHz. The circuit schematic of the co-designed PA-filter module is plotted in Figure 11.5. A 17 nH inductor is connected to the end of the stub, and a 250-Ω resistor is in series with the inductor as the amplifier stabilizer. From simulations using manufacturer's models, this network provides an impedance of $|Z| > 350\ \Omega$ at a frequency higher than 3 GHz, which presents an effective open. An additional 17 nH inductor is connected to the end of the output bias line to further prevent RF power leakage.

### 11.1.5 Experimental Results

The fabricated circuit, shown in Figure 11.6(a), is first measured using a network analyzer. Figure 11.6(b) shows the measured small-signal frequency response. In this measurement, the transistor gate is biased at a Class-AB point, leading to a quiescent current of 260 mA at $V_{DS}$ = 25 V. The curves of $S_{21}$ and $S_{22}$ indicate the desired Chebyshev filter shape, similar to the stand-alone filters, while the pass-band has a small-signal gain of around 15 dB. The simulated equal-ripple return loss is around 14.5 dB, slightly lower than the design value of 15 dB. This

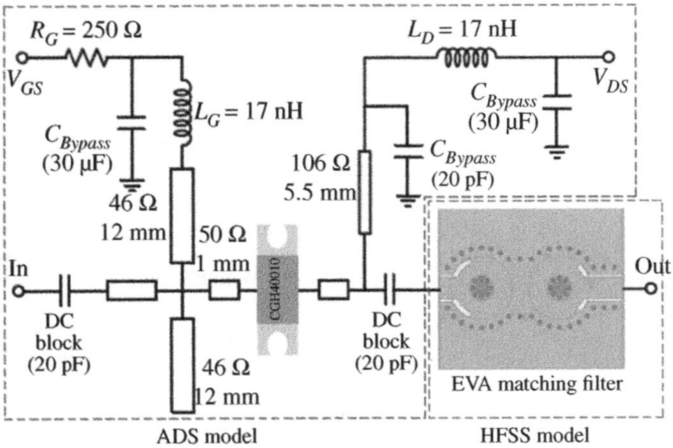


**Figure 11.5** Circuit schematic of the co-designed PA-filter module.

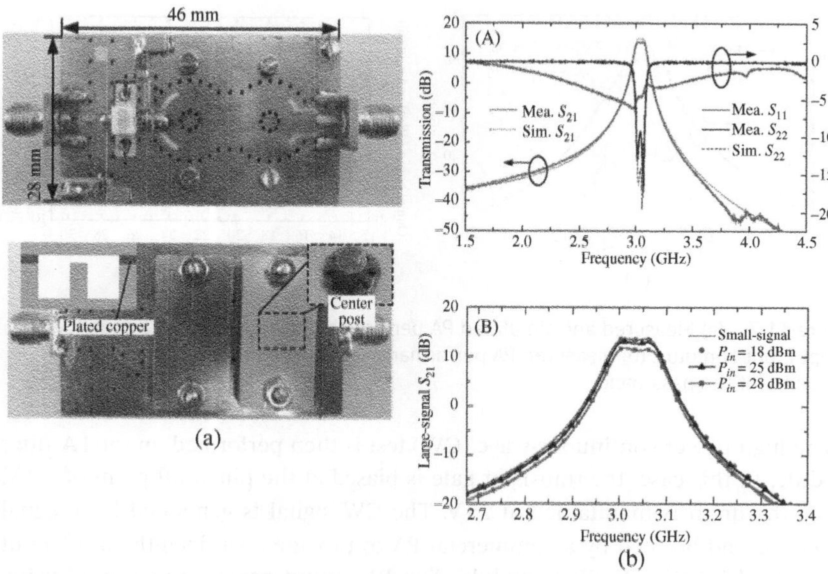


**Figure 11.6** Fabricated PA-filter co-design module. Measured and simulated small-signal frequency response of the co-designed PA and filter.

is basically caused by the load-line matching condition imposed in the PA design, introducing a slight mismatch in the small-signal case.

The $S_{11}$ shape is similar to that of a regular PA design, as the high-$Q$ filter is completely isolated by the transistor seen from the input port. The measured reflection coefficient, $S_{11}$, is around −3 to −7 dB, slightly lower than that of the regular PA design, such as those reported in [145]. This is mainly due to the fact that the input matching network in this design does not include any additional stabilization resistance (e.g., series and parallel resistors to the gate), since this amplifier has already been stable. In practical applications, $S_{11}$ might need to be further reduced by improving the input matching. Simulated results are also plotted for comparison, matching well with the measurements. It is important to highlight that the simulated $S_{22}$ presents a 14.2-dB equi-ripple, slightly lower than the target value of this design. This is mainly due to the slight mismatch in the input port, induced by the load-line matching imposed in the PA design.

Figure 11.6(b) shows the large-signal frequency response of the co-designed PA-filter circuit. It can be seen that the gain drops as the input power level increases from 18 to 28 dBm, which is due to the PA saturation. At the low input power of 18 dBm, the large-signal frequency response is consistent with the small-signal results as expected.

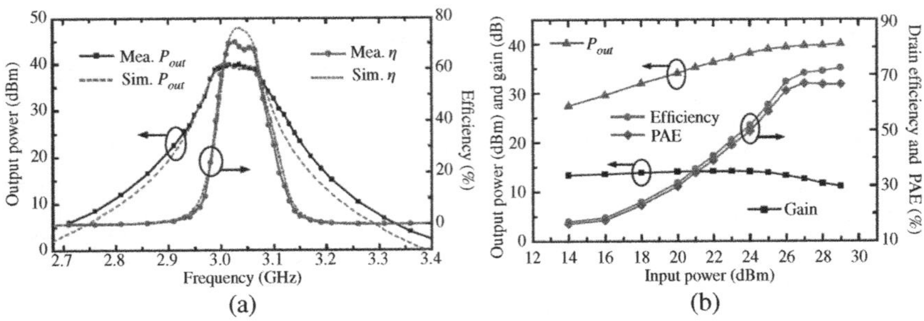


**Figure 11.7** (a) Measured and simulated PA performance with a 28-dBm frequency-swept input stimulus. (b) Measured PA performance versus input power at the center frequency of the pass-band.

The high-power continuous-wave (CW) test is then performed on the PA-filter module. In this case, the transistor gate is biased at the pinch-off point of 3.3 V, while the drain is maintained at 25 V. The CW signal is generated by a signal generator and boosted by a commercial PA to provide a sufficiently large input power to drive this PA-filter module. The PA output power is measured using a power meter. Figure 11.7(a) shows the measured output power and efficiency across the frequency range from 2.7 to 3.4 GHz with a constant input power of 28 dBm, indicating a measured efficiency of 71% at the center frequency of 3.05 GHz. An efficiency of >50% and output power of 40 dBm were measured across the entire passband.

Subsequently, the dynamic response of the PA is measured by varying the input power from 14 to 29 dBm, as shown in Figure 11.7(b). Gain compression occurs when the input power reaches 25 dBm and the highest PAE of 68% is achieved when the input power is greater than 27 dBm.

### 11.1.6 Co-Design of PA and Three-Pole High-*Q* Tunable Filter

In this section, the co-design method is further extended to the case of a PA integrated with a three-pole tunable filter. The synthesis method is similar to the one presented earlier for the two-pole implementation. Figure 11.8 shows this concept, while Figure 11.9(a) and (b) show the implemented PA.

Small-signal measurement is first conducted with the fabricated PA-filter module to determine its frequency response and tuning range. Figure 11.9(c) shows the measured S-parameters when the circuit is operating at 3.08 GHz. A near-perfect Chebyshev filter shape is achieved with 15.5-dB small-signal gain in the passband and 3-dB bandwidth of 54 MHz (1.7%). Simulation results also indicate good agreement with the measurements. The achieved equi-ripple return loss is increased to 10-dB primarily because of the load-line matching condition applied in the

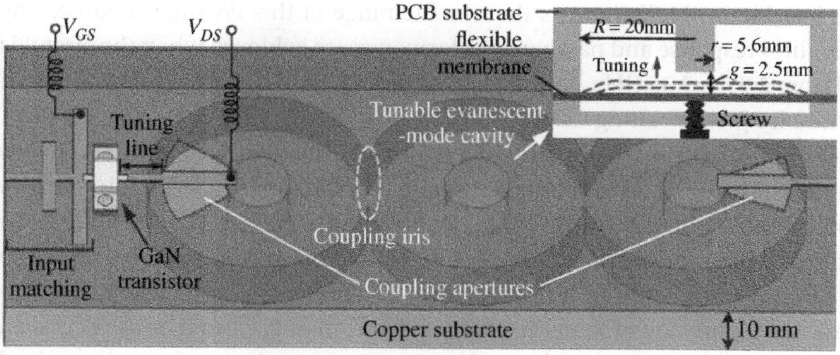


**Figure 11.8** 3-D illustration of the PA with a tunable 3-pole evanescent-mode filter. Load-pull results and input impedance trajectory frequency response of the designed filter.

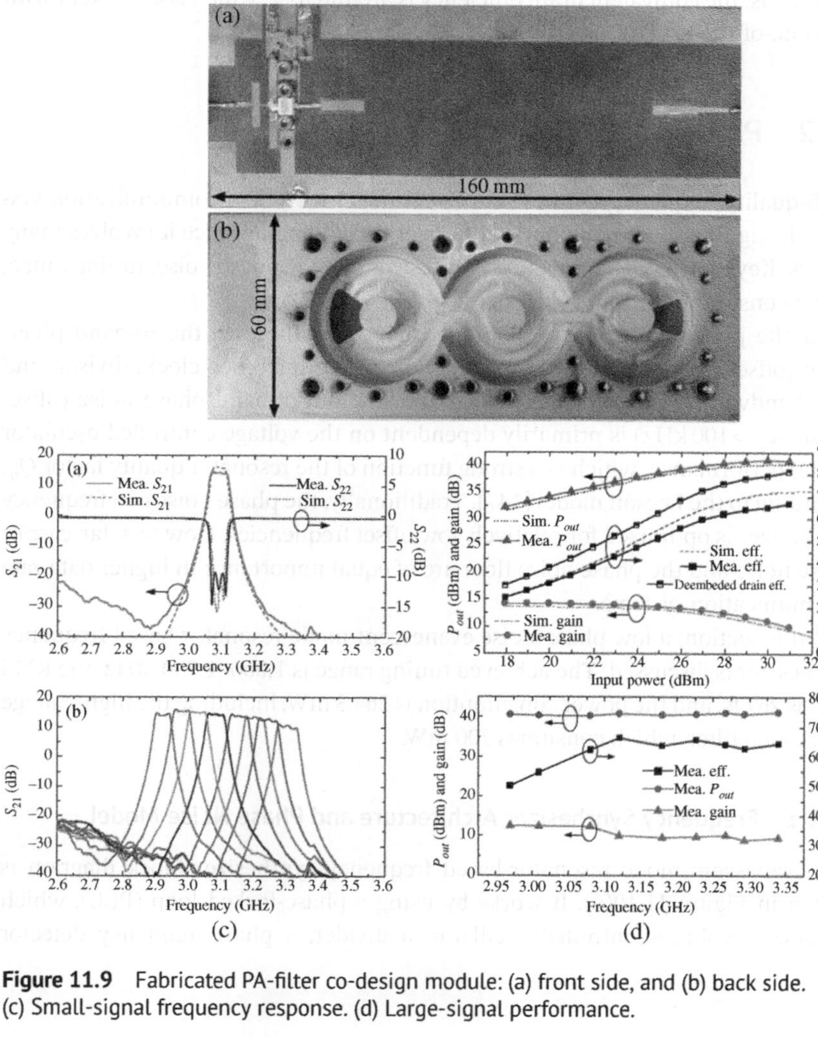


**Figure 11.9** Fabricated PA-filter co-design module: (a) front side, and (b) back side. (c) Small-signal frequency response. (d) Large-signal performance.

PA design. Figure 11.9(c) shows the tunable range of this PA-filter module. The desired filter response and passband performance are achieved when the circuit is tuned from 2.95 to 3.35 GHz. A small-signal gain of >10 dB is obtained throughout the entire frequency tuning range.

Figure 11.9(c) shows the measured PA performance under various input powers when the circuit operates at the center frequency of 3.1 GHz. The PA gain starts to compress at an input power of 24 dBm. The highest PA efficiency of 67% is achieved at a gain compression of approximately 2 dB, with a maximum output power of 40 dBm. Simulation results are also plotted for comparison. The measured PA performance is slightly lower than the simulated one, mainly caused by the degraded quality factor of the resonators due to the inaccuracies and imperfections involved in the fabrication process. The measured efficiency can be de-embedded to the transistor drain plane by subtracting the filter loss (simulated) from the original data. Thus, the equivalent drain efficiency is around 82%, which is consistent with the state-of-the-art PAs.

## 11.2 Phase-Locked Loop

High-quality frequency synthesizers are critical in wireless communication systems. Designing such frequency synthesizer is challenging since it involves many factors. Key metrics of a frequency synthesizer include phase noise, tuning range, power consumption, and circuit size.

For the phase noise model of the frequency synthesizer, the in-band phase noise (offset frequency <100 kHz) depends on the reference clock, divisor, and loop bandwidth [146]. On the other hand, the out-of-band phase noise (offset frequency >100 kHz) is primarily dependent on the voltage-controlled oscillator (VCO) performance, which is a strong function of the resonator quality factor $Q_u$, according to the Leeson model [147]. Traditionally, the phase noise of a frequency synthesizer is optimized for relatively low offset frequencies. However, far-carrier phase noise and the phase noise floor are of equal importance in higher data rate communication systems.

In this section, a low phase-noise evanescent-mode resonator-based frequency synthesizer is discussed. The achieved tuning range is 1,000–1,700 MHz, the RMS jitter is 267 fs, and the power consumption is 264.3 mW, including the high-voltage active loop filter, which consumes 200 mW.

### 11.2.1 Frequency Synthesizer Architecture and Phase Noise Model

The evanescent-mode resonator-based frequency synthesizer block diagram is shown in Figure 11.10(a). It works by using a phase-locked loop (PLL), which includes a voltage-controlled oscillator, a divider, a phase-frequency detector

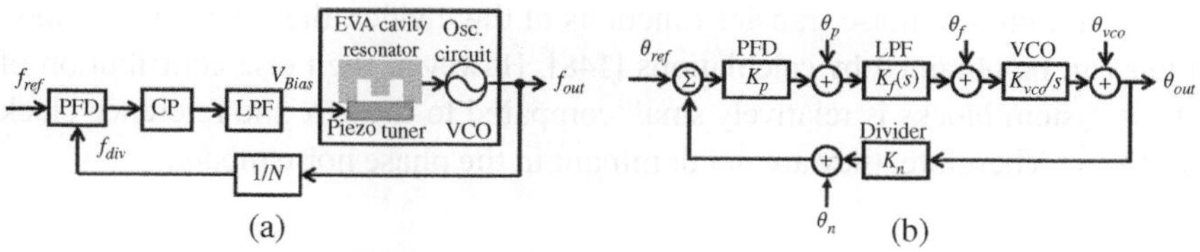


**Figure 11.10** (a) A frequency synthesizer block diagram using an evanescent-mode resonator, and (b) its phase noise model.

(PFD), a charge pump (CP), and a loop lowpass filter (LPF). By a proper co-design of both the resonator and the VCO, the phase noise can be optimized.

The divider reduces the output oscillation frequency ($f_{out}$) down to the level of the reference clock frequency ($f_{ref}$). The divisor ($N$) is fed into a programmable frequency divider to be able to generate a tunable output frequency. The output of the PFD is sent to the CP and the active loop filter, which in turn controls the tuner of the resonator (piezoelectric actuator). As a result, the PLL output frequency is adjusted to $N \times f_{ref}$.

The input-referred phase noise of each system block is modeled, as shown in Figure 11.10(b). The phase noise of a frequency synthesizer is calculated from the transfer function of each noise source. $K_p$, $K_f(s)$, $K_{vco}$, and $K_n$ are the gain constants of the PFD, the loop filter, the VCO, and the divider, respectively. Note that $K_f$ is a function of $s$, which plays the main role in providing the system type and order.

From the noise model in Figure 11.10(b), the transfer function of the reference clock phase noise is

$$H_{ref}(s) = \frac{\theta_{out}}{\theta_{in}} = \frac{K_p K_f(s) K_{vco}}{s + \dfrac{K_p K_f(s) K_{vco}}{N}} \tag{11.4}$$

Also, the transfer function of the VCO phase noise is

$$H_{vco}(s) = \frac{\theta_{out}}{\theta_{in}} = \frac{s}{s + \dfrac{K_p K_f(s) K_{vco}}{N}} \tag{11.5}$$

From (11.4), we see that the noise transfer function of the reference clock ($H_{ref}(s)$) is a lowpass function. Also, from (11.5), the transfer function of the VCO noise ($H_{vco}(s)$) is a highpass function. Consequently, the output phase noise at low frequency offset ($f_m < f_{0,loopbandwidth}$) is primarily dependent on the input reference phase noise ($\theta_{ref}$). On the other hand, at high frequency offset, the output phase noise is primarily dependent on the phase noise of the VCO ($\theta_{vco}$). As a result, choosing a low $N$ and a low phase noise reference clock helps reduce the PLL phase noise at low offset frequency range. On the other hand, the phase noise at high offset frequency can be improved by reducing the VCO phase noise.

In addition, the noise transfer functions of the divider, the PFD, and the loop filter can be obtained by calculations [148]. However, the noise contribution of those system blocks is relatively small compared to those of the reference clock and VCO. Therefore, they are not dominant in the phase noise model.

## 11.2.2 Circuit Design and Optimizations

### 11.2.2.1 Evanescent-Mode Cavity Resonator

In order to couple energy to and from an EVA cavity, a coupling slot is used as discussed in Section 2.3. However, the improvement on $Q_u$ is limited by the current crowding and radiation losses introduced by the slots. Coaxial coupling, on the other hand, can maintain the tuning capabilities of the EVA cavity without those limitations, resulting in a higher $Q_u$. This is shown in Figure 11.11, where $Q_u$ is simulated for the resonator using the two coupling methods. Consequently, coaxial coupling is chosen to improve $Q_u$.

### 11.2.2.2 Voltage-Controlled Oscillator Design

Due to their simplicity and low parasitics, single-ended oscillators are used in this work. Figure 11.12(a) demonstrates the oscillator configuration. In this configuration, a feedback capacitor ($C_{FB}$) is added to increase the negative resistance associated with the reflection coefficients $\Gamma_{in}$ and $\Gamma_{out}$. Therefore, only one lumped element is used, reducing parasitic resistance, which maintains high $Q$ for the oscillator. The $Q$ of the feedback capacitance also needs to be high to minimize noise.

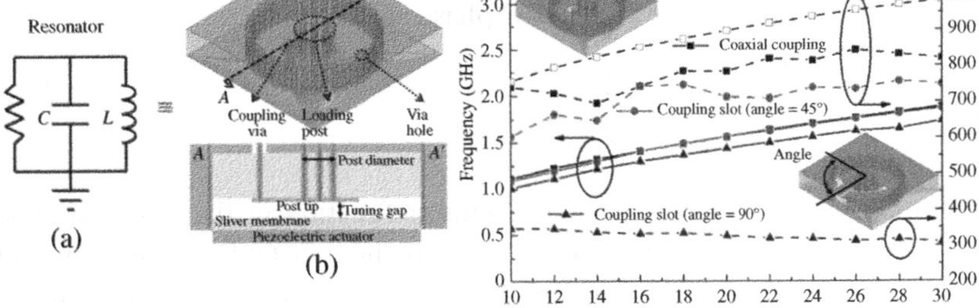


**Figure 11.11** (a) The equivalent circuit model of the resonator, (b) the physical structure of the resonator with the coaxial coupling structure, and (c) simulated frequency (solid lines) and $Q_u$ (dashed lines) performance of EVA resonators with coupling slot, coaxial coupling structure, and theory.

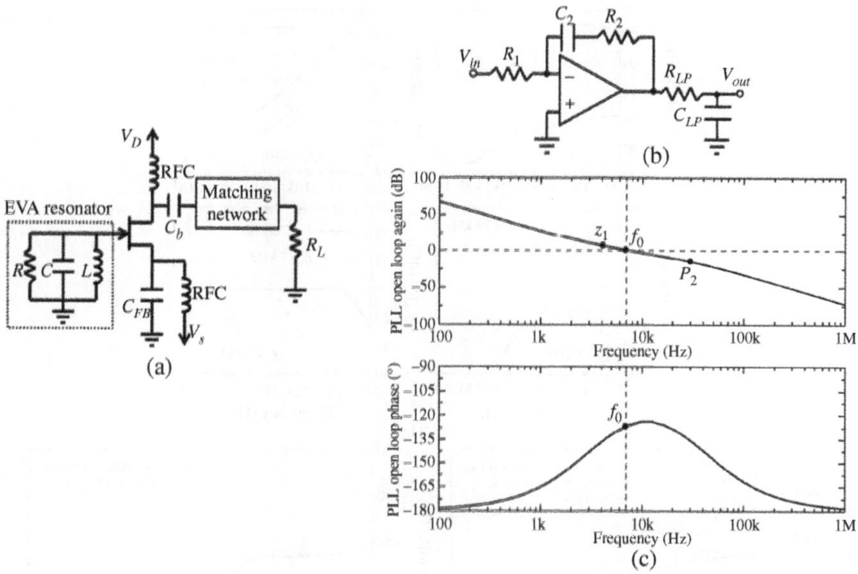


**Figure 11.12** (a) The VCO circuit schematic used in this work, where the EVA cavity is used as a resonator. (b) The schematic of a two-pole and one-zero active loop filter. (c) Simulated open loop magnitude and phase frequency response with the PLL operating frequency of 1,000 MHz.

The transistor is biased at the lowest phase noise. In Figure 11.12(a), $V_s$ is biased allowing 0 V bias at the gate, which is compatible with the resonator being a DC-short.

#### 11.2.2.3 Phase-Locked Loop Design

The function of the lowpass filter (LPF) in Figure 11.10(a) and high voltage amplification can both be provided by an active loop filter circuit. Figure 11.12(b) shows the circuit of the active loop filter, which can provide the required reference spur suppression [146, 149], in addition to the bias voltage for the piezoelectric actuator. This circuit also needs to satisfy the phase margin requirements to guarantee a stable operation of the PLL.

The phase noise from the VCO and the reference clock are the main sources of phase noise. As a result, the total phase noise of the PLL can be estimated by

$$PN_{total} = PN_{ref}H_{ref}(s) + PN_{vco}H_{vco}(s) \tag{11.6}$$

where $PN_{ref}$ and $PN_{vco}$ are the reference clock phase noise and the VCO phase noise, respectively. The noise sources along with their corresponding transfer functions are plotted in Figure 11.13(a) based on (11.4) and (11.5). The loop bandwidth affects the amount of the total phase noise into the system. This is shown

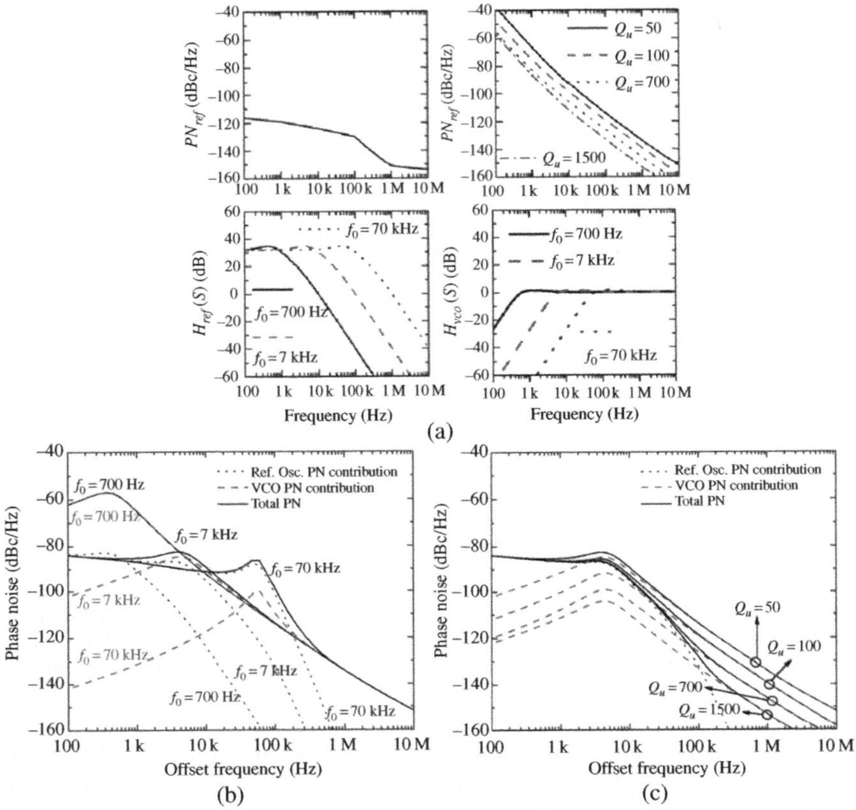


**Figure 11.13** (a) Measured reference clock phase noise ($PN_{ref}$) and simulated VCO phase noise ($PN_{vco}$) and their noise transfer function, $H_{ref}$ and $H_{vco}$, and simulated total system phase noise and the contributions from the reference clock and the VCO with (b) different loop bandwidths ($f_0$) at $Q_u = 50$, and (c) different $Q_u$ at $f_0 = 7$ kHz.

in Figure 11.13(b), where the total phase noise response is plotted at different loop bandwidths. As we can observe, the area under the phase noise curve (which is equivalent to jitter) is minimized at a certain loop bandwidth. As a result, the feedback impedance is chosen to provide that loop bandwidth.

From the phase noise response shown in Figure 11.13(a), the overall PLL phase noise response can be derived for different $Q_u$. Figure 11.13(c) shows the effect of $Q_u$ on the phase noise, which is prominent at high-frequency offsets. The divisor $N$ of the divider is chosen based on the operating frequency and the available reference clock. The value of $N$ also affects the phase noise at low frequency offsets.

#### 11.2.2.4 Measured Results

The left side of Figure 11.14(a) shows the implemented VCO, which has a footprint of 18 mm × 18 mm. A Rogers TMM3 material with 3.175 mm thickness is used as the cavity substrate. A silver membrane seals the top of the cavity to make an adjustable capacitance gap with the post.

The PLL board is made on a Rogers 4350B substrate with 0.508 mm thickness. An additional surface mount power splitter is placed to divide the VCO power between the output and the feedback loop and match the load impedance for both the output terminal and the divider input. The reference clock frequency ($f_{ref}$), 50 MHz, is provided by a crystal oscillator, where A divide-by-2 prescaler (ADF4156) is placed before the reference clock as a buffer. The APEX PA97DR power op-amp is used to amplify the DC bias voltage to the range of ±200 V, which is required to drive the piezoelectric actuator. It also serves as the active loop filter.

For system stability, the simulated phase margin $\Phi_M$ and loop bandwidth $f_0$ are 54° and 7 kHz at the operating frequency of 1000 MHz ($N = 40$ with $f_{ref} = 25$ MHz), and $\Phi_M$ and $f_0$ are 48° and 5 kHz at the operating frequency of 1700 MHz ($N = 68$ with $f_{ref} = 25$ MHz).

Figure 11.14(b) shows the measured synthesizer output spectrum. The transistor dc bias is optimized to minimize the phase noise for each operating frequency. The tuning range is from 1,000 to 1,700 MHz (1.7:1) and the output power is between −2.8 and 4.1 dBm after the power splitter. The overall tuning range is limited by the piezoelectric bending capability, which is approximately ±20 μm.

The left- and right-hand sides of the reference spurs across the tuning frequency are shown in Figure 11.15(a). The measured reference spurs are around −57 to −78 dBc.

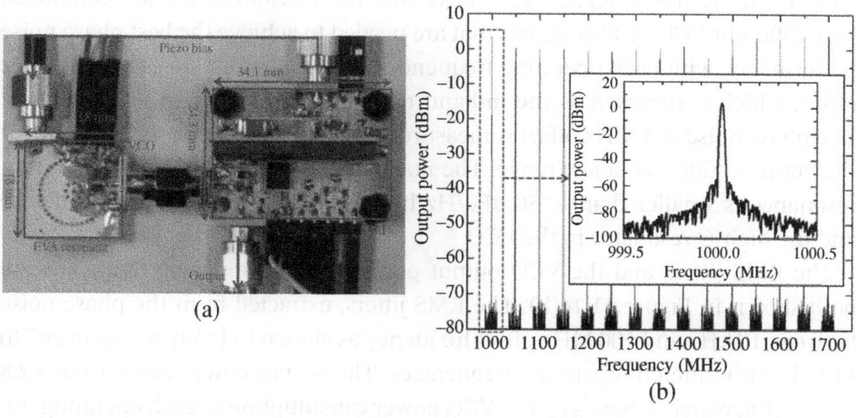


**Figure 11.14** (a) The fabricated PLL. (b) Measured output signal power spectrum at different output frequencies.

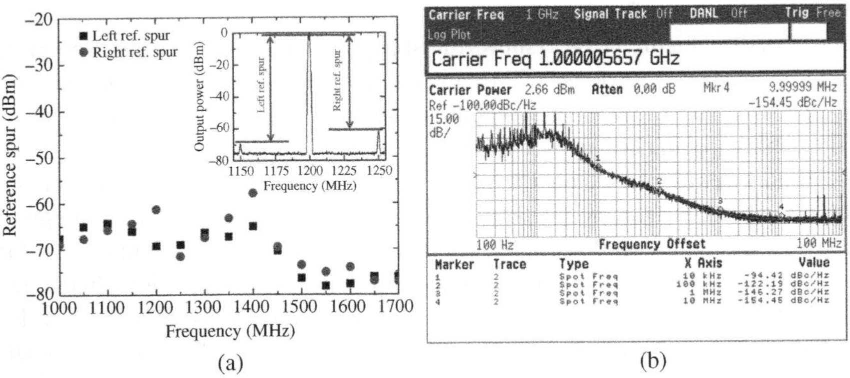


**Figure 11.15** (a) Measured reference spur across oscillatory frequencies. (b) Measured phase noise versus offset frequency at oscillation frequency of 1,000 MHz.

Figure 11.15(b) shows the measured phase noise versus offset frequency at an oscillation frequency of 1,000 MHz. At this oscillation frequency and at offsets below 4 kHz, the phase noise is dominated by the normalized reference clock phase noise. Therefore, the phase noise below 4 kHz remains at the same level at about −70 dBc/Hz. At offsets between 4 kHz and 1 MHz, the VCO noise dominates the phase noise performance. The phase noise has a slope of −20 dB/decade, which is the same as the VCO noise performance. At offsets beyond 1 MHz, the phase noise reaches the noise floor, which is around −154 dBc/Hz. The measured phase noise is shown in Figure 11.16(a) along with the oscillation frequency from 1,000 to 1,700 MHz with offset frequencies of 10 kHz, 100 kHz, 1 MHz, and 10 MHz (noise floor), respectively. Note that the measurements are conducted with different VCO dc bias signals that are needed to achieve the best phase noise performance with each operating frequency. The phase noise at 10 and 100 kHz offset, which is regarded as the in-band noise, is around −85 to −115 dBc/Hz. The phase noise at 1 MHz offset is measured between −137 and −147 dBc/Hz for the entire tuning frequency range. The noise floor at around 10 MHz away from resonance is smaller than −150 dBc/Hz because of the low noise active device and the high-$Q$ resonator in the VCO.

The RMS jitters and the VCO output power across the tuning frequency are both shown in Figure 11.16(b). The RMS jitters, extracted from the phase noise between 10 kHz and 100 MHz offset frequency as shown in [150], are from 267 to 1173 fs with different operating frequencies. The output power varies from −2.8 to 4.1 dBm, which depends on the VCO power consumption of each operating frequency. The power consumption for each circuit is: (1) 14 mA from a 1.7 V supply (23.8 mW) for the VCO, (2) 15 mA from a 2.7 V supply (40.5 mW) for the PLL board,

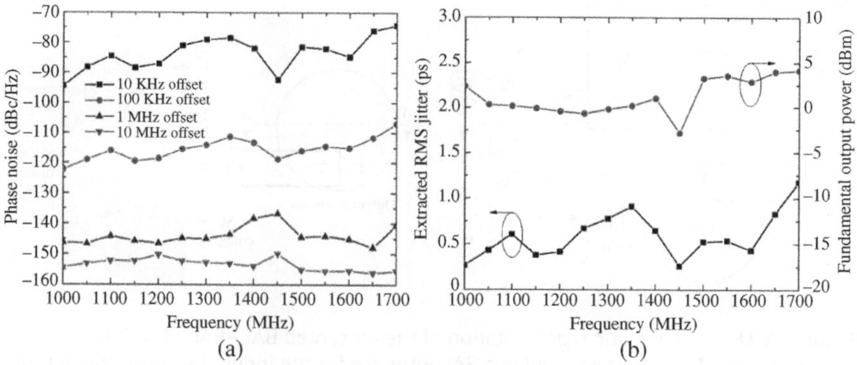


**Figure 11.16** (a) Measured phase noise versus oscillation frequency. (b) Extracted RMS jitters (10 kHz to 100 MHz) and measured output power versus oscillation frequency.

and (3) 0.5 mA from a ±200 V supply (200 mW) for the active loop filter. The total power consumption is 264.3 mW.

## 11.3 Balanced–Balanced Tunable Filtering LNA

A BAL–BAL tunable filtering LNA using cavity resonators and a single npn transistor are presented in this section. This concept, shown in Figure 11.17, can operate with differential sources (e.g., loop antennas). Using a single transistor reduces the noise sources and power consumption. The evanescent-mode cavity resonators at the input and output have a balanced interface externally and a single-ended interface internally with the transistor. They also provide continuously tunable filtering. The presented design is fabricated and measured results are presented.

### 11.3.1 Cavity Resonator-LNA Co-Design

The schematic of the presented BAL–BAL LNA is shown in Figure 11.18. The RF input signal is coupled differentially into the input resonator. The input resonator

**Figure 11.17** A fully balanced LNA can interface with other balanced components in RF front-ends.

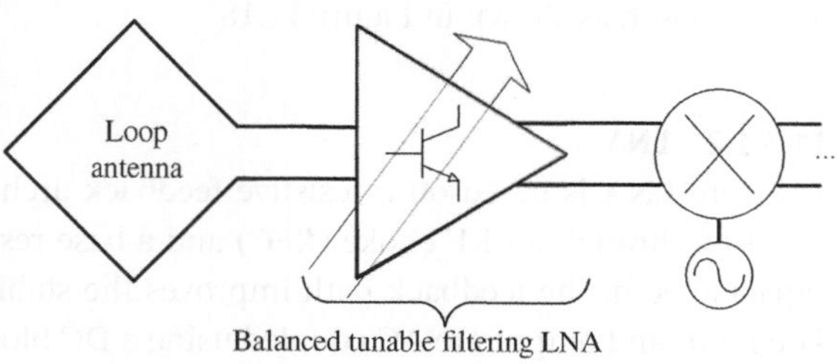

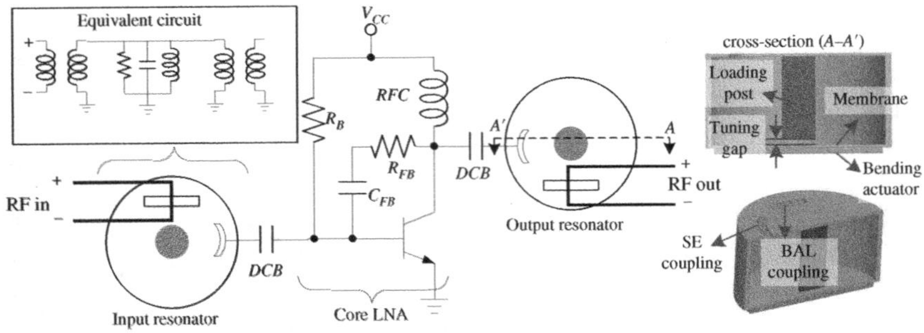


**Figure 11.18** A schematic representation of the presented BAL–BAL LNA. The evanescent-mode resonators provide a BAL interface for the input (RFin) and the output (RFout), and an SE interface with the core LNA.

has a SE coupling attached to the core LNA. This structure is replicated at the RF output. The SE terminal of the input resonator is designed at the optimal noise matching point provided by the transistor model. This also serves in minimizing the need for external discrete components for noise matching. The RF input and output ports are matched to a balanced 100 Ω. The details of the cavity resonators and the core LNA are discussed in this section.

#### 11.3.1.1 Evanescent-Mode Resonators

Evanescent-mode resonators have a narrow-band response, which provides a filtering response. The high quality factor ($Q_u$) of such resonators results in low insertion loss.

The coupling slots, shown in Figure 11.18, excite the magnetic fields in the cavity resonators. As a result, they can be modeled as a transformer. The RLC tank in the equivalent circuit model represents the evanescent-mode resonator. Since the capacitance formed by the post and the membrane is tunable, the frequency response of the resonator is also tunable.

The SE coupling is realized using a single shorted transmission line above a coupling slot to maximize the magnetic field coupling, while the BAL coupling is realized using a U-shape transmission line with the coupling aperture below its middle point, as shown in Figure 11.18.

#### 11.3.1.2 LNA

The core LNA is based on a resistive feedback architecture. The npn transistor is biased through an RF choke (*RFC*) and a base resistance ($R_B$). The additional capacitance in the feedback path improves the stability of the overall amplifier. The input and output are AC coupled using a DC blocking capacitors (*DCB*).

### 11.3.2 Implementation and Measured Results

The BAL–BAL LNA is fabricated using a PCB process. The fabrication requires two laminated substrates. The first one is a 10-mil thick RO-4350B, which holds the discrete components and the coupling microstrip lines. The second substrate, which hosts the cavity, is a 60-mil thick RO-4003. The cavities on this substrate result in an unloaded quality factor of 350 in simulation. The detailed structure of the BAL–BAL LNA is shown in Figure 11.19.

The npn transistor used in this work is the BFU730F from NXP Semiconductors, biased at 30 mA from a 3 V source. The feedback resistance is 500 Ω and the feedback capacitance is 0.1 pF. The piezoelectric actuators are biased from a ±100 V DC source to provide the sought tuning. It is worth mentioning here that piezoelectric actuators consume practically zero current for biasing.

The frequency response of the BAL–BAL LNA is first tested using a four-port network analyzer. The mixed-mode $S$-parameters for the gain and matching

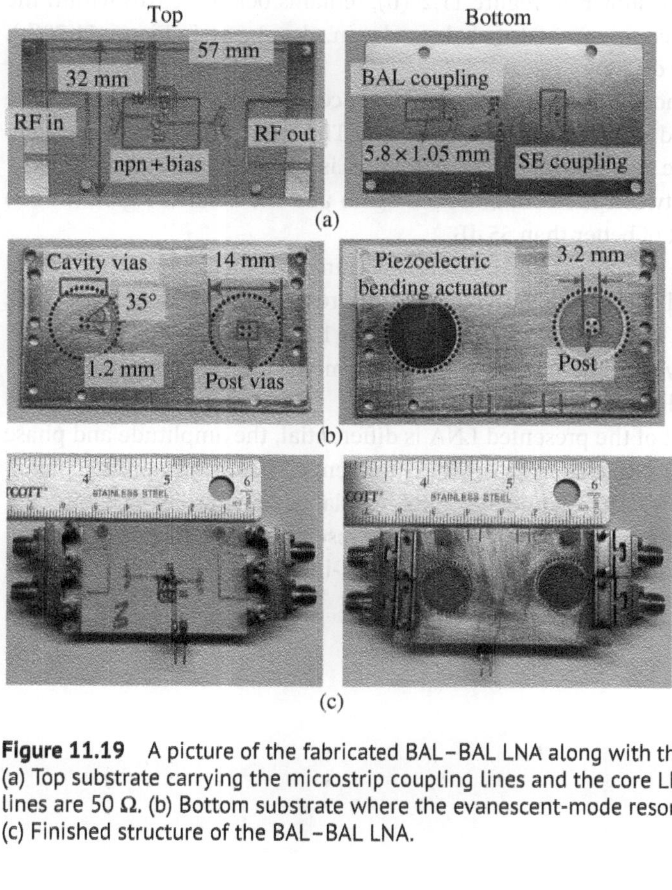


**Figure 11.19** A picture of the fabricated BAL–BAL LNA along with the dimensions. (a) Top substrate carrying the microstrip coupling lines and the core LNA. The microstrip lines are 50 Ω. (b) Bottom substrate where the evanescent-mode resonators are built. (c) Finished structure of the BAL–BAL LNA.

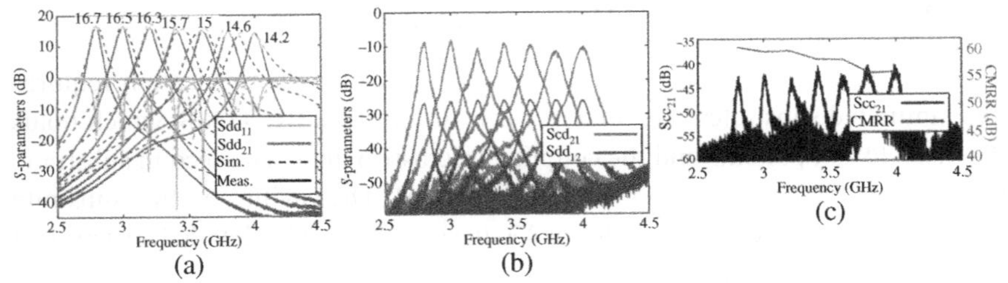


**Figure 11.20** (a) Measured and simulated results showing the balanced matching ($Sdd_{11}$) and the balanced gain ($Sdd_{21}$) with the 2.8–4 GHz tuning range with 0.2 GHz steps. (b) Measured common-mode-to-differential gain ($Scd_{21}$) of the amplifier along with the reverse gain ($Sdd_{12}$). (c) Measured common-mode gain and the CMRR.

are calculated from the measurement and simulation results and are shown in Figure 11.20(a). The balanced gain of the LNA varies between 16.7 and 14.2 dB throughout the tuning range of 2.8–4 GHz.

The reverse gain, shown in Figure 11.20(b), remains below −25 dB within the tuning range. The common-mode to balanced gain, also shown in Figure 11.20(b), remains below −9 dB.

The common-mode to common-mode gain ($Scc_{21}$) is also extracted from the measurements and is shown in Figure 11.20(c). These data are used to calculate the common-mode rejection ratio (CMRR), which is also shown in Figure 11.20(c). The difference between the common-mode gain and the balanced gain (in dB) results in a CMRR of better than 55 dB.

The noise figure for the BAL–BAL LNA is measured within the tuning range and is between 3.4 and 4.6 dB. The 3-dB bandwidth across the tuning range is 50 MHz at 2.8 GHz and 76 MHz at 4 GHz. This represents 1.8–1.9% fractional bandwidth. Two-tone linearity measurements are also performed with $\Delta f = 1$ MHz. The IIP3 is found to be 11 dBm at 3GHz.

Since the output of the presented LNA is differential, the amplitude and phase imbalance are also extracted from the measurements. The LNA amplitude imbalance remains below 0.9 dB, while the phase imbalance remains below 6°.

Other active applications of evanescent-mode resonators include non-reciprocal filters [151], linearity measurements [38], and self-interference cancellation [152].

# 12

# Monitoring and Control

## 12.1 Monitoring and Control of PCB-based Resonators: Diplexer Example

Diplexers are essential in many full-duplex radio frequency (RF) front-ends. Many modern and evolving wireless systems are adopting multiband operation, demanding tunable diplexers in their front-ends. Since the difference in power levels between transmit (Tx) and receive (Rx) paths is typically large, tunable diplexers have to be both highly linear and with high Tx–Rx isolation.

In this section, we discuss monitoring and control of evanescent-mode-based diplexer with spectrum-aware automatic tuning, as shown in Figure 12.1. Evanescent-mode cavity-based resonators are employed in this particular implementation. The tunable diplexer in this work is designed in the 0.75–1 GHz range. The bandwidths of the diplexer channels are 20 MHz at the low-frequency band and 30 MHz at the high-frequency band. The receiver port includes a band-stop filter (BSF) and a coupler, to sample the RF power and determine the existence of a potential jammer. The rejection frequency of the BSF can be adjusted to accommodate for that or, in the absence of jammers, can be adjusted to improve the isolation between the transmitter and receiver of the diplexer. Tuning the diplexer and suppressing a jammer are all done automatically and in real time without the need of manual tuning or any laboratory-scale test equipment such as network or spectrum analyzers.

### 12.1.1 System Architecture

In this section, the system-level concepts are detailed. The RF path and diplexer structure are explained first. Subsequently, we discuss the method to monitor and control the diplexer, followed by the RF spectrum sampling method used to detect the existence of a jammer within the spectrum of interest.

*Tunable Evanescent-Mode Filters: Principles, Implementation, and Applications*,
First Edition. Dimitrios Peroulis and Mohammad Abu Khater.

Companion website: www.wiley.com/go/Peroulis/TunableEvanescent-ModeFilters

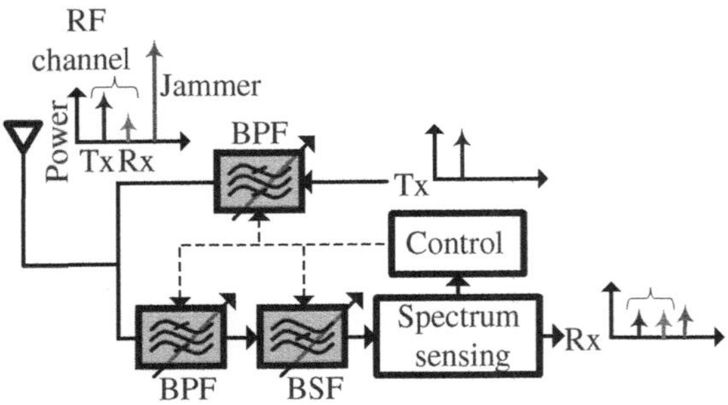


**Figure 12.1** The tunable diplexer can automatically adjust the frequency response to suppress a jammer or improve the isolation between the transmitter and the receiver depending on the condition of the spectrum.

#### 12.1.1.1 Diplexer

In order to achieve a tunable diplexer, two identical second-order tunable band pass filters (BPFs) can be combined. A BSF is introduced at the receiver port and is used for sensing the spectrum. It can also be used to improve the RF performance in terms of isolation or jammer suppression. Figure 12.2 demonstrates this concept. The tunability is achieved by adjusting the resonance frequency of each resonator individually.

In order to combine two BPFs, the diplexer is designed such that, near the source, the BPFs are as close as possible to an open circuit outside the passband. This allows the filters to share the same common port (antenna) with minimal effect on each other. Since the BPFs in the Tx and Rx paths appear near a short circuit outside the passband, a 90° transmission line is used to produce the sought open-circuit response to minimize the loading effect between the ports. As a result, a matching transmission line is included between the common port (source) and each one of the other two ports as shown in Figure 12.2.

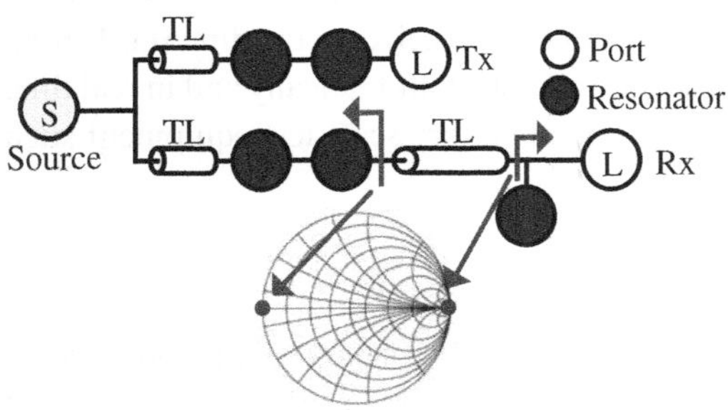


**Figure 12.2** A coupling diagram of the diplexer. The transmitting path has a two-pole bandpass filter, while the receiving path has a two-pole bandpass filter and a bandstop filter. Transmission lines are used to provide wideband matching between the Tx and Rx near the source.

To achieve the desired bandpass and bandstop responses in the Rx port, the interface between them has to be designed properly as well. The problem arises from the fact that the BPF appears as a near short circuit outside the passband, where the BSF appears as a near open at the rejection frequency (see Figure 12.2). This results in an undesired resonance (peaking in the response). This issue is resolved by introducing a 180° transmission line between the BPF and the BSF. As a result, the bandpass and bandstop responses are both achieved with minimal effect on the overall performance.

Piezoelectric actuator disks are used for tuning. Under different bias voltages, a piezoelectric disk bends the membrane in the resonator. Figure 12.3 shows how the biasing of the piezoelectric disk can tune the resonator.

#### 12.1.1.2 Resonators Monitoring and Control

Monitoring and control of resonators is needed to ensure each resonator in the diplexer is tuned to the required resonant frequency in real time. Monitoring each resonator separately is necessary since some fabrication non-idealities might result in nonidentical resonators. Also, tuning devices, such as piezoelectric tuners, may show hysteretic and creep behaviors during actuation. Such mechanical non-idealities result in nondeterministic tuning on the frequency.

In order to achieve reliable tuning, a monitoring system should be able to read the resonant frequency of each cavity. The resonant frequency of an evanescent-mode cavity can be approximated as

$$f_{Resonant} \approx \frac{1}{2\pi\sqrt{LC(V_{BIAS})}} \tag{12.1}$$

where $f_{Resonant}$ is the cavity resonant frequency, $L$ is the inductance from the evanescent-mode structure, and $C(V_{BIAS})$ is the capacitance between the post and the membrane as a function of the bias of the piezoelectric actuator. The approximation comes from the fact that (12.1) ignores parasitic effects. The capacitance $C(V_{BIAS})$ can be approximated as a parallel plate capacitor or

$$C(V_{BIAS}) \approx \frac{\epsilon A}{g(V_{BIAS})} \tag{12.2}$$

where $\epsilon$ is the permittivity, $A$ is the area of the tip of the post, and $g$ is the gap between the post and the membrane. From (12.1) to (12.2),

$$f_{Resonant} \approx \frac{\sqrt{g(V_{BIAS})}}{2\pi\sqrt{L\epsilon A}} \tag{12.3}$$

From the last equation we notice that $f_{Resonant}$ is a hysteretic function of $V_{BIAS}$, but a deterministic function of $g$ (or $C$). As a result, this relationship can be used for monitoring the resonator. This is demonstrated in Figure 12.3.

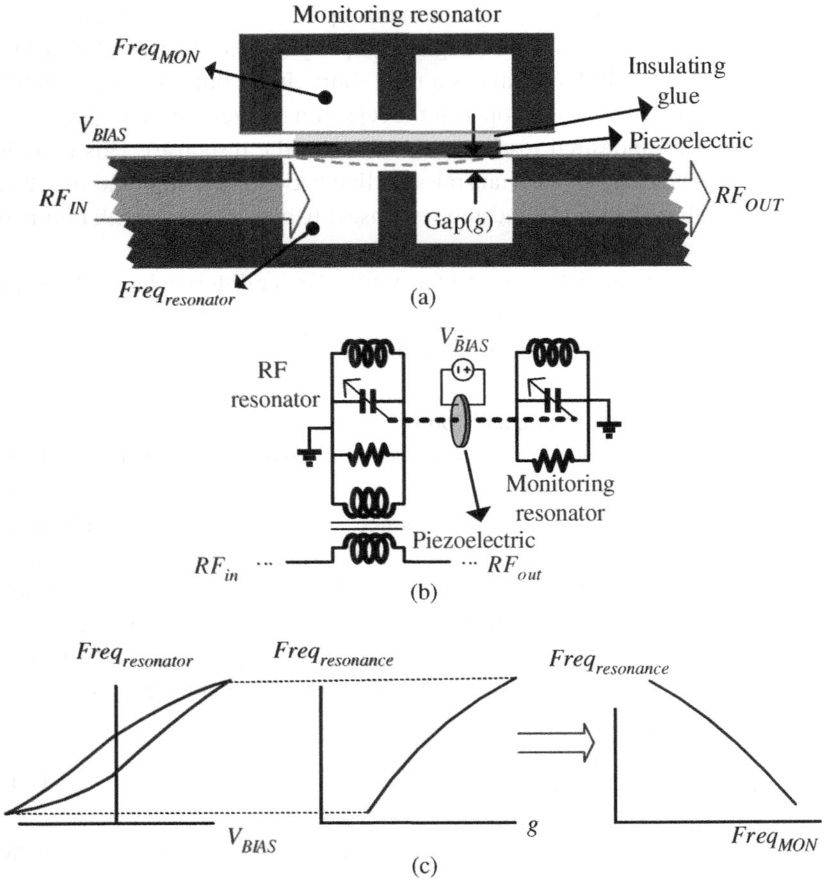


**Figure 12.3** (a) Cross section of the monitoring structure showing how the gap can change by changing the bias voltage [153, 154], (b) circuit model of the resonators in the monitoring structure in (a), and (c) conceptual relationships between the resonant frequencies, bias voltage, and the gap. The relationship between resonant frequency and bias voltage is hysteretic. On the other hand, the gap gives a non-hysterestic relationship with the resonant frequency.

To detect the change in the gap of the cavity, the monitoring structure in Figure 12.3(a) is used, where a monitoring resonator is mounted on top of the RF resonator. This way, the RF path is completely independent from the monitoring path, which is necessary for limiting RF leakage and preserving the high $Q$ of the resonator. Furthermore, the gaps of the monitoring and the RF resonators are inversely proportional to each other (one-to-one relationship) since increasing one results in reducing the other.

Detecting the frequency ($Freq_{MON}$) of the monitoring resonator is realized by embedding it in an oscillator loop. The oscillator generates a sinusoid with a frequency proportional to that of the resonant frequency of the monitoring resonator or

$$f_{oscillator} \propto Freq_{MON} \tag{12.4}$$

where $f_{oscillator}$ is the output frequency of the oscillator. A frequency counter can read frequency $f_{oscillator}$. A frequency divider, however, is required to lower this frequency to a point where it can be read by practically any digital controller (a microcontroller or a field programmable gate array [FPGA]). Since the output of the frequency counter is proportional to $Freq_{MON}$, the relationship between the output of the frequency counter and the frequency of the RF resonator can be now established in a calibration stage. Figure 12.4 shows this.

Since the relationship between $Freq_{CounterOutput}$ and the frequency of a monitored resonator in the diplexer ($Freq_{resonator}$) can be developed, we can determine the resolution of the system (the smallest frequency step the system can achieve). The average resolution can be defined as

$$Resolution_{Avg} = \left| \frac{\Delta Freq_{resonator}}{\Delta Freq_{CounterOutput}} \right| Hz/step \tag{12.5}$$

The output of the frequency counter can be expressed by

$$Freq_{CounterOutput} = (f_{Osc}/D) \times T \tag{12.6}$$

where $f_{Osc}$ is the frequency of the oscillator, $D$ is the divisor of the frequency divider, and $T$ is the control cycle time. As a result, the average resolution can be rewritten as

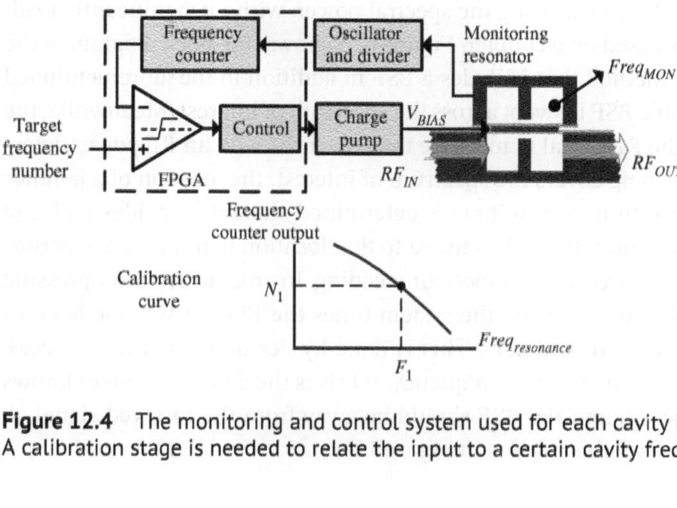


**Figure 12.4** The monitoring and control system used for each cavity in the diplexer. A calibration stage is needed to relate the input to a certain cavity frequency.

$$Resolution_{Avg} = \left| \frac{\Delta Freq_{resonator} D}{\Delta f_{Osc} T} \right| Hz/step \tag{12.7}$$

From (12.7), we observe that we can improve by increasing $T$ (at the cost of slowing down the response of the system) or decreasing $D$ (provided that the divided clock is not too fast to read). Also, the resolution can be improved by increasing the frequency range of the monitoring resonator, which is normally dictated by the tuning mechanism. The parameters in (12.7) are typically known at the system-level design. Consequently, the resolution can be specified then.

With the information available from the monitoring technique, a resonator can now be controlled for tuning. To control a resonator, the number obtained from the frequency counter is compared to a number that represents the target frequency (which is acquired from the calibration stage mentioned earlier). The output of the comparator triggers a charge pump controller. The controller converts the output of the comparator to an increasing (or decreasing) analog voltage, such that the bias voltage keeps increasing (or decreasing) until the frequency of the resonator is at the sought location (in Figure 12.4, $Freq_{CounterOutput} = TargetFrequency$ Number). In that case, the frequency of the resonator is at its target and the output of the charge pump stays fixed. Obviously, this monitoring and control methodology can determine whether the RF resonator is part of a BPF or a BSF.

#### 12.1.1.3 Spectrum Sensing

In order for the diplexer, and the control loop, to be spectrum-aware, a method for sensing the spectrum is required. The presented method employs the monitoring and control from the previous section to tune the resonator, along with RF power detection circuit.

A typical method for measuring the spectral power, without significantly loading the RF path, is based on a coupler. Figure 12.5 shows the block diagram of the spectrum sensing method that includes a BSF in addition to the aforementioned coupler. This tunable BSF is swept across the spectrum of interest. Meanwhile, the coupler samples the RF signal to measure the power through an RF power detector. When the sweeping covers the spectrum of interest, the location of a jammer (or self-interference from transmitter) is determined since it coincides with the peak power. At this point, the BSF is tuned to that location to improve the performance of the diplexer according to spectrum reading. In order to avoid suppressing the wanted signal at the receiver, the system tunes the BSF outside the band of interest in the absence of jammers. This is done by deciding whether the peak power coincides with the receiver frequency. If this is the case, the system knows that no jammers exist, and the BSF should be away from the received signal. A flow chart summarizes this process in Figure 12.6.

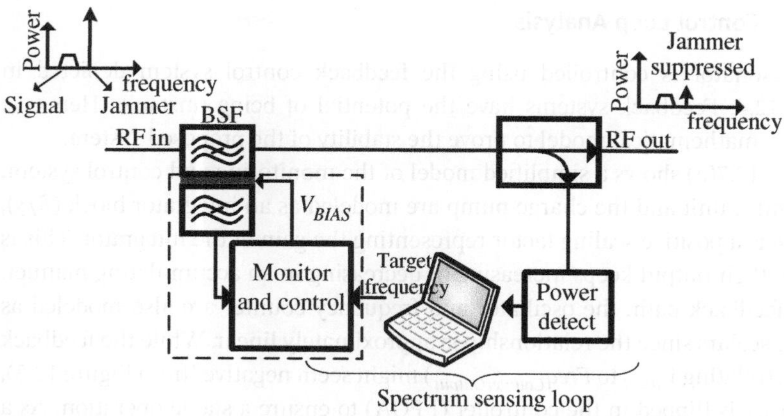


**Figure 12.5** The spectrum sensing method used in this work is capable of scanning the spectrum and deciding the location of possible jammer. Once found, the jammer is suppressed by the BSF.

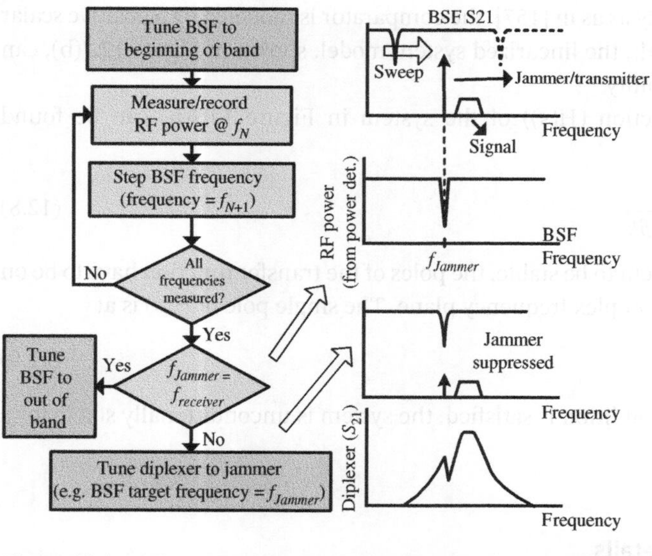


**Figure 12.6** A flowchart, along with conceptual spectral plots, of the spectrum sensing method.

Recently, a more advanced spectrum and jamming detection methods were presented [155, 156]. These methods can still integrate with evanescent-mode device such as the diplexer discussed in this section.

### 12.1.2 Control Loop Analysis

Each resonator is controlled using the feedback control system depicted in Figure 12.4. Feedback systems have the potential of being unstable. Here, we discuss a mathematical model to prove the stability of the proposed system.

Figure 12.7(a) shows a simplified model of the monitoring and control system. The control unit and the charge pump are modeled as an integrator block ($\beta/s$), where $\beta$ is a positive scaling factor representing the gain of the integrator. This is because their output keeps increasing or decreasing in an accumulating manner. In the feedback path, the oscillator and frequency counter are also modeled as positive scalars since the relationship is approximately linear. While the feedback scalar $k$ (relating $V_{BIAS}$ to $Freq_{CounterOutput}$) might seem negative (from Figure 12.3), its polarity is flipped in the controller (FPGA) to ensure a stable operation. As a result, $k$ is a positive scalar in Figure 12.7.

The comparator, however, is a nonlinear device. This means that linear system analysis cannot be directly applied to it. As a result, a linearization method is used.

The comparator is modeled using the describing function method, which studies the behavior of the system in the fundamental mode in the frequency domain. Using a similar analysis as in [157], the comparator is modeled by a positive scalar factor ($\alpha$). As a result, the linearized system model, shown in Figure 12.7(b), can be analyzed for stability.

The transfer function (H(s)) of the system in Figure 12.7(b) can be found to be

$$H(s) = \frac{\alpha\beta}{s + \alpha\beta k} \tag{12.8}$$

In order for the system to be stable, the poles of the transfer function have to be on the left half of the complex frequency plane. The single pole of $H(s)$ is at

$$s = -\alpha\beta k \tag{12.9}$$

Since the stability condition is satisfied, the system is unconditionally stable for a bounded input.

### 12.1.3 Design Details

#### 12.1.3.1 Diplexer Structure

The diplexer consists of five resonators: two in the transmitter side (synthesized as a second-order BPF) and three in the receiver side (two of which are synthesized as a second-order BPF, followed by a first-order BSF). Most of the fabrication process is compatible with conventional printed circuit board (PCB) fabrication processes. A drawing of the diplexer is shown in Figure 12.8.

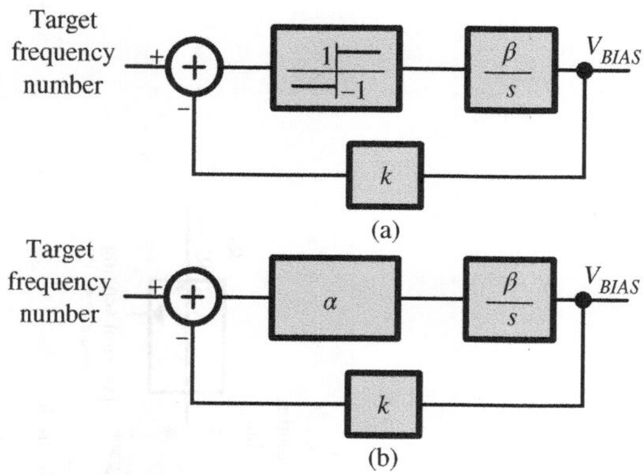


**Figure 12.7** (a) A model of the tuning control system, and (b) linearization of the control system in (a) using a describing function method.

The BPFs are synthesized using the coupling matrix method. The coupling coefficients are based on a Chebyshev filter response with external coupling of 1.037 and inter-resonator coupling of 1.287 (for 15 dB equi-ripple response). The external coupling is realized with a coupling aperture to the cavity. Also, the inter-resonator coupling is realized with coupling apertures connected with a transmission line, in addition to an opening between the cavities, as shown in Figure 12.8. The transmission line between the inter-resonator coupling is needed to achieve the correct coupling value and does not need to be 50 Ω.

The cylindrical shape of each resonator is realized by through-hole vias that define its side walls. The bottom side is implemented by a copper PCB layer. Since the top side of the cavity is a moving membrane, it cannot be fabricated in a standard PCB process. The membrane is attached post-manufacturing by soldering a silver membrane on the cavity (shown in Figure 12.8). Silver has been chosen since it has lower tensile strength (less stiff) than copper, while conductivity is not compromised.

A piezoelectric actuator is used to control the membrane-post distance. Since the piezoelectric actuator needs to be biased on both sides, a conductive glue is used to attach it to the membrane of the diplexer resonators, while an insulating glue is used to attach it to the monitoring resonator. The bias electrode is soldered on the disk before the monitoring resonator is mounted as shown in Figure 12.3.

#### 12.1.3.2 Oscillator and Frequency Counter

An oscillator is designed to generate a sinusoidal signal, which in turn is used to determine the frequency of the monitoring resonator. The chosen oscillator should be able to oscillate with a single-ended, parallel resistor, inductor, and capacitor (RLC) tank (the model of the monitoring resonator).

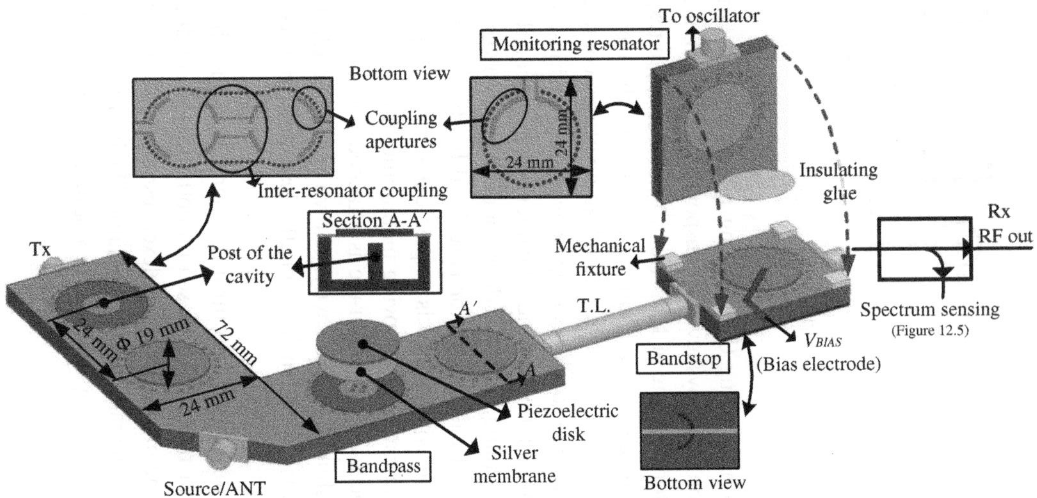


**Figure 12.8** The structure of the diplexer along with a breakout view showing the internal structure of the resonators. Monitoring resonators are mounted on each resonator of the diplexer. The bottom views show the coupling apertures for the bandpass and bandstop filters.

Since any practical oscillator has output phase noise, and since phase noise causes jitter [158], the output period of the oscillator can be expressed by

$$T_{osc}(t) = T_0 + \Phi(t) \tag{12.10}$$

where $T_{osc}$ is the actual period of oscillation, $T_0$ is the nominal constant period, and $\Phi(t)$ is a random variable that represents fluctuations in oscillation period. The system relies on counting the divided number of cycles over a control time $T$. This can be represented as

$$T = \Sigma_1^N \left(DT_0 + \Phi(t)\right) = NDT_0 + \Sigma_1^N D\Phi(t) \tag{12.11}$$

where $N$ is the overall number of cycles counted. For a sufficiently large $N$, $\Sigma_1^N D\Phi(t) \approx 0$, since $\Phi$ is a random variable. The assumption of $N$ being large can be validated by the fact that the monitoring oscillator operates at a much higher speed (about 100 MHz or higher) compared to the control loop (~1 kHz). As a result, the phase noise response of an oscillator has practically no effect on the resolution of the system.

Given the restrictions above, a Colpitts oscillator can be used since it can oscillate with the single-ended monitoring resonator using few overhead components and with no components causing inductive coupling. Since the employed Colpitts oscillator uses the monitoring resonator as an LC tank, when the monitoring resonator is tuned, the frequency of the output sinusoid changes accordingly. The divider can be an off-the-shelve device that takes a sinusoidal input and outputs a digital clock at a lower frequency. The schematic of the oscillator including an RLC model of the resonator is shown in Figure 12.9.

The frequency counter is implemented as a binary ripple counter. This eases the processing inside a digital system. The output of the counter is reset at the end

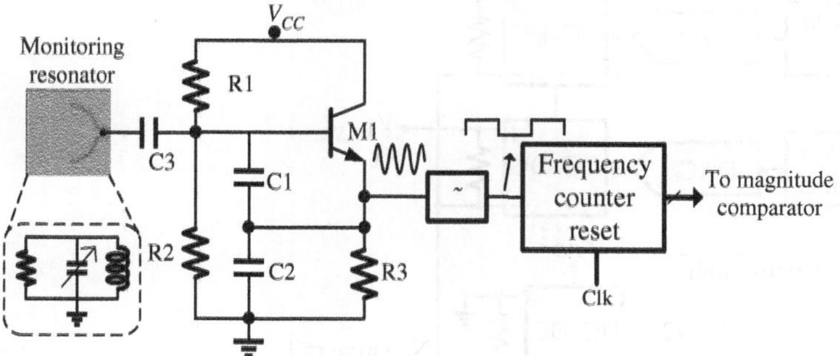


**Figure 12.9** Schematic diagram of the oscillator used with the monitoring resonator. The frequency counter provides the frequency number and then is cleared every control cycle.

of each control cycle. Once the frequency counter number is obtained, it can be compared to the target frequency number by a binary magnitude comparator. The output of the comparator dictates whether the frequency of the resonator needs to be increased or decreased.

We should note here that the coupling between the RF resonators and the monitoring resonators (or oscillator) is negligible. This is primarily because the two resonators are electrically isolated, which minimizes any conduction coupling. Also, in evanescent-mode cavity filters, the fields are almost entirely contained within the cavities, minimizing any induction or radiation coupling. In addition, the monitoring signals and the RF signals operate at different frequencies, which further reduces any effect from one resonator to the other.

#### 12.1.3.3 Control Unit and Charge Pump

The control unit has to be able to take a digital input (from the magnitude comparator) and generate an analog voltage to drive the charge pump.

The control unit in Figure 12.10 consists of an up/down counter and a digital-to-analog converter (DAC). If the input from the comparator implies that the voltage needs to be increased, the up/down comparator counts up, thus increasing the output voltage of the DAC (and vice versa). If the frequency of the resonator is at the target frequency, the up/down counter is configured such that it stops counting to maintain the voltage constant.

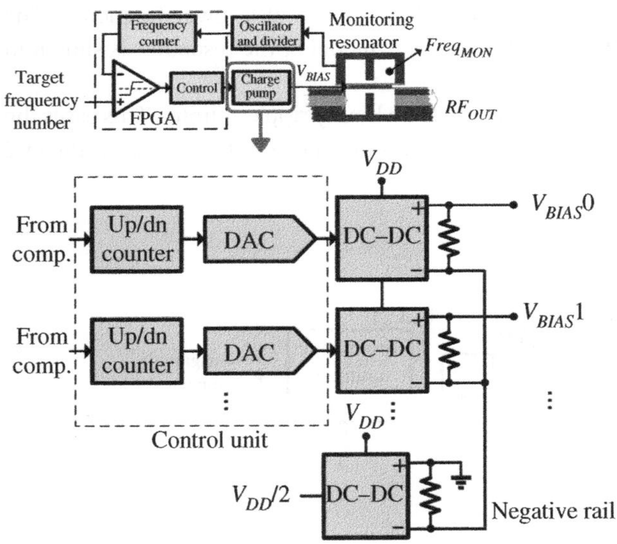


**Figure 12.10** The schematic of the control unit and the charge pump (in Figure 12.4). Each piezoelectric actuator is driven independently by one of the outputs ($V_{BIAS}$).

The charge pump needs to provide a bipolar high voltage in order to drive the piezoelectric actuators (typically ±180 V). While there are bipolar DC–DC converters, the one used here is unipolar to simplify the control and design. The commercially available DC–DC converter used here (UMHV0505) is biased from a single source and can be controlled using an analog voltage input from the controller discussed earlier.

In order to achieve the negative voltage, one DC–DC converter is configured such that it generates a negative voltage rail. This negative voltage rail is used as a reference by the tunable DC–DC converters. This way, the tunable DC–DC converter can generate voltages between the negative rail to a positive maximum. This is possible since the DC–DC converter has an isolated output.

Since the output voltage can be either positive or negative, the output node can source or sink current. This can be problematic for DC–DC converters since they can only source currents. As a result, the resistors at the output of each DC–DC converter are needed to ensure that each DC–DC converter is sourcing current all the time. The resistor values, however, can be relatively high (in the order of 1 MΩ), because the piezoelectric actuators are high impedance loads, so the power consumption of those resistors is relatively low.

### 12.1.4 Implementation and Measurements

#### 12.1.4.1 Implementation

The presented system is built using the components listed in Table 12.1. The diplexer and the control loop components are all shown in Figure 12.11.

**Table 12.1** Building components of the presented system.

| | |
|---|---|
| PCB substrate | TMM3 |
| Oscillator transistor (M1) | BFP420 |
| Oscillator capacitors: C1, C2, C3 (respectively) | 39 pF, 6.8 pF, 8.2 pF |
| Oscillator resistors: R1, R2, R3 (respectively) | 930 Ω, 4.2 kΩ, 150 Ω |
| Frequency divider | MC100E137 (÷128) |
| Level shifter | SN65ELT23 |
| Logic (frequency counter, magnitude comp, and controller) | ALTERA DE2 FPGA |
| Digital-to-analog converter (DAC) | DAC7565IAPW |
| DC–DC converter | UMHV0505 |
| Coupler | DBTC-17-5+ |
| RF power detector | HMC713MS8E |

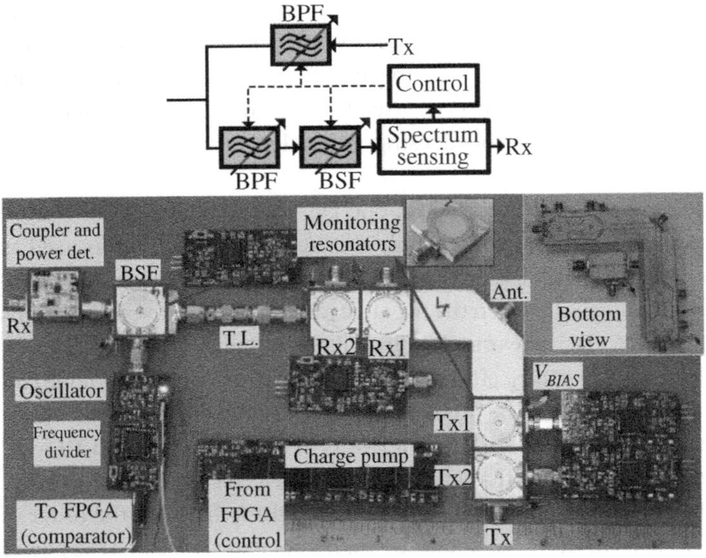


**Figure 12.11** Photograph of the system in Figure 12.1 including the diplexer, monitoring resonator, and monitoring and control printed circuit boards (PCBs).

The diplexer was designed in the 0.75–1 GHz frequency range, which covers several modern communication standards. The dimensions of the structure are shown in Figure 12.8.

Since the monitoring resonator is a part of an oscillator, lowering its resonant frequency simplifies the oscillator design. The frequency of the monitoring resonator, however, cannot be arbitrarily decreased. This is because the quality factor ($Q$) would degrade accordingly, unless the size of the resonator is increased, which conflicts with the size of the diplexer. Given the dimensions of the diplexer in Figure 12.8, it has been found that resonators operating around 400 MHz would provide sufficiently stable performance in the oscillator. As piezoelectric actuators typically tune a resonator around one octave, monitoring resonators were designed in the 350–650 MHz range.

The frequency divider operates in the emitter-coupled logic (ECL) voltage level. As a result, a level shifter is required to interface with the rest of the logic.

#### 12.1.4.2 Monitoring Performance

The monitoring and control systems are first measured with the diplexer. Each pole is tuned separately by changing the input *TargetFrequency* Number. The $Freq_{CounterOutput}$ and the $Freq_{resonator}$ (from the network analyzer [NA]) are recorded. The setup for that is shown in Figure 12.12(a). Although the resonators in an RF path (e.g., Tx path) are inseparable, the frequency of each resonator

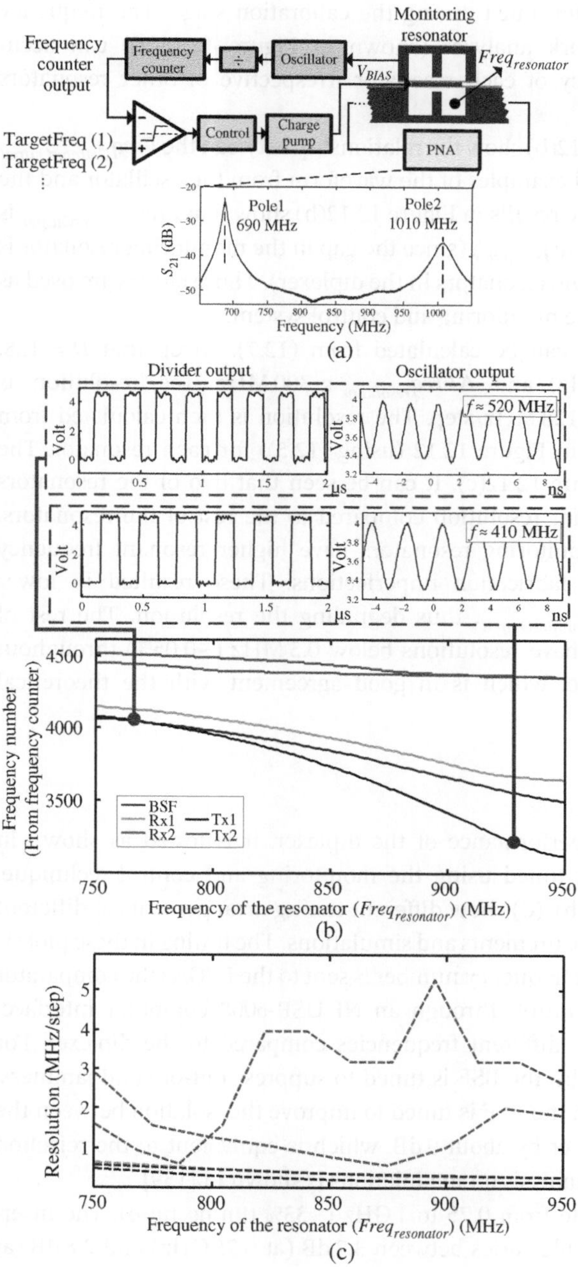


**Figure 12.12** (a) Connection diagram to measure the performance of the monitoring and control systems. The frequency of each pole (resonator) in the BPF can be accurately determined, even at maximum separation as shown in the measurements. (b) Measured output from the frequency counter for each resonator with examples from the oscillator and divider outputs. (c) The frequency resolution of the system across the tuning range.

can still be accurately determined during the calibration stage. The frequency response from the network analyzer (shown in Figure 12.12(a)) can accurately reveal the frequency of each resonator irrespective of other resonators frequencies.

The results in Figure 12.12(b) show the relationship between the $Freq_{CounterOutput}$ and $Freq_{resonator}$. Measured examples of the waveform from the oscillator and the divider are also shown. The results in Figure 12.12(b) show that $Freq_{CounterOutput}$ is inversely proportional to $Freq_{resonator}$ (since the gap in the monitoring resonator is complimentary to that of the resonators in the diplexer). Those curves are used as the calibration stage for the monitoring and control system.

The average resolution can be calculated from (12.7). Given that $D = 128$, $T = 1$ ms, $\Delta f_{Osc} \approx 300$ MHz, and $\Delta Freq_{resonator} \approx 300$ MHz, the resolution is predicted to be around 0.128 MHz/step. The resolution is then calculated from the measurement results in Figure 12.12 (using (12.5)) for each resonator. The results are plotted in Figure 12.12(c). It can be seen that two of the resonators (Rx2 and Tx2) have inferior resolution compared to the rest of the resonators. This is because those monitoring resonators have higher resonant frequency than the design due to fabrication imperfections. These resulted in lower dynamic range for $Freq_{CounterOutput}$, thus degrading the resolution. The rest of the resonators, however, have resolutions below 0.5 MHz (~0.05%) throughout most of the tuning range, which is in good agreement with the theoretical calculations.

#### 12.1.4.3 RF Performance

In order to test the RF performance of the diplexer, it is tested as shown in Figure 12.13(a) and then tuned using the monitoring and control technique. The plots in Figure 12.13(b)–(d) show different tuning arrangements at different frequencies from both measurements and simulations. The tuning in these plots is fully automatic. The target frequency number is sent to the FPGA (the comparator in the monitoring and control) through an NI USB-6008 computer interface. The BSF is also tuned to different frequencies compared to the diplexer. For example, in Figure 12.13(b), the BSF is tuned to suppress out-of-band jammers. In Figure 12.13(c) and (d), the BSF is tuned to improve the isolation between the transmitter and the receiver by about 9 dB, which is equivalent to the rejection level of the BSF. Further analysis on modulations is shown in [159].

The diplexer can operate from 0.75 to 1 GHz (~33% tuning ratio). The insertion loss on the receiver side varies between 3.7 dB (at 0.75 GHz) and 2.8 dB (at 1 GHz), while on the transmitter side it is between 2.2 dB (at 0.7 GHz) and 1.8 dB (at 1 GHz), as shown in Figure 12.14. The difference in insertion loss between the transmitter and the receiver is due to the matching transmission line, coupler,

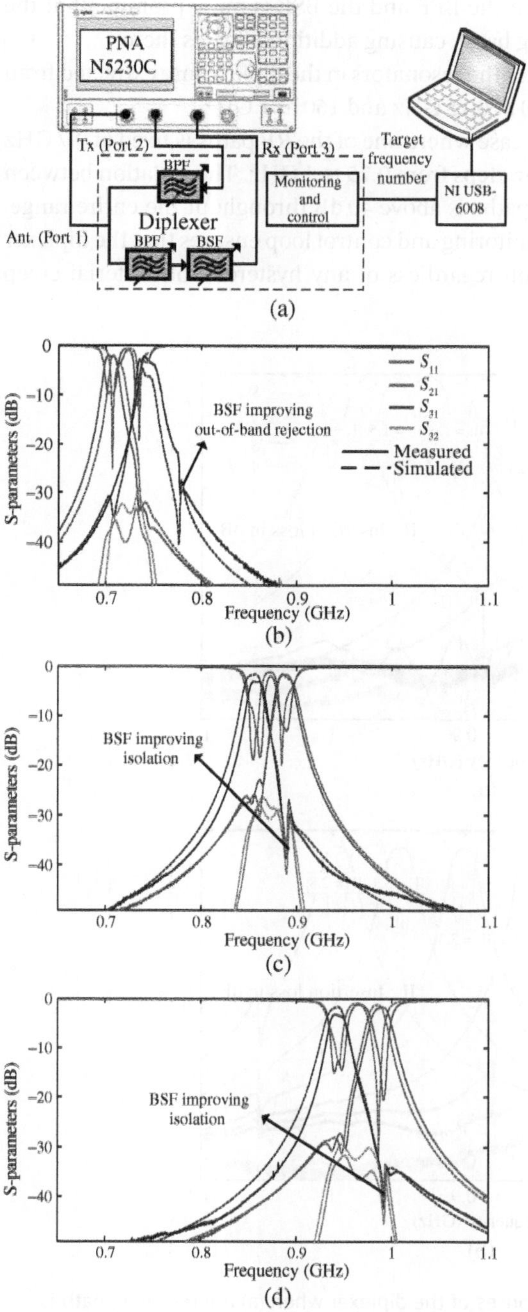


**Figure 12.13** (a) Measurement setup used to measure the RF performance of the diplexer. The monitoring and control is used to tune the diplexer. (b)–(d) The measured (solid) and simulated (dashed) frequency responses of the diplexer under different tuning configurations.

and BSF. The mismatch between the BPF and the BSF is more prominent at the low-frequency part of the tuning band, causing additional losses there.

The unloaded quality factors for the resonators in the diplexer are extracted from simulations and found to be 150 at 0.75 GHz and 160 at 1 GHz.

Figure 12.14 also presents the case where one of the RF paths is fixed at 0.7 GHz and the other is tuned in 50 MHz steps from 0.75 to 1 GHz. The isolation between the transmitting and receiving paths is above 40 dB throughout the entire range. For all measured curves, the monitoring and control loop ensures that the diplexer remains at that tuning condition regardless of any hysteresis or material creep over time.

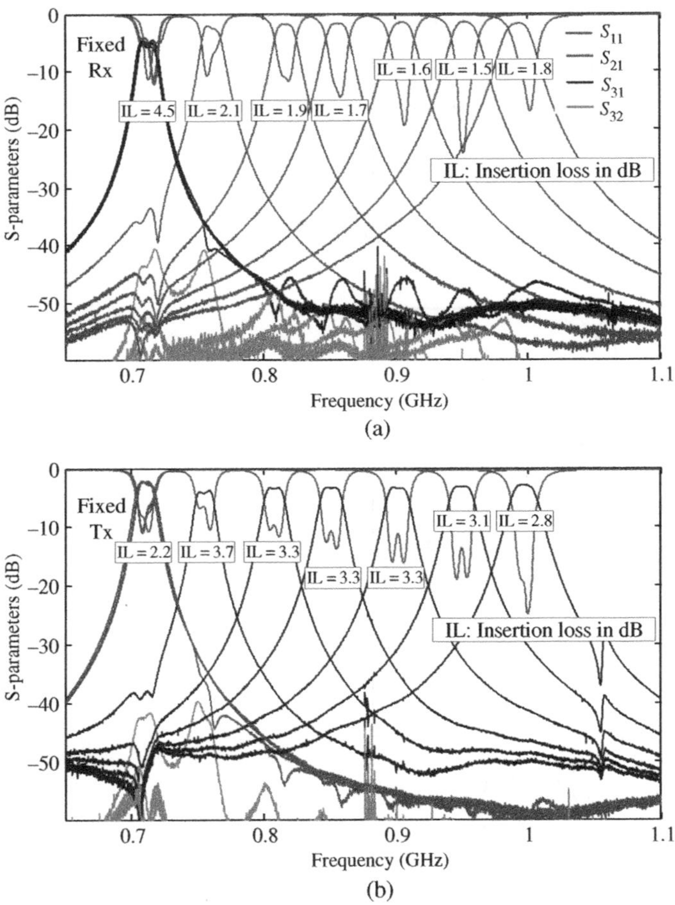


**Figure 12.14** The measured responses of the diplexer when (a) the receiving path is fixed at 0.7 GHz and when (b) the transmitting path is fixed at 0.7 GHz, along with the insertion loss (IL) for each band (in dB). The BSF is tuned outside the window.

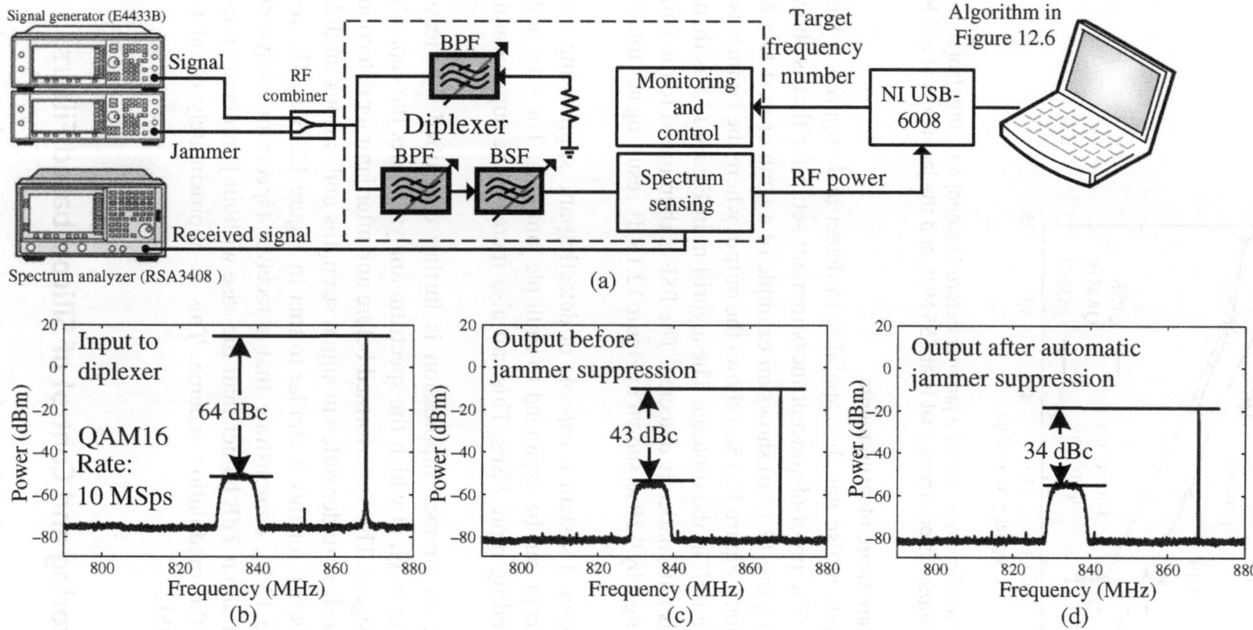


**Figure 12.15** (a) Measurement setup of the presented system during jammer detection and suppression test. (b) The input spectrum of a QAM16 signal (at 835 MHz) along with a jammer (at 868 MHz). (c) The output spectrum from the diplexer. (d) The spectrum after the jammer was automatically detected, and the BSF is tuned to suppress it further.

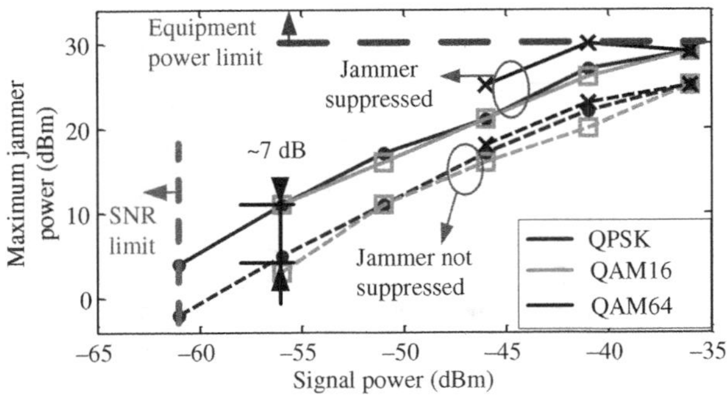


**Figure 12.16** The maximum power of a jammer before (dashed) and after (solid) the automatic jammer suppression. The signal is at 855 MHz, and the jammer is at 835 MHz.

#### 12.1.4.4 Spectrum-aware Measurements

The presented system also has the capability to detect and suppress a jammer in the spectrum. The proof-of-concept measurement setup for that is shown in Figure 12.15(a). Figure 12.15(b) shows an example of a jammer next to a signal (33 MHz separation). Figure 12.15(c) shows the output where the jammer is only suppressed by the BPF in the diplexer. The algorithm in Figure 12.6 is then run, and the jammer is automatically detected. The BSF is then tuned to the jammer, further suppressing it by 9 dB as shown in Figure 12.15(d), resulting in a total rejection of 30 dB.

While the presented system is capable of detecting and suppressing a single jammer, the concept can be expanded to multiple jammers. The only addition required is appending more BSFs. This can also increase the suppression of a single jammer.

The advantage of jammer suppression is further quantified by measuring the jammer power level, at which the spectrum analyzer loses the capability to demodulate the signal. This is performed before and after jammer detection and suppression, as well as under different signal strengths and various modulation techniques. The setup for this is similar to that in Figure 12.15(a). The results, shown in Figure 12.16, demonstrate that a receiver (spectrum analyzer) can handle a jammer about 7 dB higher than the case without jammer suppression, independent of the modulation scheme. This is approximately equal to the rejection of the BSF.

## 12.2 Monitoring and Control of Silicon-based Filters

This section presents a fully electronic monitoring and control system for electrostatically tunable micro-electro-mechanical systems (MEMS) cavity filters.

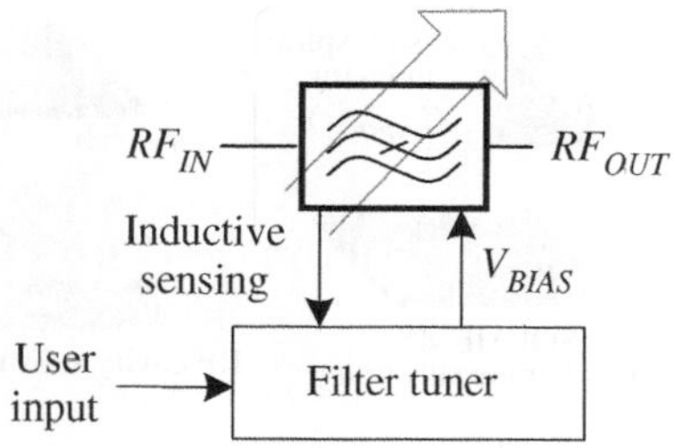


**Figure 12.17** Inductive proximity sensing is used to monitor MEMS tunable filters. The monitoring is the feedback in a control loop, which tunes the filter to a desired frequency response based on an input from the user.

The monitoring of the MEMS membrane is based on inductive proximity sensing for each cavity membrane [160, 161]. The sensing inductor doubles as the bias electrode, minimizing the presence of metallic objects in the vicinity of the magnetic field. The sensing method measures the absolute deflection of the MEMS membrane and adjusts the MEMS DC bias voltage until the membrane is at the desired deflection. Figure 12.17 depicts a conceptual drawing of the system.

### 12.2.1 Monitoring Concepts and Optimizations

Figure 12.18(a) shows a more detailed schematic of the system. When a DC bias voltage ($V_{BIAS}$) is applied to the sensing inductor ($L_{SPIRAL}$), the MEMS membrane bends toward $L_{SPIRAL}$, which increases the resonant frequency of the RF cavity (due to a decrease in the loading capacitance). This results in reducing the effective value of $L_{SPIRAL}$ (due to the counteracting eddy currents in the membrane), which in turn increases the resonant frequency of the series LC circuit. The capacitors C1 and C2 are fixed external components. The shift in the LC resonance frequency changes the amplitude response of the monitoring circuit. As a result, by measuring $V_{SENSE}$ and relating it to the resonant frequency of the corresponding RF cavity, the filter response can be monitored. This information is used to adjust the bias voltage $V_{BIAS}$ until the desired RF response is achieved.

A spiral inductor is chosen due to its easily manufacturable planar structure. Additionally, since the spiral inductor covers most of the area above the membrane, we can conveniently engage it as the MEMS bias electrode.

The circuit is driven by an AC source, referenced to a crystal oscillator. This achieves a stable excitation, which is necessary in detecting the motion of thin membranes (typically ~1 μm) with small deflection range (typically ~10 μm).

#### 12.2.1.1 Inductive Proximity Sensing

A circuit model for the monitoring structure is shown in Figure 12.18(b). Figure 12.18(c) shows the AC equivalent circuit. $R_S$ is the output impedance of

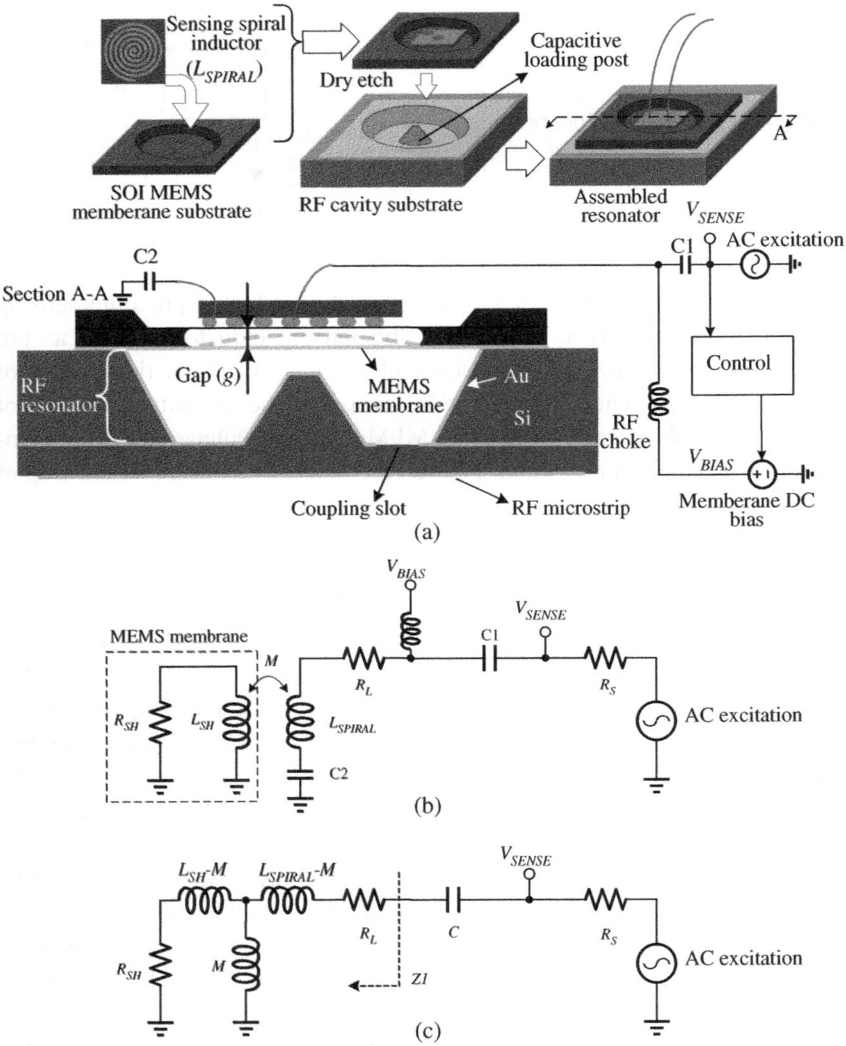


**Figure 12.18** (a) A schematic structure of the inductive sensing method where a spiral inductor also functions as the biasing electrode and the readout from the spiral is used as feedback to bias the MEMS membrane to a desired location. (b) A circuit model for the inductive sensing in (a). (c) Equivalent AC circuit model of the circuit in (b).

the AC source, and $R_L$ is the effective resistance of the spiral inductor. $L_{SH}$ and $R_{SH}$ are the effective sheet inductance and resistance of the membrane, discussed in more detail later in this section.

The value of $L_{SPIRAL}$ is calculated using [162]

$$L_{Spiral} = \frac{\mu_0 N^2}{(r_{out} - r_{in})^2} \times \int_0^{\pi} \int_{r_{in}}^{r_{out}} \int_{r_{in}}^{r_{out}} \frac{\cos(\theta) r_1 r_2 dr_1 dr_2 d\theta}{\sqrt{r_1^2 + r_2^2 - 2r_1 r_2 \cos(\theta)}} \tag{12.12}$$

where $\mu_0$ is the free space permeability, $N$ is the number of turns, and $r_{in}$ and $r_{out}$ are the inner and outer radii of the spiral structure, respectively.

Since the eddy currents in the membrane run in a circular fashion, and since the metal of the membrane has a finite resistance, the membrane can be modeled as an RL circuit as shown in Figure 12.18(b) [163]. The effective inductance of the sheet membrane ($L_{SH}$) can also be calculated using (12.12), while multiplying it by 0.5 to accommodate for the absence of gaps between the tracks and setting $N = 1$.

When the spiral inductor is in close proximity to the metallic membrane, the mutual inductance between them can be calculated as [162]

$$M = \kappa(g)\sqrt{L_{SPIRAL} L_{SH}} = \frac{\mu_0 N}{(r_{out} - r_{in}) r_{out}} \times \int_0^{\pi} \int_{r_{in}}^{r_{out}} \int_0^{r_{out}} \frac{\cos(\theta) r_1 r_2 dr_1 dr_2 d\theta}{\sqrt{g^2 + r_1^2 + r_2^2 - 2r_1 r_2 \cos(\theta)}} \tag{12.13}$$

where $\kappa(g)$ is the coupling coefficient as a function of the gap between the spiral inductor and the membrane, as shown in Figure 12.18(a).

To evaluate the resultant effective inductance of $Z1$ (at the terminals of the spiral), shown in Figure 12.18(c), as a function of the gap $g$, the circuit is analyzed with (12.12) and (12.14) evaluated numerically. The details of the circuit analysis are discussed in Section 12.2.1.2. Figure 12.19 shows the calculated and simulated sensed inductance ($L$ at $Z1$) as a function of $g$. It is worth mentioning here that the practical range indicated in Figure 12.19 is due to the fact that the gap $g$ is small (typically $< 50$ μm) compared to the bias electrodes (spiral inductor in this case, typically $> 1000$ μm).

#### 12.2.1.2 Circuit Optimization

A change in the gap $g$ (whether intentional due to a DC bias or unintentional due to imperfections such as long-term creep) changes the effective value of the sensed inductance at $Z1$, which in turn affects the frequency response of the sensing node ($V_{SENSE}$). The circuit model in Figure 12.18(c) is analyzed, and each parameter in the circuit is optimized to improve the sensitivity of the circuit to changes in $g$.

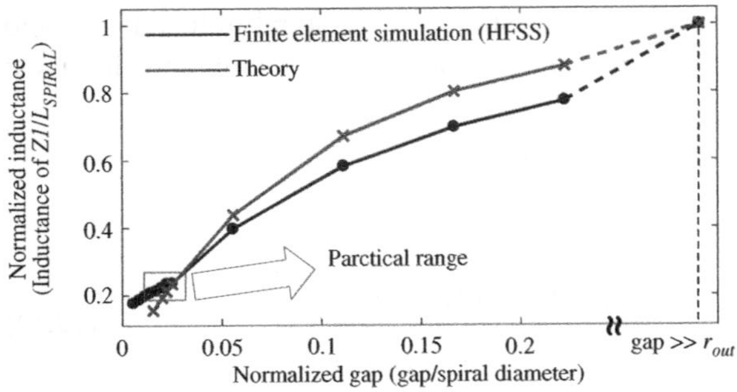


**Figure 12.19** Simulated and calculated equivalent inductance of $Z1$ versus the gap $g$.

The sensitivity of the circuit primarily depends on how much $Z1$ changes per unit change in g. To simplify the analysis, an impedance sensitivity is defined as

$$\begin{aligned} ImpedanceSensitivity &= \left|\frac{\partial Z1}{\partial g}\right| \\ &= \left|\frac{\partial M}{\partial g} \times \frac{\partial Z1}{\partial M}\right| \end{aligned} \tag{12.14}$$

In order to maximize the sensitivity, each term in (12.14) is maximized separately.

To optimize the contribution of the first term in (12.14), $\frac{\partial M}{\partial g}$ is evaluated from (12.13)

$$\frac{\partial M}{\partial g} = \frac{\mu_0 N}{(r_{out} - r_{in})r_{out}} \times \int_0^{\pi} \int_{r_{in}}^{r_{out}} \int_0^{r_{out}} \frac{-g\cos(\theta) r_1 r_2 dr_1 dr_2 d\theta}{(g^2 + r_1^2 + r_2^2 - 2r_1 r_2 \cos(\theta))^{\frac{3}{2}}} \tag{12.15}$$

As we observe from (12.15), the sensitivity of the mutual inductance to the gap increases by decreasing g. Decreasing g arbitrarily, however, can limit the tuning range and can complicate the fabrication process. Also, increasing the number of turns $N$ and decreasing $(r_{out} - r_{in})$ can further increase $\frac{\partial M}{\partial g}$. On the other hand, these changes may require small feature dimensions that complicate the fabrication of the spiral inductor.

To optimize the contribution of the second term in (12.15), $Z1$ is found by analyzing the circuit shown in Figure 12.18(c) as

$$Z1 = R_L + j\omega L_{SPIRAL} + \frac{\omega^2 M^2}{R_{SH} + j\omega L_{SH}} \tag{12.16}$$

where $\omega$ is the operating angular frequency of the AC excitation. From (12.16), we get

$$\frac{\partial Z1}{\partial M} = \frac{2\omega^2 M}{R_{SH} + j\omega L_{SH}} = \frac{2\omega^2 \kappa \sqrt{L_{SPIRAL} L_{SH}}}{R_{SH} + j\omega L_{SH}} \tag{12.17}$$

From (12.17), we observe that we can improve by increasing the mutual inductance. This can be achieved by increasing the coupling coefficient and by increasing $L_{SPIRAL}$, which is consistent with the conclusions drawn from (12.15). Increasing the operating frequency is also an efficient way to increase the sensitivity, as long as the frequency is below the inductor self-resonant frequency.

While reducing sheet resistance and inductance of the membrane, $R_{SH}$ and $L_{SH}$, by increasing the metal thickness improves the sensitivity, it might result in a much higher actuation voltage to bend the membrane. As a result, $R_{SH}$ and $L_{SH}$ are typically dictated by the MEMS design and fabrication process.

For further circuit optimizations, the sensing voltage $V_{SENSE}$ is evaluated. $V_{SENSE}$ is found by analyzing the circuit in Figure 12.18(c), resulting in

$$V_{SENSE} = V_{IN}\left(1 - \frac{j\omega CR_S}{1 + j\omega CR_S + j\omega CZ1}\right) \tag{12.18}$$

Due to the complexity of the expressions in (12.18) and (12.16), the effect of each design parameter might not be obvious. As a result, the amplitude of $V_{SENSE}$ is evaluated numerically and verified with Simulation Program with Integrated Circuit Emphasis (SPICE) simulations. Figure 12.20 shows the effect of each design parameter on $|V_{SENSE}|$.

As shown in Figure 12.20(a), when driving the circuit at a frequency where $V_{SENSE}$ has the highest slope $\left(\left|\frac{\partial V_{SENSE}}{\partial \omega}\right|\right)$, the voltage sensitivity to a change in g is maximized. In other words, if the circuit is operating at the maximum slope frequency, and the gap g changes, the resonant frequency shifts, causing the maximum change in $V_{SENSE}$. The evaluation of (12.18) shows that the maximum slope occurs just below the resonant frequency as shown in the figure. It is important here to realize that, since a MEMS membrane deflection is typically very small (10–20 μm) compared to the size of the actuating inductor, the deviation in the frequency of maximum slope is also small. As a result, a linear approximation of the maximum slope region is used across the tuning range.

We should also note here that choosing the frequency of the highest slope in $V_{SENSE}$ does not contradict the requirement of choosing high frequency as concluded from (12.17). To satisfy both conditions, the frequency of highest slope should be placed as high as possible by reducing $C$. The effect of reducing $C$, or increasing $L_{SPIRAL}$, which improves the sensitivity, is shown in Figure 12.20(b).

The effect of the spiral resistance ($R_L$) is shown in Figure 12.20(c). Lower inductor resistance causes a sharper frequency response, resulting in a better voltage sensitivity. In order to reduce $R_L$, the spiral inductor should be fabricated with thick metal on a high-quality substrate.

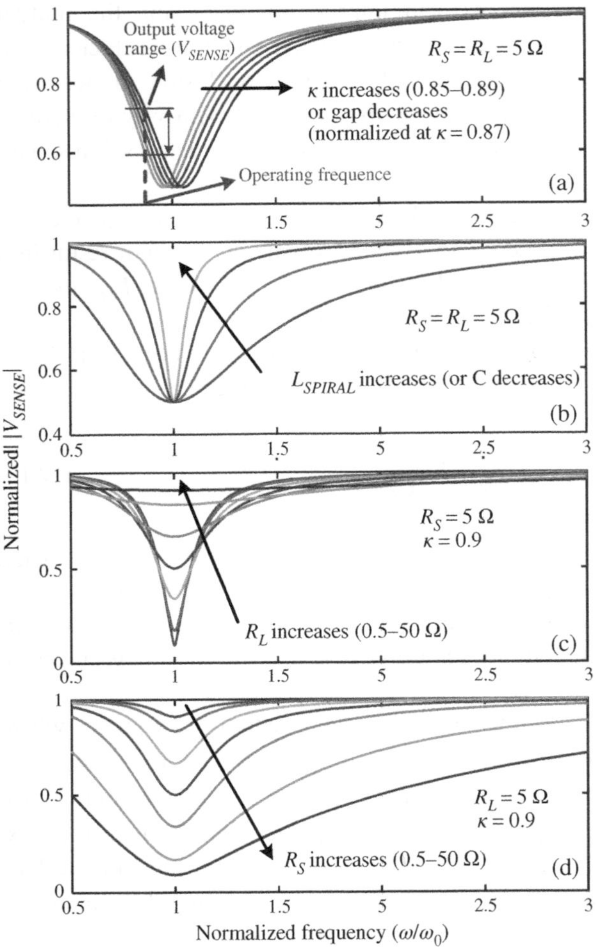


**Figure 12.20** The amplitude of $V_{SENSE}$ versus frequency and how it is affected by (a) the coupling coefficient (or the gap), (b) the spiral inductance and the external capacitance, (c) the inductor resistance $R_L$, and (d) the source resistance $R_S$.

If there is design flexibility to control the source impedance ($R_S$), its value can further optimize the sensitivity, as plotted in Figure 12.20(d). When $R_S \ll R_L$ or $R_S \gg R_L$, the output shows no significant slope near the resonant frequency. As a result, for each $R_L$, there is an optimal value for $R_S$ that produces the highest sensitivity. The optimal value of $R_S$ can be found numerically by sweeping its value and calculating the maximum slopes of the curves in Figure 12.20(d) for a given value of $R_L$. An example of the relationship between $R_L$ and optimal $R_S$ is shown in Figure 12.21.

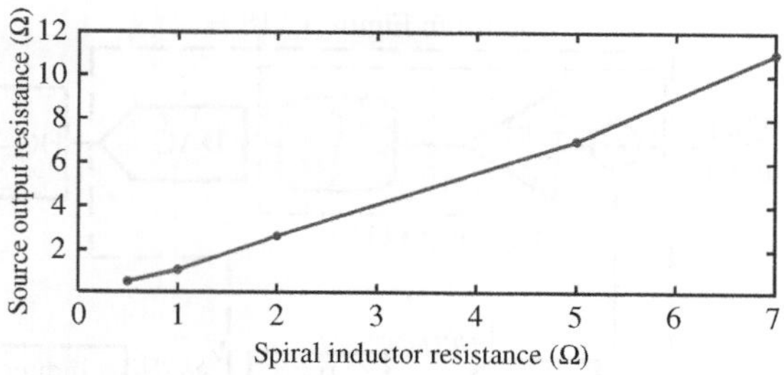


**Figure 12.21** For each value of the inductor resistance $R_L$, there is a specific value for the source resistance $R_S$ such that the sensitivity is maximized.

#### 12.2.1.3 Spiral Inductor

The analysis shown previously assumes the general case where $L_{SPIRAL}$ and $R_L$ can vary independently. In realistic implementations, however, a higher inductance typically results in a higher $R_L$. The increase in $L_{SPIRAL}$ and the increase in $R_L$ have opposite effects on sensitivity. As a result, a tradeoff exists when using a spiral inductor.

Figure 12.22 shows the normalized highest slope on the amplitude of $V_{SENSE}$ versus the number of turns in the spiral inductor. Since $r_{in}$ and $r_{out}$ are kept constant (1700 μm and 2000 μm, respectively, which fits the MEMS design), the tracks are fitted to the same area by reducing the track width and the spacing between them. As shown in Figure 12.22, our design results in the highest voltage sensitivity when the spiral inductor has six or seven turns.

### 12.2.2 Control System Design

The control loop utilizes the monitoring system discussed earlier for feedback to ensure that the MEMS membrane sheet is at the desired deflection, which results

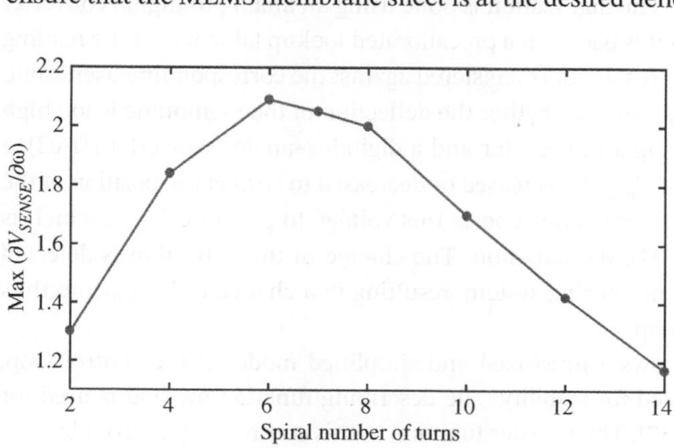


**Figure 12.22** The tradeoff between the number of turns in the spiral inductor and its effect on the sensitivity of the inductive monitoring.

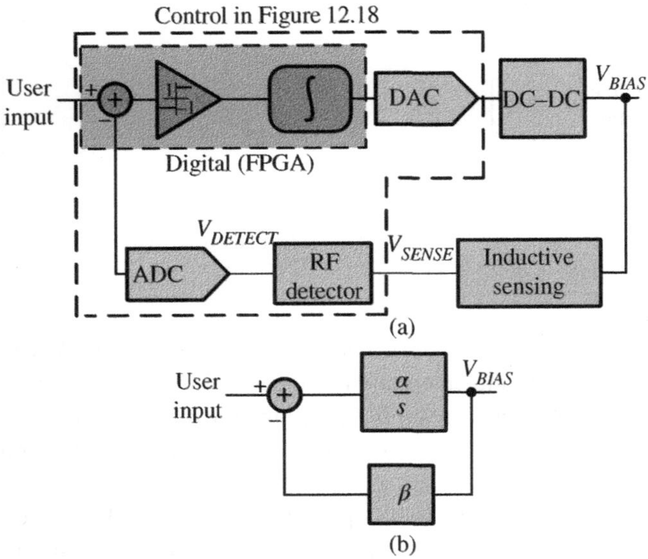


**Figure 12.23** (a) A block diagram of the presented monitoring and control system and (b) a linearized and simplified block diagram of the control loop in (a).

in the desired filter frequency response. A separate independent control loop is applied to each resonator in the filter.

The control loop, shown in Figure 12.23(a), takes a digital input from the user and compares it to a digitized version of $V_{DETECT}$, which is the measure of the amplitude of $V_{SENSE}$. The digitization is done using an analog-to-digital converter (ADC). The user input is based on a precalibrated lookup table where the reading from the NA for each resonator is registered against the corresponding user input. This comparison determines whether the deflection of the membrane is too high or too low. Then, using an integrator and a digital-to-analog converter (DAC), a low-voltage replica of $V_{BIAS}$ is increased or decreased to correct the position of the membrane. A DC–DC converter boosts this voltage to generate $V_{BIAS}$, which is sufficiently high for MEMS actuation. The change in the deflection is detected using the inductive monitoring system, resulting in a change in $V_{SENSE}$ and thus closing the control loop.

Figure 12.23(b) shows a linearized and simplified model of the control loop, which can be analyzed for stability. The describing function method is used for linearization [128, 157]. The transfer function of this control loop is given by

$$H(s) = \frac{\alpha}{s + \alpha\beta} \tag{12.19}$$

where $\alpha$ and $\beta$ are positive scalars resembling the forward and feedback path gains, respectively. $H(s)$ has a single pole in the left-hand plane (at $-\alpha\beta$), which denotes that the system is unconditionally stable for bounded inputs.

### 12.2.3 Testbed Filter Design

This monitoring and control system is implemented on an absorptive BSF operating in the 30–42 GHz frequency range. The design of such filters is discussed in Chapter 6.

### 12.2.4 Implementation and Measurements

#### 12.2.4.1 Fabrication

The spiral inductors are fabricated using a PCB fabrication process on a 4-mil-thick 370HR substrate, with 8.9 μm thick copper. The spiral inductor has seven turns, and the measured $L_{SPIRAL}$ is 42 nH compared to the simulated 51 nH. This discrepancy is due to fabrication and simulation modeling errors.

The MEMS membranes and filter cavities are fabricated on a silicon substrate in a process similar to the one in [101], with the exception that each biasing electrode is replaced with a spiral inductor as shown in Figure 12.24(a). The spiral inductors are glued to the MEMS substrate. Each gold membrane is 1 μm thick and has 5-μm-deep corrugations to improve its stability and reduce the required actuation voltage. The membranes are released using $XeF_2$ dry etching. Small holes (5 μm in radius) are etched to provide access for the etchant to remove the silicon layer beneath it. The MEMS membrane substrate is then attached on the cavity substrate to form the BSF.

The BSF is placed on a connectorized PCB to simplify testing. The connectors are 2.4 mm end launch. The connectorized filter structure is shown in Figure 12.24(b).

The connectorized filter is then attached to the inductive sensing and control board as shown in Figure 12.24(c). The inductive sensing and control PCB is fabricated on a 4-layer, 62-mil-thick Rogers 4350B substrate. The spiral inductors' terminals are attached to the monitoring circuits and are driven by the AC signal and the bias signal. The AC signal is generated by an FPGA and an operational amplifier, which is used for conditioning and harmonic filtering. The components used to realize the control system in Figure 12.23 are summarized in Table 12.2.

The system runs from a single 5 V power supply and consumes a total of 350 mA for the control of both resonators: 130 mA is consumed in the FPGA development board and 28 mA in the DC–DC converters. Since most of the monitoring and control components are electronic, embedding the system in an on-chip solution with high-efficiency charge pumps [164] can significantly reduce power consumption.

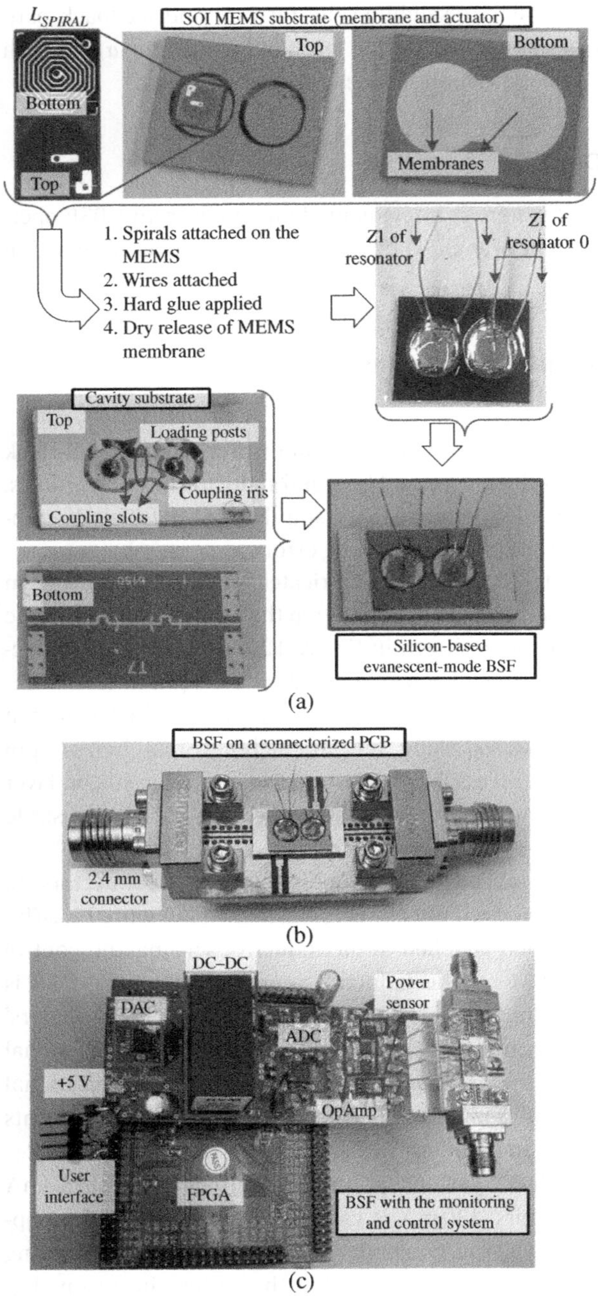


**Figure 12.24** (a) The assembly process of the silicon-based RF MEMS filter with the monitoring spiral inductor. (b) The BSF is placed on a connectorized PCB to simplify RF testing. (c) The PCB of the filter attached to the rest of the monitoring and control system.

**Table 12.2** List of components used in the monitoring and control.

| | | | |
|---|---|---|---|
| DAC | DAC8565 | ADC | ADS8353 |
| FPGA | Altera Cyclone IV | RF power sensor | LTC5507 |
| DC–DC | UMHV0505 | Conditioning Opamp | MAX4203 |
| C1, C2 | 47 pF | RF choke | 1.5 μH |

#### 12.2.4.2 Sensing Feedback

The inductive sensing blocks are measured separately to verify their operation. When the filter is attached to the PCB substrate, as shown in Figure 12.24(b), the spiral inductor leads are connected to a network analyzer to measure the inductance versus the membrane bias voltage $V_{BIAS}$. Since increasing the bias voltage decreases the gap g, the inductance is expected to decrease, which is observed in the measurements in Figure 12.25. The value of the inductance is then used to extract the value of the gap g from simulation results. The corresponding values of g are also plotted in Figure 12.25.

With the filter connected to the sensing circuit as shown in Figure 12.24(c), the output of the power detector $V_{DETECT}$ is measured across a range of excitation frequencies. This measurement, shown in Figure 12.26(a), is used to find the frequency of the highest sensitivity as discussed in Section 12.2.1.2. The fabricated inductive monitoring system shows highest sensitivity at 142 MHz, which is chosen as the AC monitoring frequency. While increasing this frequency, by decreasing C1 and C2, can improve the sensitivity, the values of the capacitors are chosen such that the operating frequency is well below the resonance of the spiral.

Figure 12.26(b) shows the change in $V_{DETECT}$ versus the bias voltage. As shown in the figure, the majority of the sensing occurs when the bias voltage exceeds 50 V, which is consistent with the inductance measurement shown in Figure 12.25. This measured range of $V_{DETECT}$ dictates the operating sensing

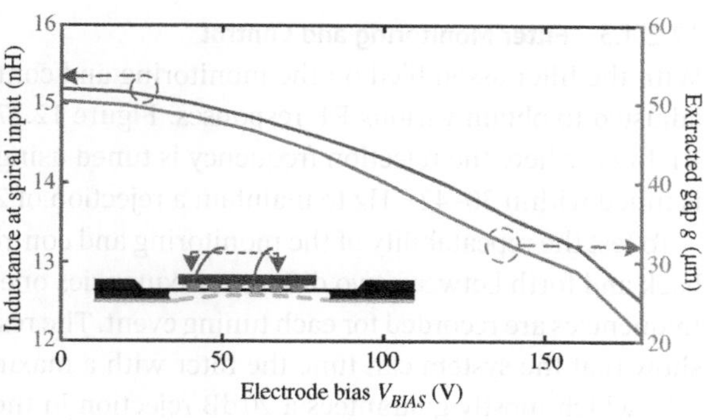


**Figure 12.25** The measured sensed inductance at *Z*1 versus the bias voltage, and the membrane gap *g* is extracted from the simulated inductance values shown in Figure 12.19.

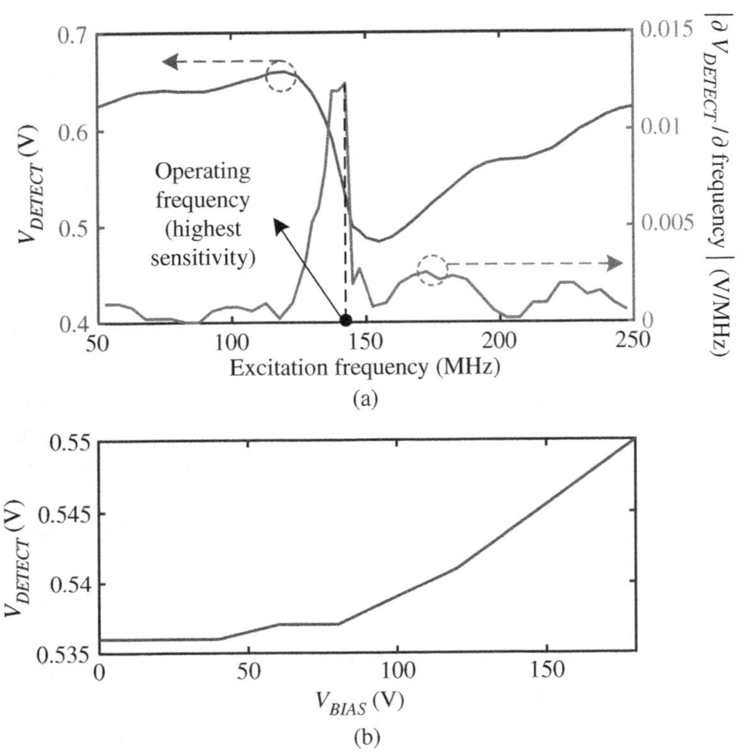


**Figure 12.26** (a) The measured frequency response of $V_{DETECT}$ and its derivative, which dictates the operating frequency. (b) The measured $V_{DETECT}$ versus the bias voltage $V_{BIAS}$, with the system operating at the highest sensitivity frequency. This relationship is the readout of the inductive sensing.

range. In other words, the user input should be the digital code for the voltage levels in this range. It should be noted that the measurement in Figure 12.26(b) shows a monotonic relationship, which is necessary to sustain stable control.

#### 12.2.4.3 Filter Monitoring and Control

With the filter assembled on the monitoring and control board, the user input is adjusted to obtain various RF responses. Figure 12.27 shows the response of the RF filter, where the rejection frequency is tuned using the control loop. The filter is tuned within 30–42 GHz to maintain a rejection of 25–51 dB.

To test the repeatability of the monitoring and control system, the filter is tuned back and forth between two different frequencies one hundred times. The center frequencies are recorded for each tuning event. The results, shown in Figure 12.28, show that the system can tune the filter with a maximum error of 1% (spread of 2%), which mostly guarantees a 20 dB rejection in the target filter response. The

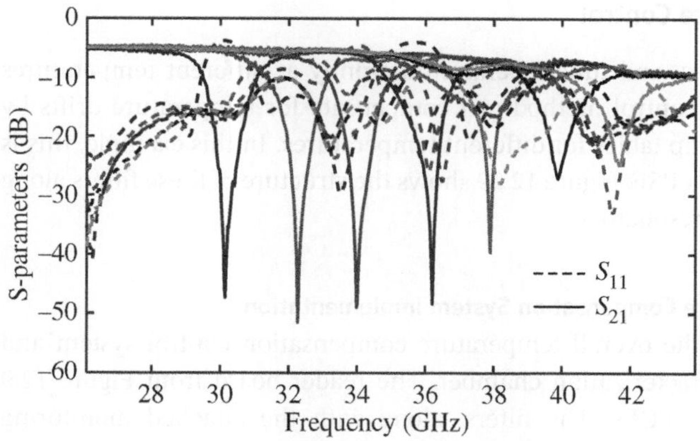


**Figure 12.27** The monitoring and control system is used to tune the BSF notch to various frequencies.

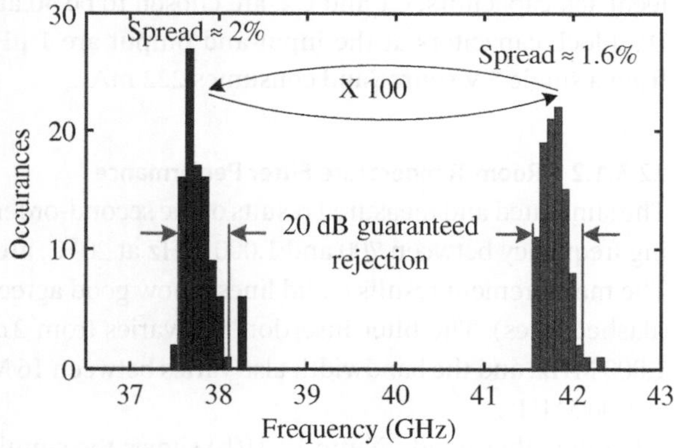


**Figure 12.28** The filter is tuned between two different frequencies one hundred times, and the center frequency is recorded for each tuning.

observed spread in the measurements is primarily due to MEMS fabrication issues such as residual stress. Well-calibrated fabrication processes can further improve this.

## 12.3 Monitoring and Control Applications: Temperature Compensation

The monitoring and control concepts presented earlier in this chapter can take a user input and translate it to a well-tuned filter in the field, without requiring test equipment. This feature can also help operate the filter in a more complex environment. In this section, we present an example of stabilizing a filter's response against temperature variations.

### 12.3.1 Temperature Control

Electronics, substrates, and tuners behave differently at different temperatures [165]. The feedback control method can compensate for temperature drifts by having different lookup tables for different temperatures. In this example, this is applied to a BPF and a BSF. Figure 12.29 shows the structure of these filters along with the monitoring resonators.

#### 12.3.1.1 Temperature Compensation System Implementation

Figure 12.30 shows the overall temperature compensation control system and the temperature characterization chamber. The reader board from Figure 12.9 is implemented with PCBs. The filters, along with the attached monitoring resonators and the reader board (including the oscillation circuit and the divider), are all placed in a hot/cold chamber for temperature measurements. The Infineon BFP420 low noise SiGe bipolar RF transistor is used in the oscillation circuit. The feedback capacitors, C1 and C2, are chosen to be 40 and 6.8 pF, respectively. The DC block capacitors at the input and output are 1 μF. The control system runs from a single 5 V source and consumes 222 mA.

#### 12.3.1.2 Room Temperature Filter Performance

The simulated and measured results of the second-order Chebyshev BPF, with tuning frequency between 700 and 1,000 MHz at 20 °C, are shown in Figure 12.31(a). The measurement results (solid lines) show good agreement with the simulations (dashed lines). The filter insertion loss varies from 2 dB at 700 MHz to 1.2 dB at 1,000 MHz, and the bandwidth also varies between 16 MHz at 700 MHz to 32 MHz at 1,000 MHz.

On the other hand, Figure 12.31(b) shows the simulated and measured results of the second-order Butterworth BSF at different frequencies between 1,300 and 1,600 MHz at 20 °C. The stopband rejections vary between 30 dB and 34 dB, and the insertion loss is around 0.16 dB. The 3 dB bandwidth is 21 MHz at 1,300 MHz and 39 MHz at 1,600 MHz.

#### 12.3.1.3 Temperature Compensation Performance

As mentioned in the beginning of this section, the temperature calibration table is first constructed in this part. Then, the closed-loop RF fillter measurement results with the control system are presented and compared to the results from the open-loop (without the control system) filters.

Figure 12.32(a) shows the measured temperature-dependent calibration curve for one resonator of the BPF. Specifically, this figure depicts the measured RF frequency of the bandpass resonator as a function of the input data within the

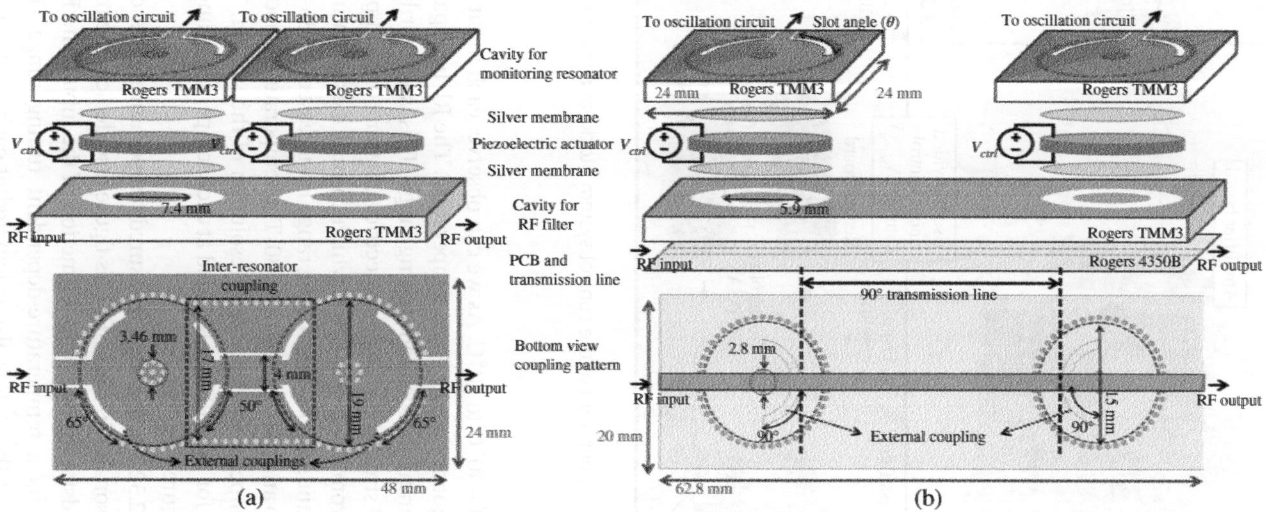


**Figure 12.29** The physical structures of (a) the second-order BPF and (b) the second-order BSF with monitoring resonators.

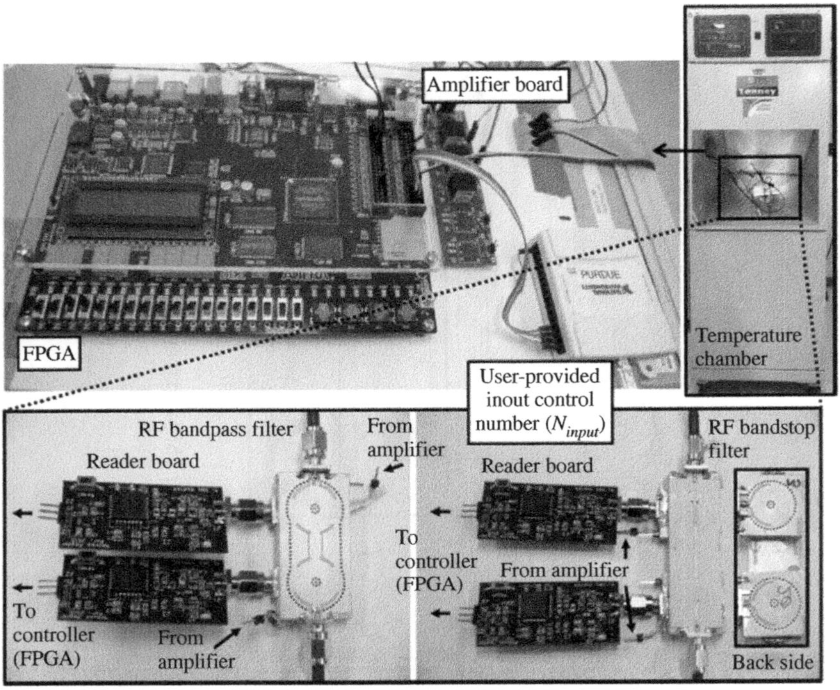


**Figure 12.30** The implementation of the control system blocks and the measurement setup.

temperature range of −40 °C to 80 °C. As we can observe, each input is related to one unique filter frequency at any given temperature. The RF bandpass resonator also exhibits a temperature-dependent tuning range. This is primarily due to the temperature-induced stresses on the piezoelectric actuator. For example, at −40 °C the tuning range is from 555 to 793 MHz, which corresponds to an input change of 27. On the other hand, at 20 °C the tuning range is increased to 536–1,022 MHz, which corresponds to an input change of 1,580. The temperature-dependent input range results in a temperature-dependent resolution of the system. The average frequency resolution for the BPF is calculated at each temperature, and the results are also shown in Figure 12.32(a).

Similarly, Figure 12.32(b) shows the measured temperature-dependent calibration curve for one resonator in the BSF. This figure also depicts the measured RF frequency of the bandstop resonator as a function of the input. The RF bandstop resonator also exhibits a temperature-dependent tuning range. The average frequency resolution of the BSF is also calculated at each temperature, and the results are also shown in Figure 12.32(b).

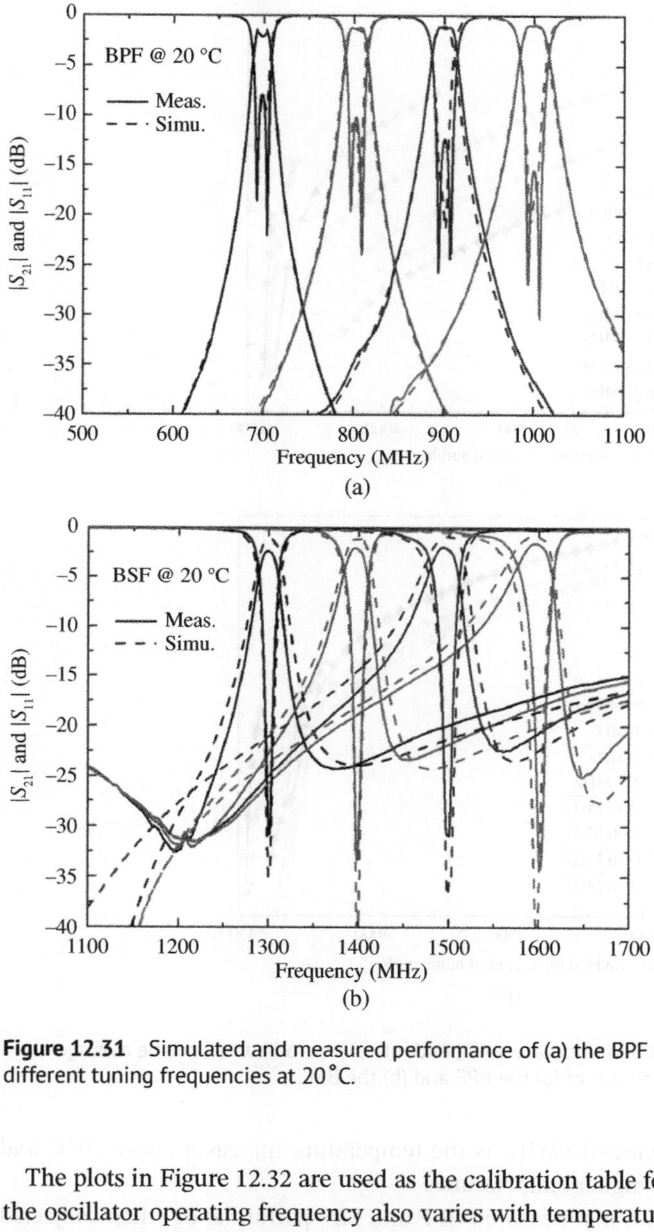


**Figure 12.31** Simulated and measured performance of (a) the BPF and (b) the BSF at different tuning frequencies at 20 °C.

The plots in Figure 12.32 are used as the calibration table for the filters. While the oscillator operating frequency also varies with temperature, these variations are compensated for during the calibration procedures.

For both calibration tables, the resolution is relatively low at −40 °C because of the low bending ability of the piezoelectric actuator. For the same reason, the frequency tuning range decreases at lower temperatures. On the other hand,

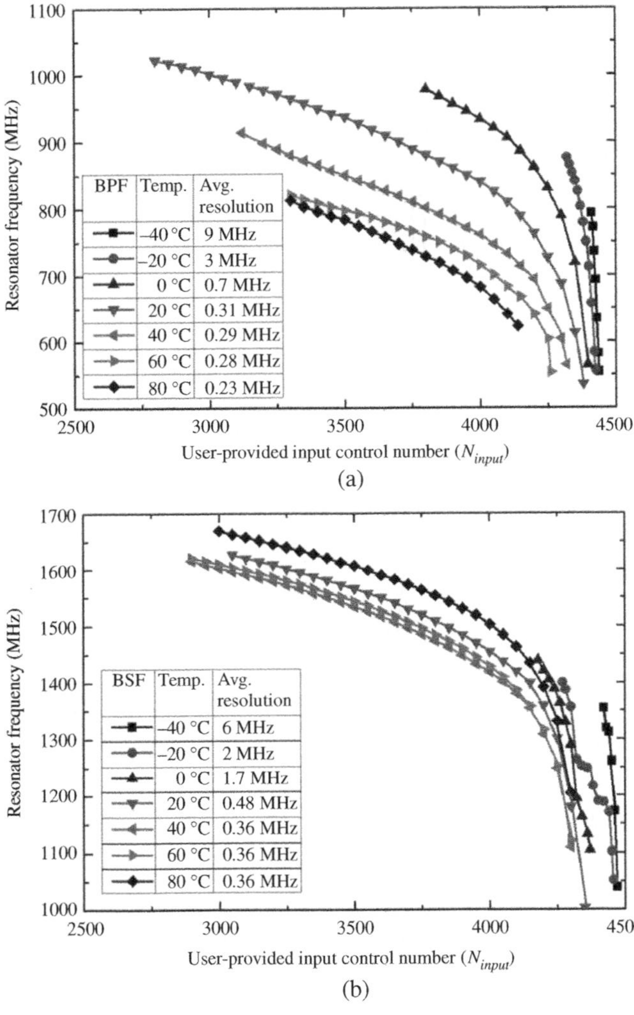


**Figure 12.32** Measured temperature-dependent calibration curve and the average resolutions of the resonator of (a) the BPF and (b) the BSF.

the resolution is below 0.5 MHz as the temperature increases above 20°C and improves further at higher temperatures.

Figure 12.33 shows the measured BPF and BSF performance when subjected to temperatures between −40°C and 80°C, both under open-loop and real-time closed-loop operation. The dashed lines of Figure 12.33(a) show the open-loop BPF response for various temperatures with constant bias voltage. The voltage is chosen to bias the BPF response to 700 MHz at 20°C. The solid lines show the BPF

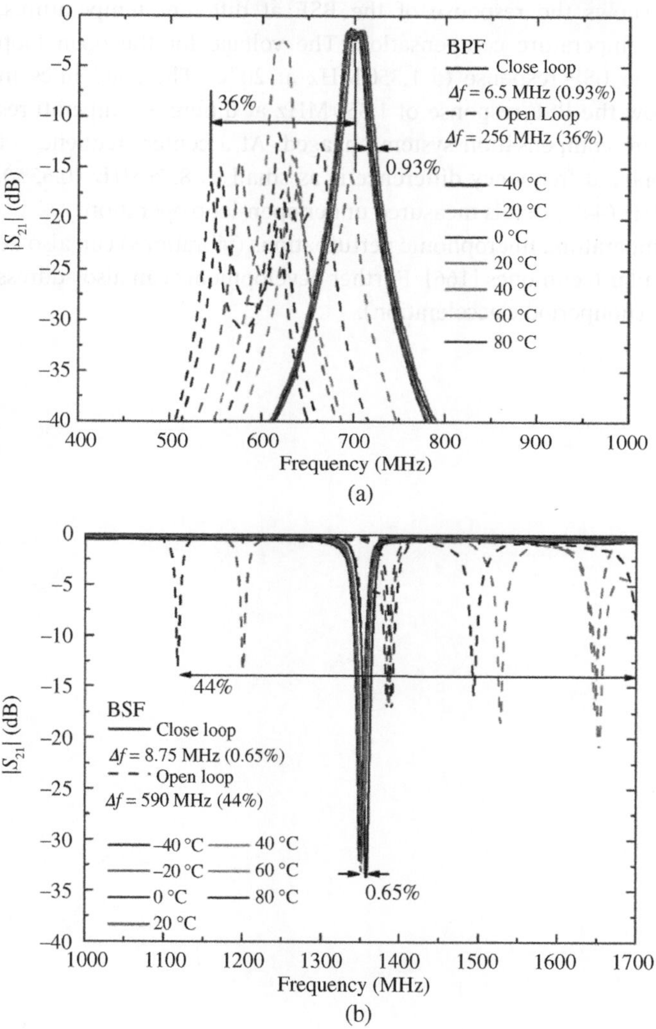


**Figure 12.33** Measured open-loop and closed-loop RF frequency response of (a) the BPF and (b) the BSF at different temperatures.

response of 700 MHz at different temperatures with the temperature compensation control system engaged. With the real-time control system, the filter operating frequency can be automatically maintained at the desired frequency regardless of temperature. For example, at a center frequency of 700 MHz, the passband frequency difference is as small as 6.5 MHz (0.93%), compared to 256 MHz (36%) that is measured under open-loop operation.

Figure 12.33(b) shows the response of the BSF at different temperatures, without and with temperature compensation. The voltage for the open loop is chosen to bias the BSF response to 1,350 MHz at 20°C. The solid lines in Figure 12.33(b) show the BSF response of 1,350 MHz at different temperatures with the temperature compensation system engaged. At a center frequency of 1,350 MHz, the stopband frequency difference is as small as 8.75 MHz (0.65%), compared to 590 MHz (44%) that is measured under open-loop operation.

In addition to temperature, microphonic perturbations (vibrations) can also be mitigated using similar techniques [166]. Further developments can also address shock performance (nonperiodic acceleration).

## References

1 A. T. Wegener and W. J. Chappell, “Simultaneous transmit and receive with a small planar array,” in *2012 IEEE/MTT-S International Microwave Symposium Digest*, 2012, pp. 1–3.

2 C. Andrews and A. C. Molnar, “A passive-mixer-first receiver with baseband-controlled rf impedance matching, < 6db nf, and > 27dbm wideband iip3,” in *2010 IEEE International Solid-State Circuits Conference – (ISSCC)*, 2010, pp. 46–47.

3 C. Andrews and A. C. Molnar, “A passive mixer-first receiver with digitally controlled and widely tunable rf interface,” *IEEE Journal of Solid-State Circuits*, vol. 45, no. 12, pp. 2696–2708, 2010.

4 C. Andrews and A. C. Molnar, “Implications of passive mixer transparency for impedance matching and noise figure in passive mixer-first receivers,” *IEEE Transactions on Circuits and Systems I: Regular Papers*, vol. 57, no. 12, pp. 3092–3103, 2010.

5 A. Mirzaei, H. Darabi, and D. Murphy, “Architectural evolution of integrated m-phase high-q bandpass filters,” *IEEE Transactions on Circuits and Systems I: Regular Papers*, vol. 59, no. 1, pp. 52–65, 2012.

6 A. Ghaffari, E. Klumperink, and B. Nauta, “8-path tunable rf notch filters for blocker suppression,” in *2012 IEEE International Solid-State Circuits Conference*, 2012, pp. 76–78.

7 C. Andrews, C. Lee, and A. Molnar, “Effects of lo harmonics and overlap shunting on n-phase passive mixer based receivers,” in *2012 Proceedings of the ESSCIRC (ESSCIRC)*, 2012, pp. 117–120.

8 T. Snow, C. Fulton, and W. J. Chappell, “Transmit–receive duplexing using digital beamforming system to cancel self-interference,” *IEEE Transactions on Microwave Theory and Techniques*, vol. 59, no. 12, pp. 3494–3503, 2011.

*Tunable Evanescent-Mode Filters: Principles, Implementation, and Applications*,
First Edition. Dimitrios Peroulis and Mohammad Abu Khater.

Companion website: www.wiley.com/go/Peroulis/TunableEvanescent-ModeFilters

**9** R. J. Cameron, "Advanced filter synthesis," *IEEE Microwave Magazine*, vol. 12, no. 6, pp. 42–61, 2011.

**10** A. Atia, A. Williams, and R. Newcomb, "Narrow-band multiple-coupled cavity synthesis," *IEEE Transactions on Circuits and Systems*, vol. 21, no. 5, pp. 649–655, 1974.

**11** R. J. Cameron, "General coupling matrix synthesis methods for chebyshev filtering functions," *IEEE Transactions on Microwave Theory and Techniques*, vol. 47, no. 4, pp. 433–442, 1999.

**12** R. J. Cameron, "Advanced coupling matrix synthesis techniques for microwave filters," *IEEE Transactions on Microwave Theory and Techniques*, vol. 51, no. 1, pp. 1–10, 2003.

**13** D. M. Pozar, *Microwave Engineering*, 4th ed. Wiley, 2011.

**14** J.-S. Hong, *Microstrip Filters for RF/Microwave Applications*, 2nd ed. John Wiley & Sons, 2011.

**15** R. J. Cameron, C. M. Kudsia, and R. R. Mansour, *Microwave Filters for Communication Systems: Fundamentals, Design and Applications*, 1st ed. Wiley-Interscience, 2007.

**16** J. Lee, E. J. Naglich, H. H. Sigmarsson, D. Peroulis, and W. J. Chappell, "Tunable inter-resonator coupling structure with positive and negative values and its application to the field-programmable filter array (fpfa)," *IEEE Transactions on Microwave Theory and Techniques*, vol. 59, no. 12, pp. 3389–3400, 2011.

**17** D. R. Jachowski, "Passive enhancement of resonator q in microwave notch filters," in *2004 IEEE MTT-S International Microwave Symposium Digest (IEEE Cat. No.04CH37535)*, vol. 3, 2004, pp. 1315–1318.

**18** D. Swanson and G. Macchiarella, "Microwave filter design by synthesis and optimization," *IEEE Microwave Magazine*, vol. 8, no. 2, pp. 55–69, 2007.

**19** E. J. Naglich, J. Lee, D. Peroulis, and W. J. Chappell, "Bandpass–bandstop filter cascade performance over wide frequency tuning ranges," *IEEE Transactions on Microwave Theory and Techniques*, vol. 58, no. 12, pp. 3945–3953, 2010.

**20** S. Amari, "Synthesis of cross-coupled resonator filters using an analytical gradient-based optimization technique," *IEEE Transactions on Microwave Theory and Techniques*, vol. 48, no. 9, pp. 1559–1564, 2000.

**21** K. Wu and W. Meng, "A direct synthesis approach for microwave filters with a complex load and its application to direct diplexer design," *IEEE Transactions on Microwave Theory and Techniques*, vol. 55, no. 5, pp. 1010–1017, 2007.

**22** M. Meng and K. Wu, "Direct synthesis of general chebyshev bandpass filters with a frequency variant complex load," in *2010 IEEE MTT-S International Microwave Symposium*, 2010, pp. 433–436.

**23** K. Yoshida, T. Takahashi, H. Kanaya, T. Uchiyama, and Z. Wang, "Superconducting slot antenna with broadband impedance matching circuit," *IEEE Transactions on Applied Superconductivity*, vol. 11, no. 1, pp. 103–106, 2001.

**24** L. Y. G. Matthaei and E. M. T. Jones, *Microwave Filters, Impedance-Matching Networks, and Coupling Structures*. McGraw-Hill, 1980.

**25** A. R. Brown and G. M. Rebeiz, "A varactor-tuned rf filter," *IEEE Transactions on Microwave Theory and Techniques*, vol. 48, no. 7, pp. 1157–1160, 2000.

**26** M. Koochakzadeh and A. A. Tamijani, "Multi-scale tunable filter covering a frequency range of 6.5:1," in *2008 IEEE MTT-S International Microwave Symposium Digest*, 2008, pp. 1023–1026.

**27** H. Dayal, "An overview of miniaturization of varactor tuned printed filter technology," in *IEEE International Microwave Symposium Workshop*, 2007.

**28** A. Abbaspour-Tamijani, L. Dussopt, and G. M. Rebeiz, "Miniature and tunable filters using mems capacitors," *IEEE Transactions on Microwave Theory and Techniques*, vol. 51, no. 7, pp. 1878–1885, 2003.

**29** D. Peroulis, S. Pacheco, K. Sarabandi, and P. B. Katehi, "Mems devices for high isolation switching and tunable filtering," in *2000 IEEE MTT-S International Microwave Symposium Digest (Cat. No.00CH37017)*, vol. 2, 2000, pp. 1217–1220.

**30** A. Pothier, J. C. Orlianges, G. Zheng, C. Champeaux, A. Catherinot, D. Cros, P. Blondy, and J. Papapolymerou, "Low-loss 2-bit tunable bandpass filters using mems dc contact switches," *IEEE Transactions on Microwave Theory and Techniques*, vol. 53, no. 1, pp. 354–360, 2005.

**31** F. A. Miranda, G. Subramanyam, F. W. van Keuls, R. R. Romanofsky, J. D. Warner, and C. H. Mueller, "Design and development of ferroelectric tunable microwave components for kuand k-band satellite communication systems," *IEEE Transactions on Microwave Theory and Techniques*, vol. 48, no. 7, pp. 1181–1189, 2000.

**32** B. Acikel, T. R. Taylor, P. J. Hansen, J. S. Speck, and R. A. York, "A new high performance phase shifter using ba/sub x/sr/sub 1-x/tio3 thin films," *IEEE Microwave and Wireless Components Letters*, vol. 12, no. 7, pp. 237–239, 2002.

**33** A. Tombak, J. P. Maria, F. T. Ayguavives, Z. Jin, G. T. Stauf, A. I. Kingon, and A. Mortazawi, "Voltage-controlled rf filters employing thin-film barium-strontium-titanate tunable capacitors," *IEEE Transactions on Microwave Theory and Techniques*, vol. 51, no. 2, pp. 462–467, 2003.

**34** J. Nath, D. Ghosh, J. P. Maria, A. I. Kingon, W. Fathelbab, P. D. Franzon, and M. B. Steer, "An electronically tunable microstrip bandpass filter using thin-film barium-strontium-titanate (bst) varactors," *IEEE Transactions on Microwave Theory and Techniques*, vol. 53, no. 9, pp. 2707–2712, 2005.

**35** Y.-C. Hu, P.-Z. Chang, and W.-C. Chuang, "An approximate analytical solution to the pull-in voltage of a micro bridge with an elastic boundary," *Journal of Micromechanics and Microengineering*, vol. 17, no. 9, pp. 1870–1876, 2007.

**36** C. L. Goldsmith, Z. Yao, S. Eshelman, and D. Denniston, "Performance of low-loss rf mems capacitive switches," *IEEE Microwave and Guided Wave Letters*, vol. 8, no. 8, pp. 269–271, 1998.

**37** M. F. Hagag, M. A. Khater, M. D. Hickle, and D. Peroulis, "Tunable siw cavity-based dual-mode diplexers with various single-ended and balanced ports," *IEEE Transactions on Microwave Theory and Techniques*, vol. 66, no. 3, pp. 1238–1248, 2018.

**38** M. A. Khater and D. Peroulis, "Dual-band dual-mode filter-enhanced linearity measurement," *IEEE Microwave and Wireless Components Letters*, vol. 31, no. 9, pp. 1083–1085, 2021.

**39** C. A. Balanis, *Antenna Theory: Analysis and Design*, 4th ed. John Wiley & Sons, 2016.

**40** M. A. Khater, "High-speed printed circuit boards: A tutorial," *IEEE Circuits and Systems Magazine*, vol. 20, no. 3, pp. 34–45, 2020.

**41** P. R. Scheeper, W. Olthuis, and P. Bergveld, "The design, fabrication, and testing of corrugated silicon nitride diaphragms," *Journal of Microelectromechanical Systems*, vol. 3, no. 1, pp. 36–42, 1994.

**42** E. Higurashi, D. Chino, T. Suga, and R. Sawada, "Au–au surface-activated bonding and its application to optical microsensors with 3-d structure," *IEEE Journal of Selected Topics in Quantum Electronics*, vol. 15, no. 5, pp. 1500–1505, 2009.

**43** G. M. Rebeiz, *RF MEMS: Theory, Design, and Technology*. John Wiley & Sons, 2003.

**44** J. Hong and W. Tang, "Dual-band filter based on non-degenerate dual-mode slow-wave open-loop resonators," in *2009 IEEE MTT-S International Microwave Symposium Digest*, 2009, pp. 861–864.

**45** M. A. Khater, M. Abdelfattah, Y.-C. Wu, W. Allen, and D. Peroulis, "Transient response enhancement of rf mems tuners using digital signal processing," in *2017 IEEE International Symposium on Circuits and Systems (ISCAS)*, 2017, pp. 1–4.

**46** J. Sorocki, I. Piekarz, S. Gruszczynski, K. Wincza, and J. Papapolymerou, "Application of 3-d printing technology for the realization of high-performance directional couplers in suspended stripline technique," *IEEE Transactions on Components, Packaging and Manufacturing Technology*, vol. 9, no. 8, pp. 1652–1658, 2019.

**47** K. Zhao and D. Psychogiou, "Monolithic SLA-based capacitively-loaded high-Q coaxial resonators and bandpass filters," in *2020 Proceedings 50th European Microwave Conference (EuMC)*, 2021, pp. 471–474.

**48** Z. Yang, R. Zhang, and D. Peroulis, "Design and optimization of bidirectional tunable mems all-silicon evanescent-mode cavity filter," *IEEE Transactions on Microwave Theory and Techniques*, vol. 68, no. 6, pp. 2398–2408, 2020.

**49** M. S. Arif, W. Irshad, X. Liu, W. J. Chappell, and D. Peroulis, "A high-Q magnetostatically-tunable all-silicon evanescent cavity resonator," in *2011 IEEE MTT-S International Microwave Symposium Digest (IMS 2011)*, 2011, pp. 3–6.

**50** E. J. Naglich, J. Lee, D. Peroulis, and W. J. Chappell, "High-q tunable bandstop filters with adaptable bandwidth and pole allocation," in *2011 IEEE MTT-S International Microwave Symposium Digest (IMS 2011)*, 2011, pp. 1–4.

**51** E. J. Naglich, M. Sinani, S. Moon, and D. Peroulis, "High-q mems-tunable w-band bandstop resonators," in *2014 IEEE MTT-S International Microwave Symposium Digest (IMS 2014)*, 2014, pp. 1–3.

**52** R. Karania and D. Kazmer, "Low volume plastic manufacturing strategies," *ASME Journal of Mechanical Design*, vol. 129, no. 12, pp. 1225–1233, 2007.

**53** MTD Micromolding, "The 6 sciences of medical micro molding: How to achieve better long-term results and faster speed to market," https://directory.qmed.com/p-free-white-paper-on-the-6-sciences-of-medical-file068602.html, (Accessed on: July 14, 2021).

**54** PTI Engineered Plastics, "The benefits of injection molded plastics in an EV/AV world," https://directory.qmed.com/over-the-years-plastic-hasdeveloped-into-one-of-file117915.html, (Accessed on: July 14, 2021).

**55** S. Olivera, H. B. Muralidhara, K. Venkatesh, K. Gopalakrishna, and C. S. Vivek, "Plating on acrylonitrile-butadiene-styrene (ABS) plastic: a review," *Journal of Materials Science*, vol. 51, no. 8, pp. 3657–3674, 2016.

**56** Xometry, "The ultimate guide to injection molding," https://www.xometry.com/ultimate-guide/injection-molding, (Accessed on: July 14, 2021).

**57** P. Adhikari, W. Yang, Y. C. Wu, and D. Peroulis, "A pcb technology-based 22-42-ghz quasi-absorptive bandstop filter," *IEEE Microwave and Wireless Components Letters*, vol. 28, no. 11, pp. 975–977, 2018.

**58** P. Leisner, M. Fredenberg, and I. Belov, "Pulse and pulse reverse plating of copper from acid sulphate solutions," *Transactions of the Institute of Metal Finishing*, vol. 88, no. 5, pp. 243–247, 2010.

**59** Z. Yang, D. Psychogiou, and D. Peroulis, "Design and optimization of tunable silicon-integrated evanescent-mode bandpass filters," *IEEE Transactions on Microwave Theory and Techniques*, vol. 66, no. 4, pp. 1790–1803, 2018.

60 M. Abdelfattah and D. Peroulis, “High-q tunable evanescent-mode cavity siw resonators and filters with contactless tuners,” *IEEE Transactions on Microwave Theory and Techniques*, vol. 67, no. 9, pp. 3661–3672, 2019.

61 P. Adhikari, W. Yang, and D. Peroulis, “A 20-26.5-GHz PCB bandpass filter tuned with contactless tuners,” *IEEE Microwave and Wireless Components Letters*, vol. 29, no. 8, pp. 513–515, 2019.

62 J. S. Hong, *Microstrip Filters for RF/Microwave Applications*, 2nd ed. Hoboken, NJ: John Wiley & Sons, Inc., 2011.

63 R. J. Cameron, C. M. Kudsia, and R. R. Mansour, *Microwave Filters for Communication Systems: Fundamentals, Design, and Applications*, 2nd ed. Hoboken, NJ: Wiley, 2018.

64 D. M. Pozar, *Microwave Engineering*, 4th ed. Hoboken, NJ: John Wiley & Sons, Inc., 2012.

65 P. Blondy and D. Peroulis, “Handling RF power: The latest advances in RF-MEMS tunable filters,” *IEEE Microwave Magazine*, vol. 14, no. 1, pp. 24–38, 2013.

66 K. Chen, “High-power microwave/ radio-frequency components, circuits, and-subsystems for next-generation wireless radio front-ends,” PhD dissertation, Purdue University, 2013.

67 A. Semnani, K. Chen, and D. Peroulis, “Microwave gas breakdown in tunable evanescent-mode cavity resonators,” *IEEE Microwave and Wireless Components Letters*, vol. 24, no. 5, pp. 351–353, 2014.

68 S. Sirici, M. A. Sanchez-Soriano, J. D. Martinez, V. E. Boria, F. Gentili, W. Bosch, and R. Sorrentino, “Design and multiphysics analysis of direct and cross-coupled SIW combline filters using electric and magnetic couplings,” *IEEE Transactions on Microwave Theory and Techniques*, vol. 63, no. 12, pp. 4341–4354, 2015.

69 S. Saeedi, J. Lee, and H. Sigmarsson, “Prediction of power handling in tunable, high-Q, substrate-integrated, evanescentmode cavity bandpass filters,” *Electronics Letters*, vol. 52, no. 10, pp. 846–848, 2016.

70 C. Ernst, V. Postoyalko, and N. G. Khan, “Relationship between group delay and stored energy in microwave filters,” *IEEE Transactions on Microwave Theory and Techniques*, vol. 49, no. 1, pp. 192–196, 2001.

71 S. J. Park, I. Reines, C. Patel, and G. M. Rebeiz, “High *q* rf-mems 4-6-ghz tunable evanescent-mode cavity filter,” *IEEE Transactions on Microwave Theory and Techniques*, vol. 58, no. 2, pp. 381–389, 2010.

72 K. Chen, H. H. Sigmarsson, and D. Peroulis, “Power handling of high-q evanescent-mode tunable filter with integrated piezoelectric actuators,” in *2012 IEEE MTT-S International Microwave Symposium Digest (IMS 2012)*, 2012, pp. 1–3.

**73** K. Sadasivan and D. Psychogiou, "Tunable 3D-printed coaxial-cavity filters with mixed electromagnetic coupling," in *2019 IEEE International Symposium on Antennas and Propagation and USNC-URSI Radio Science Meeting (APSURSI 2019)*, 2019, pp. 1703–1704.

**74** D. Psychogiou and K. Sadasivan, "Tunable coaxial cavity resonator-based filters using actuated liquid metal posts," *IEEE Microwave and Wireless Components Letters*, vol. 29, no. 12, pp. 763–766, 2019.

**75** K. Zhao and D. Psychogiou, "A monolithic vertical integration concept for compact coaxial-resonator-based bandpass filters using additive manufacturing," *IEEE Microwave and Wireless Components Letters*, vol. 31, no. 6, pp. 689–692, 2021.

**76** M. D. Sinanis, M. Abdelfattah, M. Cakmak, and D. Peroulis, "A 2.2-4.2 ghz low-loss tunable bandpass filter based on low cost manufacturing of abs polymer," in *IEEE Wireless and Microwave Technology Conference (WAMICON)*, 2018, pp. 1–4.

**77** F. Cai, W. Khan, and J. Papapolymerou, "A low loss X-band filter using 3-D Polyjet technology," in *2015 IEEE MTT-S International Microwave Symposium Digest (IMS 2015)*, 2015, pp. 8–11.

**78** S. Moon, H. H. Sigmarsson, H. Joshi, and W. J. Chappell, "Substrate integrated evanescent-mode cavity filter with a 3.5 to 1 tuning ratio," *IEEE Microwave and Wireless Components Letters*, vol. 20, no. 8, pp. 450–452, 2010.

**79** D. R. Jachowski, "Compact, frequency-agile, absorptive bandstop filters," in *IEEE MTT-S International Microwave Symposium Digest, 2005*, 2005, pp. 513–516.

**80** A. C. Guyette, I. C. Hunter, R. D. Pollard, and D. R. Jachowski, "Perfectly-matched bandstop filters using lossy resonators," in *IEEE MTT-S International Microwave Symposium Digest, 2005*, 2005, pp. 517–520.

**81** M. D. Hickle and D. Peroulis, "Theory and design of frequency-tunable absorptive bandstop filters," *IEEE Transactions on Circuits and Systems I: Regular Papers*, vol. 65, no. 6, pp. 1862–1874, 2018.

**82** E. J. Naglich, J. Lee, D. Peroulis, and W. J. Chappell, "Extended passband bandstop filter cascade with continuous 0.85–6.6-ghz coverage," *IEEE Transactions on Microwave Theory and Techniques*, vol. 60, no. 1, pp. 21–30, 2012.

**83** P. Pal, M. Gosálvez, and K. Sato, "Etched profile control in anisotropic etching of silicon by tmah+triton," *Journal of Micromechanics and Microengineering*, vol. 22, no. 5, p. 065013, 2012.

**84** W. Yang, M. A. Khater, E. J. Naglich, D. Psychogiou, and D. Peroulis, "Frequency-selective limiters using triple-mode filters," *IEEE Access*, vol. 8, pp. 114 854–114 863, 2020.

**85** G. L. Matthaei, "Narrow-band, fixed-tuned, and tunable bandpass filters with zig-zag hairpin-comb resonators," *IEEE Transactions on Microwave Theory and Techniques*, vol. 51, no. 4, pp. 1214–1219, 2003.

**86** M. Sanchez-Renedo, R. Gomez-Garcia, J. I. Alonso, and C. Briso-Rodriguez, "Tunable combline filter with continuous control of center frequency and bandwidth," *IEEE Transactions on Microwave Theory and Techniques*, vol. 53, no. 1, pp. 191–199, 2005.

**87** J. S. Sun and T. Itoh, "Dual-mode tunable filter with simple bandwidth control scheme," in *2009 IEEE MTT-S International Microwave Symposium Digest*, 2009, pp. 497–500.

**88** E. Fourn, C. Quendo, E. Rius, A. Pothier, P. Blondy, C. Champeaux, J. C. Orlianges, A. Catherinot, G. Tanne, C. Person, and F. Huret, "Bandwidth and central frequency control on tunable bandpass filter by using mems cantilevers," in *IEEE MTT-S International Microwave Symposium Digest, 2003*, vol. 1, 2003, pp. 523–526.

**89** E. J. Naglich, J. Lee, and D. Peroulis, "Tunable bandstop filter with a 17-to-1 upper passband," in *2012 IEEE/MTT-S International Microwave Symposium Digest*, 2012, pp. 1–3.

**90** K. W. Wong, R. R. Mansour, and G. Weale, "Compact tunable bandstop filter with wideband balun using ipd technology for frequency agile applications," in *2015 IEEE MTT-S International Microwave Symposium*, 2015, pp. 1–3.

**91** I. Reines, S. Park, and G. M. Rebeiz, "Compact low-loss tunable *x*-band bandstop filter with miniature rf-mems switches," *IEEE Transactions on Microwave Theory and Techniques*, vol. 58, no. 7, pp. 1887–1895, 2010.

**92** A. Anand, Y. Liu, and X. Liu, "Substrate-integrated octave-tunable combline bandstop filter with surface mount varactors," in *2014 IEEE International Wireless Symposium (IWS 2014)*, 2014, pp. 1–4.

**93** A. C. Guyette, "Intrinsically switched varactor-tuned filters and filter banks," *IEEE Transactions on Microwave Theory and Techniques*, vol. 60, no. 4, pp. 1044–1056, 2012.

**94** D. R. Jachowski, "Octave tunable lumped-element notch filter," in *2012 IEEE/MTT-S International Microwave Symposium Digest*, 2012, pp. 1–3.

**95** T. Lee, J. Lee, E. J. Naglich, and D. Peroulis, "Octave tunable lumped-element notch filter with resonator-q-independent zero reflection coefficient," in *2014 IEEE MTT-S International Microwave Symposium (IMS2014)*, 2014, pp. 1–4.

**96** D. R. Jachowski and A. C. Guyette, "Sub-octave-tunable microstrip notch filter," in *2009 IEEE International Symposium on Electromagnetic Compatibility*, 2009, pp. 99–102.

**97** X. Y. Zhang, C. H. Chan, Q. Xue, and B. Hu, "Rf tunable bandstop filters with constant bandwidth based on a doublet configuration," *IEEE Transactions on Industrial Electronics*, vol. 59, no. 2, pp. 1257–1265, 2012.

**98** A. C. Guyette, "Varactor-tuned bandstop filters with tunable center frequency and bandwidth," in *2010 IEEE International Conference on Wireless Information Technology and Systems*, 2010, pp. 1–4.

**99** A. Anand and X. Liu, "Capacitively tuned electrical coupling for reconfigurable coaxial cavity bandstop filters," in *2015 IEEE MTT-S International Microwave Symposium*, 2015, pp. 1–3.

**100** M. D. Hickle and D. Peroulis, "Octave-tunable constant absolute bandwidth bandstop filter utilizing a novel passively-compensated coupling method," in *2016 IEEE MTT-S International Microwave Symposium (IMS)*, 2016, pp. 1–4.

**101** M. D. Hickle, M. D. Sinanis, and D. Peroulis, "Tunable high-isolation w-band bandstop filters," in *2015 IEEE MTT-S International Microwave Symposium*, 2015, pp. 1–4.

**102** M. D. Hickle, M. D. Sinanis, and D. Peroulis, "Design and implementation of an intrinsically-switched 22–43 ghz tunable bandstop filter," in *2016 IEEE 17th Annual Wireless and Microwave Technology Conference (WAMICON)*, 2016, pp. 1–3.

**103** A. C. Guyette and E. J. Naglich, "Short-through-line bandstop filters using dual-coupled resonators," *IEEE Transactions on Microwave Theory and Techniques*, vol. 64, no. 2, pp. 459–466, 2016.

**104** K. Chen, J. Lee, W. J. Chappell, and D. Peroulis, "Co-design of highly efficient power amplifier and high-*q* output bandpass filter," *IEEE Transactions on Microwave Theory and Techniques*, vol. 61, no. 11, pp. 3940–3950, 2013.

**105** T. Snow, J. Lee, and W. J. Chappell, "Tunable high quality-factor absorptive bandstop filter design," in *2012 IEEE/MTT-S International Microwave Symposium Digest*, 2012, pp. 1–3.

**106** E. J. Naglich, D. Peroulis, and W. J. Chappell, "Wide spurious free range positive-to-negative inter-resonator coupling structure for reconfigurable filters," in *2013 IEEE MTT-S International Microwave Symposium Digest (MTT)*, 2013, pp. 1–4.

**107** X. Chen and K. Wu, "Substrate integrated waveguide cross-coupled filter with negative coupling structure," *IEEE Transactions on Microwave Theory and Techniques*, vol. 56, no. 1, pp. 142–149, 2008.

**108** M. A. Sanchez-Soriano, S. Sirci, J. D. Martnez, and V. E. Boria, "Compact bandpass filters based on a new substrate integrated waveguide coaxial cavity," in *2016 IEEE MTT-S International Microwave Symposium (IMS)*, 2016, pp. 1–4.

**109** C. Liao, P. Chi, and C. Chang, "Microstrip realization of generalized chebyshev filters with box-like coupling schemes," *IEEE Transactions on Microwave Theory and Techniques*, vol. 55, no. 1, pp. 147–153, 2007.

**110** M. Abdelfattah and D. Peroulis, "A novel independently-tunable dual-mode siw resonator with a reconfigurable bandpass filter application," in *2018 IEEE/MTT-S International Microwave Symposium - IMS*, 2018, pp. 1091–1094.

**111** M. Sanchez-Soriano, S. Sirci, J. D. Martnez, and V. E. Boria, "Compact dual-mode substrate integrated waveguide coaxial cavity for bandpass filter design," *IEEE Microwave and Wireless Components Letters*, vol. 26, no. 6, pp. 386–388, 2016.

**112** S. Nam, B. Lee, B. Koh, and J. Lee, "Reconfigurable bandpass filter with resonators in cul-de-sacs for producing notches," *IEEE Transactions on Components, Packaging and Manufacturing Technology*, vol. 7, no. 9, pp. 1531–1542, 2017.

**113** M. Ohira and Z. Ma, "A parameter-extraction method for microwave transversal resonator array bandpass filters with direct source/load coupling," *IEEE Transactions on Microwave Theory and Techniques*, vol. 61, no. 5, pp. 1801–1811, 2013.

**114** H. Joshi, H. H. Sigmarsson, S. Moon, D. Peroulis, and W. J. Chappell, "High-*q* fully reconfigurable tunable bandpass filters," *IEEE Transactions on Microwave Theory and Techniques*, vol. 57, no. 12, pp. 3525–3533, 2009.

**115** S. Amari and F. Seyfert, "Design of dual-mode ridge cavity filters," in *2009 IEEE MTT-S International Microwave Symposium Digest*, 2009, pp. 1625–1628.

**116** J. Lee, E. J. Naglich, and W. J. Chappell, "Frequency response control in frequency-tunable bandstop filters," *IEEE Microwave and Wireless Components Letters*, vol. 20, no. 12, pp. 669–671, 2010.

**117** S. Amari, "Direct synthesis of folded symmetric resonator filters with source-load coupling," *IEEE Microwave and Wireless Components Letters*, vol. 11, no. 6, pp. 264–266, 2001.

**118** J. Lee and K. Sarabandi, "A miniaturized conductor-backed slot-line resonator filter with two transmission zeros," *IEEE Microwave and Wireless Components Letters*, vol. 16, no. 12, pp. 660–662, 2006.

**119** J. B. Muldavin and G. M. Rebeiz, "Novel dc-contact mems shunt switches and high-isolation series/shunt designs," in *2001 31st European Microwave Conference*, 2001, pp. 1–3.

**120** R. V. Snyder, "Inverted-resonator evanescent mode filters," in *1996 IEEE MTT-S International Microwave Symposium Digest*, vol. 2, 1996, pp. 465–468.

**121** V. Walker and I. C. Hunter, "Design of cross-coupled dielectric-loaded waveguide filters," *IEE Proceedings - Microwaves, Antennas and Propagation*, vol. 148, no. 2, pp. 91–96, 2001.

**122** J. Lee, M. S. Uhm, and I.-B. Yom, "A dual-passband filter of canonical structure for satellite applications," *IEEE Microwave and Wireless Components Letters*, vol. 14, no. 6, pp. 271–273, 2004.

**123** R. Levy, "Filters with single transmission zeros at real or imaginary frequencies," *IEEE Transactions on Microwave Theory and Techniques*, vol. 24, no. 4, pp. 172–181, 1976.

**124** G. Pfitzenmaier, "Synthesis and realization of narrow-band canonical microwave bandpass filters exhibiting linear phase and transmission zeros," *IEEE Transactions on Microwave Theory and Techniques*, vol. 30, no. 9, pp. 1300–1311, 1982.

**125** H. Meng and K. Wu, "Direct optimal synthesis of a microwave bandpass filter with general loading effect," *IEEE Transactions on Microwave Theory and Techniques*, vol. 61, no. 7, pp. 2566–2573, 2013.

**126** M. A. Khater, M. Awajan, and D. Peroulis, "A k-band resonant impedance tuner with a solid-state hybrid tuning," in *2022 IEEE 22nd Annual Wireless and Microwave Technology Conference (WAMICON)*, 2022, pp. 1–4.

**127** K. Saitou and K. Kageyama, "Tunable duplexer having multilayer structure using ltcc," in *IEEE MTT-S International Microwave Symposium Digest, 2003*, vol. 3, 2003, pp. 1763–1766.

**128** M. A. Khater, Y. Wu, and D. Peroulis, "Tunable cavity-based diplexer with spectrum-aware automatic tuning," *IEEE Transactions on Microwave Theory and Techniques*, vol. 65, no. 3, pp. 934–944, 2017.

**129** E. E. Djoumessi and K. Wu, "Electronically tunable diplexer for frequency-agile transceiver front-end," in *2010 IEEE MTT-S International Microwave Symposium*, 2010, pp. 1472–1475.

**130** H. Ezzeddine, P. Mazet, S. Bila, S. Verdeyme, F. Seyfert, D. Pacaud, J. Puech, and L. Estagerie, "Design of a compact dual-band diplexer with dual-mode cavities," in *2012 42nd European Microwave Conference*, 2012, pp. 455–458.

**131** V. Radonić, V. Crnojević-Bengin, A. Baskakova, and I. Vendik, "Multilayer microwave diplexers based on dual-mode resonators for ism/wifi bands," in *Proceedings of 2014 Mediterranean Microwave Symposium (MMS2014)*, 2014, pp. 1–4.

**132** T. Duong, W. Hong, Z. Hao, W. Huang, J. Zhuang, and V. Vo, "A millimeter wave high-isolation diplexer using selectivity-improved dual-mode filters," *IEEE Microwave and Wireless Components Letters*, vol. 26, no. 2, pp. 104–106, 2016.

**133** X. Guan, F. Yang, H. Liu, and L. Zhu, "Compact and high-isolation diplexer using dual-mode stub-loaded resonators," *IEEE Microwave and Wireless Components Letters*, vol. 24, no. 6, pp. 385–387, 2014.

**134** E. A. Ogbodo, Y. Wu, and Y. Wang, "Microstrip diplexers with dual-mode patch resonant junctions," in *2016 46th European Microwave Conference (EuMC)*, 2016, pp. 1155–1158.

**135** M. D. Hickle and D. Peroulis, "A widely-tunable substrate-integrated balun filter," in *2017 IEEE MTT-S International Microwave Symposium (IMS)*, 2017, pp. 274–277.

**136** W. Zhao, Y. Zhang, and M. Zhan, "Design and performance of a w-band microstrip rat-race balanced mixer," in *2010 International Conference on Microwave and Millimeter Wave Technology*, 2010, pp. 713–716.

**137** J. Tsai, Y. Lee, T. Huang, C. Yu, and J. G. J. Chern, "A 90-nm cmos broadband and miniature q-band balanced medium power amplifier," in *2007 IEEE/MTT-S International Microwave Symposium*, 2007, pp. 1129–1132.

**138** J. S. G. Hong and M. J. Lancaster, *Microstrip Filters for RF / Microwave Applications*. John Wiley & Sons, 2004.

**139** P. Reynaert, "Polar modulation," *IEEE Microwave Magazine*, vol. 12, no. 1, pp. 46–51, 2011.

**140** D. R. Parveg, P. Singerl, A. Wiesbauer, H. M. Nemati, and C. Fager, "A broadband, efficient, overdriven class-j rf power amplifier for burst mode operation," in *The 5th European Microwave Integrated Circuits Conference*, 2010, pp. 424–427.

**141** E. Serebryakova, A. Samulak, K. Blau, and M. Hein, "Reconstruction filters for switched-mode power amplifier systems," in *2009 European Microwave Conference (EuMC)*, 2009, pp. 1453–1456.

**142** C. Patel, M. Yavuz, and S. Nanda, "Femtocells [industry perspectives]," *IEEE Wireless Communications*, vol. 17, no. 5, pp. 6–7, 2010.

**143** M. A. Khater, Y.-C. Wu, M. Abdelfattah, and D. Peroulis, "A compact octave tunable switched-power-combining pa," *IEEE Access*, vol. 9, pp. 15 212–15 220, 2021.

**144** P. Lin, J. Zhang, Y. Chen, and Q. Zhang, "Macro-femto heterogeneous network deployment and management: from business models to technical solutions," *IEEE Wireless Communications*, vol. 18, no. 3, pp. 64–70, 2011.

**145** P. Saad, C. Fager, H. Cao, H. Zirath, and K. Andersson, "Design of a highly efficient 2.4-ghz octave bandwidth gan-hemt power amplifier," *IEEE Transactions on Microwave Theory and Techniques*, vol. 58, no. 7, pp. 1677–1685, 2010.

**146** B. Razavi, *Design of Analog CMOS Integrated Circuits*. McGraw-Hill, 2000.

**147** D. Leeson, "A simple model of feedback oscillator noise spectrum," *Proceedings of the IEEE*, vol. 54, no. 2, pp. 329–330, 1966.

**148** P. Delos, "Phase locked loop noise transfer functions," *High Frequency Electronics*, 2016.

**149** T. Lee and T. Lee, *The Design of CMOS Radio-Frequency Integrated Circuits*. Cambridge University Press, 2004.

**150** M. J. Underhill, "Time jitter and phase noise – now and in the future?" in *2012 IEEE International Frequency Control Symposium Proceedings*, 2012, pp. 1–8.

**151** M. A. Khater, A. Fisher, and D. Peroulis, "Switch-based non-reciprocal filter for high-power applications," in *2024 IEEE International Microwave Filter Workshop (IMFW)*, 2024, pp. 180–182.

**152** M. A. Khater, J. Zhou, Y.-C. Wu, H. Krishnaswamy, and D. Peroulis, "A tunable 0.86-1.03 ghz fdd wireless communication system with an evanescent-mode diplexer and a self-interference-cancelling receiver," in *2017 IEEE MTT-S International Microwave Symposium (IMS)*, 2017, pp. 376–379.

**153** M. A. Khater and D. Peroulis, "Real-time feedback control system for tuning evanescent-mode cavity filters," *IEEE Transactions on Microwave Theory and Techniques*, vol. 64, no. 9, pp. 2804–2813, 2016.

**154** M. A. Khater and D. Peroulis, "Pulse injection for real-time monitoring of tunable high-q filters," *Microwave and Optical Technology Letters*, vol. 56, no. 3, pp. 761–764, 2014.

**155** M. A. Khater and D. Peroulis, "Multioctave interference detectors with sub-microsecond response," *IEEE Transactions on Microwave Theory and Techniques*, vol. 71, no. 6, pp. 2693–2701, 2023.

**156** M. A. Khater and D. Peroulis, "2-8 ghz interference detector with 1.1 $\mu$s response," *IEEE Microwave and Wireless Components Letters*, vol. 32, no. 6, pp. 756–759, 2022.

**157** J. Shang, H. Li, X. You, T. Q. Zheng, and S. Wang, "A novel stability analysis approach based on describing function method using for dc-dc converters," in *2015 IEEE Applied Power Electronics Conference and Exposition (APEC)*, 2015, pp. 2642–2647.

**158** A. Zanchi, A. Bonfanti, S. Levantino, and C. Samori, "General sscr vs. cycle-to-cycle jitter relationship with application to the phase noise in pll," in *2001 Southwest Symposium on Mixed-Signal Design (Cat. No.01EX475)*, 2001, pp. 32–37.

**159** M. A. Khater and D. Peroulis, "Spectrum-aware jammer suppression using evanescent-mode cavity filters," in *2015 IEEE MTT-S International Microwave Symposium*, 2015, pp. 1–4.

**160** M. A. Khater, M. Abdelfattah, M. D. Sinanis, and D. Peroulis, "An inductor-based real-time monitoring and control system for tunable cavity mems filters," in *2018 IEEE/MTT-S International Microwave Symposium - IMS*, 2018, pp. 401–404.

**161** M. A. Khater, M. Abdelfattah, M. D. Sinanis, and D. Peroulis, "Monitoring and control of mems tunable filters using inductive proximity sensing," *IEEE Transactions on Microwave Theory and Techniques*, vol. 66, no. 12, pp. 5605–5613, 2018.

**162** S. Babic and C. Akyel, "Improvement in calculation of the self- and mutual inductance of thin-wall solenoids and disk coils," *IEEE Transactions on Magnetics*, vol. 36, no. 4, pp. 1970–1975, 2000.

**163** H. Wang, W. Li, and Z. Feng, "Noncontact thickness measurement of metal films using eddy-current sensors immune to distance variation," *IEEE Transactions on Instrumentation and Measurement*, vol. 64, no. 9, pp. 2557–2564, 2015.

**164** M. A. Khater and D. Peroulis, "Variable-output charge-pump for piezoelectric and electrostatic tunable rf filters," in *2015 IEEE International Symposium on Circuits and Systems (ISCAS)*, 2015, pp. 1362–1365.

**165** M. A. Khater, K. Zeng, and D. Peroulis, "Temperature-compensated lumped element tunable bandpass filter," in *2016 IEEE MTT-S International Microwave Symposium (IMS)*, 2016, pp. 1–4.

**166** M. A. Khater and D. Peroulis, "Vibration mitigation for evanescent-mode cavity filters," in *2014 IEEE MTT-S International Microwave Symposium (IMS2014)*, 2014, pp. 1–4.

**167** M. Sinani, *High-Q tunable microwave cavity resonators and filters with scalable manufacturing technologies for 5G communications* (Doctoral dissertation, Purdue University), 2022.

**168** E. Naglich, *Filter synthesis and design techniques for highly adaptable systems* (PhD dissertation, Purdue University), 2013.

**169** T.-C. Lee, *Synthesis of microwave and acoustic wave resonator filters for wireless communication systems* (PhD dissertation, Purdue University), 2014.

# Index



*Tunable Evanescent-Mode Filters: Principles, Implementation, and Applications*,
First Edition. Dimitrios Peroulis and Mohammad Abu Khater.

Companion website: www.wiley.com/go/Peroulis/TunableEvanescent-ModeFilters

## *d*

Printed and bound by CPI Group (UK) Ltd, Croydon, CR0 4YY
17/09/2026
14964311-0003